Springer Theses

Recognizing Outstanding Ph.D. Research

Aims and Scope

The series "Springer Theses" brings together a selection of the very best Ph.D. theses from around the world and across the physical sciences. Nominated and endorsed by two recognized specialists, each published volume has been selected for its scientific excellence and the high impact of its contents for the pertinent field of research. For greater accessibility to non-specialists, the published versions include an extended introduction, as well as a foreword by the student's supervisor explaining the special relevance of the work for the field. As a whole, the series will provide a valuable resource both for newcomers to the research fields described, and for other scientists seeking detailed background information on special questions. Finally, it provides an accredited documentation of the valuable contributions made by today's younger generation of scientists.

Theses are accepted into the series by invited nomination only and must fulfill all of the following criteria

- They must be written in good English.
- The topic should fall within the confines of Chemistry, Physics, Earth Sciences, Engineering and related interdisciplinary fields such as Materials, Nanoscience, Chemical Engineering, Complex Systems and Biophysics.
- The work reported in the thesis must represent a significant scientific advance.
- If the thesis includes previously published material, permission to reproduce this must be gained from the respective copyright holder.
- They must have been examined and passed during the 12 months prior to nomination.
- Each thesis should include a foreword by the supervisor outlining the significance of its content.
- The theses should have a clearly defined structure including an introduction accessible to scientists not expert in that particular field.

More information about this series at http://www.springer.com/series/8790

Vivishek Sudhir

Quantum Limits on Measurement and Control of a Mechanical Oscillator

Doctoral Thesis accepted by
École Polytechnique Fédérale de Lausanne, Switzerland

Author
Dr. Vivishek Sudhir
LIGO Laboratory
Massachusetts Institute of Technology
Cambridge, MA
USA

Supervisor
Prof. Tobias J. Kippenberg
Institute of Physics
École Polytechnique Fédérale de Lausanne
Lausanne
Switzerland

ISSN 2190-5053 ISSN 2190-5061 (electronic)
Springer Theses
ISBN 978-3-319-69430-6 ISBN 978-3-319-69431-3 (eBook)
https://doi.org/10.1007/978-3-319-69431-3

Library of Congress Control Number: 2017955357

Printed on acid-free paper

This Springer imprint is published by Springer Nature
The registered company is Springer International Publishing AG
The registered company address is: Gewerbestrasse 11, 6330 Cham, Switzerland

The hardest thing of all to see is what is really there.

J.A. Baker, The Peregrine

To my father, who by patiently answering all
my questions, became my first science teacher,
And to my mother, who taught me everything else

Supervisor's Foreword

The study of radiation pressure coupling between light and motion in optical interferometers has a long history, one that predates its study in atomic systems. The seminal work of Braginsky in the 1960s predicted that the radiation pressure of light in an optical interferometer causes two disturbances that limit the ability to measure optical displacements: a classical—dynamic back-action—that causes parametric instability of the mirror, and a quantum mechanical limit imposed by the quantum fluctuations of the radiation pressure force. While originally formulated within the context of interferometric gravitational wave detection, both classical dynamical back-action and its quantum mechanical counterpart have become observable in experiments using cavity optomechanical systems that utilize high-Q and small mass mechanical oscillators coupled to intense optical fields stored in cavities. While dynamical back-action physics, exemplified by laser sideband cooling of massive mechanical oscillators, has been accessed by a wide range of nano- and micron-scale optomechanical systems, observing the limits imposed by the quantum nature of light has been far more challenging. The latter is compounded by the fact that the quantum fluctuations of the radiation pressure force are faint, and further, they can be easily masked by classical noises.

The present thesis from Vivishek Sudhir constitutes the first experiments carried out in our laboratory at EPFL that observes and studies the quantum nature of the radiation pressure interaction. This is achieved via advances in the ability to perform sensitive measurements and to operate novel nano-optomechanical systems at low cryogenic temperatures. These advances enable a series of experiments that demonstrate the origin and physical effects that result from radiation pressure quantum noise. A newly developed optomechanical system consisting of a high-Q small mass nanostring dispersively coupled to the evanescent field of an optical whispering gallery mode resonator enabled to achieve vacuum cooperativities near unity, translating into an ability to measure the motion of the nanostring with an imprecision at the standard quantum limit (SQL) with less than a single intracavity photon on average. In fact, by engineering the nanostring to oscillate at frequencies untouched by other measurement noises, it was possible to measure its motion with an imprecision 40dB below that at the SQL. With such record measurement sensitivity, the oscillator could be feedback cooled to an occupation of only 5 quanta (corresponding to 16% ground-state occupation). Although feedback cooling had been proposed as a method to reduce the thermal motion of mechanical oscillators since decades, its utility for reducing thermal noise to a level comparable to the zero-point motion of the oscillator had to wait till this thesis.

This experiment sets the stage for the exploration that is carried out in the rest of the thesis—a dissection of linear quantum measurements. The ability to cool the mechanical oscillator to such low occupancy implies that the corresponding quantum back-action, resulting from the radiation pressure quantum fluctuations, has been canceled. Therefore, feedback cooling to occupancies below that due to back-action should more accurately be viewed as an example of quantum feedback. In a subsequent series of experiments, this thesis goes on to precisely explore the different manifestations of this suppressed quantum back-action. Quantum back-action is shown to give rise to two effects: pondermotive squeezing of light and motional sideband asymmetry. A particularly interesting aspect of the work is that both effects are observed in one and the same experiment for the first time, depending on how one analyzes the post-measurement optical field: in case of homodyne detection, one observes that the strong measurement causes squeezing of the light, while for the case of heterodyne detection (combined with feedback cooling) the experiments reveal an asymmetry in the sidebands scattered by the mechanical oscillator. The work therefore shows that the two effects have precisely the same origin: the generation of correlations between the amplitude and the phase fluctuations of the measurement laser. Measuring such quantum correlations is a daunting task, and the experimenter has to exercise extreme care to rule out classical effects. Both manifestations of quantum correlations can be mimicked by classical noises—the difference lying only in their calibration. The present thesis achieves this in a particularly elegant way: operation with large vacuum cooperativity in the Doppler regime combined with feedback cooling eliminates a large number of classical measurement noises; in particular, and counterintuitively, the experiment is not sensitive to classical phase noise of the measurement lasers. Moreover, significant sources of systematic error are eliminated by being able to observe sideband asymmetry by only varying the electronic gain of the feedback path, and not the laser power or detuning. These advantages together allow to demonstrate a quantum mechanical sideband asymmetry at the level of 10% in agreement with the occupation of the oscillator. Together, optical squeezing and sideband asymmetry show the quantum correlations induced by an optical field by a "macroscopic" mechanical oscillator. The culmination of the thesis is to probe, using sideband asymmetry, the regime where feedback of detected quantum noise leads to noise squashing. In this regime, the sideband asymmetry disappears. On the one hand, it corroborates the trustworthiness of the calibration of classical measurement noises. On the other hand, it highlights a basic limitation of quantum feedback: although it can cancel back-action caused by quantum noise, it cannot overcome detection quantum noise.

Overall, the present thesis represents the first experiments at the laboratory at EPFL where long-predicted quantum effects of radiation pressure were probed and analyzed. This thesis provides the first glimpse into the quantum nature of mechanical oscillators, and a step toward quantum control of macroscopic mechanical systems, after that of atoms, ions, and superconducting circuits in recent decades.

Lausanne, Switzerland
August 2017

Prof. Tobias J. Kippenberg

Abstract

The precision measurement of position has a long-standing tradition in physics. Cavendish's verification of the universal law of gravitation using a torsion pendulum, Perrin's confirmation of the atomic hypothesis via the precise measurement of the Brownian motion, and the verification of the mechanical effect of electromagnetic radiation all belong to this classical heritage. Quantum mechanics posits that the measurement of position results in an uncertain momentum; an idea developed to full maturity within the context of interferometric searches for gravitational waves. Over the past decade, standing at the confluence of quantum optics and nanomechanics, cavity optomechanics has emerged as a powerful platform to study the quantum limits of position measurements.

The subject of this thesis is the precision measurement of the position of a nanomechanical oscillator, the fundamental limits of such measurements, and its relevance to measurement-based feedback control. The nanomechanical oscillator is coupled to light confined in an optical micro-cavity via radiation pressure. The fluctuations in the position of the oscillator are transduced onto the phase of the light, while quantum fluctuations in the amplitude of the light lead to a disturbance in the momentum of the oscillator. We perform an interferometric position measurement with a sensitivity, that is, 10^5 times below what is required to resolve the zero-point motion of the oscillator, constituting the most precise measurement of thermal motion yet. The resulting disturbance—measurement back-action—is observed to be commensurate with the uncertainty principle, leading to a 10% contribution to the total motion of the oscillator.

The continuous record of the measurement (performed in a 4 K cryogenic environment) furnishes the ability to resolve the zero-point motion of the oscillator within its decoherence rate—the necessary condition for measurement-based feedback control of the state of the oscillator. Using the measurement record as error signal, the oscillator is cooled toward its ground state, resulting in a factor 10^4 suppression of its total (thermal and back-action) motion, to a final occupation of 5 phonons on average.

Measurements generally proceed by establishing correlations between the system being measured and the measuring device. For the class of quantum measurements employed here—continuous linear measurements—these correlations arise due to measurement back-action. These back-action-induced correlations appear as correlations between the degrees of freedom of the measuring device. For interferometric position measurements, quantum correlations are established between the phase and amplitude of the light. In a homodyne measurement, they lead to optical squeezing, while in a heterodyne measurement, they appear as an asymmetry in the sidebands carrying information about the oscillator position. Feedback is used to enhance sideband asymmetry, a first proof-of-principle demonstration of the ability to control quantum correlations using feedback. In the regime where amplified vacuum noise dominates the feedback signal, the disappearance of sideband asymmetry visualizes a fundamental limit of linear feedback control. Using a homodyne detector, we also characterize these quantum correlations manifested as optical squeezing at the 1% level.

Keywords Quantum Measurement · Cavity Optomechanics · Quantum Feedback Quantum Correlations

Acknowledgements

It seems to me I am trying to tell you a dream—making a vain attempt, because no relation of a dream can convey the dream-sensation, that commingling of absurdity, surprise, and bewilderment in a tremor of struggling revolt, that notion of being captured by the incredible which is of the very essence of dreams...

Joseph Conrad, Heart of Darkness

Experimental physics demands a certain degree of manual dexterity, the patience to tolerate the mundane, and an inexhaustible supply of optimism. Tobias hired me to work on an immensely sophisticated experiment despite any proof of my possessing these qualities; I am indebted to him for the belief he has placed in me. I would also like to acknowledge his dedication to fostering an environment where the pursuit of science is unencumbered by other worldly concerns. Stefan, Ewold, and Samuel bore the brunt of a theorist trying to perform delicate experiments; the stern, yet encouraging, stewardship of this trio ensured that I enjoyed the culture shock. Dal Wilson was more than a colleague and a post-doc; his pragmatic approach to physics provided the ideal counterpoint in our attempts to forge a deeper understanding of things ranging from vibration isolation to the subtleties of quantum mechanics. Without the patient efforts of Amir, Ryan, and Hendrik in designing, fabricating, and testing of devices, none of the results reported here would have been possible. I hope I have been able to bequeath a better vision of the future of our experiment to Sergey. Conversations with Alexey, Daniel, and Nathan have enabled me to vicariously learn how to measure microwave "photons". John Jost once taught me how to decorate a Christmas tree; since then he has imparted some wisdom on frequency metrology, and some on atomic physics. Together with Dal, Victor and Caroline have probably spent the most time with me outside the lab; it was fun to exercise an air of social normalcy with this bunch. Christophe has been an amazing sparring partner on every subject capable of being brought under scientific scrutiny.

The personal space required for the pursuit of science comes through the sacrifice of many people. My family back home in India—parents, sister, and grandparents—have had to patiently wait for the brief interludes when I go home. No words can do justice to this and the very many other pleasures they have surrendered over the years for my sake. Before physics attracted me, I was charmed by someone else; I am lucky to have married her. Longtime friends, mentors—Ajith and Subeesh—have played pivotal roles in my being able to engage in physics; I hope to repay this enormous debt, some day.

It was a privilege to have learnt from a few great teachers during my formative years. Their vision of an indelible unity in nature continues to motivate my study of physics.

Contents

Chapter 1
Prologue

The history of science shows that even during the phase of her progress in which she devotes herself to improving the accuracy of the numerical measurement of quantities with which she has long been familiar, she is preparing the materials for the subjugation of the new regions, which would have remained unknown if she had been contented with the rough methods of her early pioneers.

James Clerk Maxwell

Electromagnetic radiation holds a unique position in man's interaction with nature. In the form of light, acting as a conduit between far away objects and the human eye, it informs us through vision. In the form of electrostatic forces between atomic-scale bodies, it forms the basis of our tactile sense. Scientific instruments that peer deeper into space and sharper into the atom, interface our feeble senses with that world through electromagnetic signals.

That electromagnetic radiation can be a causative agency took much longer to be realized. The historic anecdote of Archimedes using mirrors to focus the sun's light onto incoming enemy ships to burn them (at the Siege of Syracuse, in 212 BC[1]), illustrates the powerful potential of intense radiation. However, radiation of such intensity was typically limited to astrophysical sources. Indeed, Kepler [1] suggested that the tails of comets point away from the sun because of an outward solar *radiation pressure*. Such a mechanical effect arising from light appears to have found favour with Newton [2]—corpuscles of light, reflecting off a surface, impart a recoil force. In the absence of a quantitative theory of light, these conjectures remained unsubstantiated.

[1]As reported by Anthemius of Tralles in his *On Burning Glasses*, ca. 700 years after the event.

V. Sudhir, *Quantum Limits on Measurement and Control of a Mechanical Oscillator*, Springer Theses, https://doi.org/10.1007/978-3-319-69431-3_1

The ensuing 100 years saw a host of experiments attempting to observe the tiny recoil force due to light in a terrestrial setting [3]. Some of these experiments set out to settle the debate between corpuscular and wave theories of light, for it was widely conjectured that if light were a wave, then it would not impart any mechanical force [4]. Bolstered by Cavendish's sensitive measurements of the gravitational force (another exceptionally weak effect) between two spherical bodies [5], the experimentalists converged upon the torsion pendulum as a sufficiently sensitive apparatus to see the pressure due to light. William Crookes pioneered this experimental technique, using a pair of vanes delicately suspended on a wire, forming a torsion pendulum that would potentially be set in motion when light impinged on one of the vanes. This instrument—Crookes radiometer [3, 6]—did respond to some photo-motive force—however the direction of motion was opposite to that expected if the force were radiation pressure. With the realization that the force at play was due to convection of the surrounding air heated by the absorbed light, a brief experimental hiatus ensued.

In the meantime, several other causative effects of the electromagnetic field were observed; perhaps most profound among them were Oersted's observation that a changing electric current causes magnetic effects, and Faraday's complementary observation that a moving magnet leads to an electric current. A priori, these electric and magnetic effects bore little relation to the mechanical radiation pressure effect that was being ardently pursued. In the same year (1873) that Crookes concluded his unsuccessful experiments to reveal the mechanical effect of light, Maxwell [7] produced his theoretical synthesis of Faraday's experiments, resulting in a unified description of electrical and magnetic phenomena.

Maxwell, in the second edition of his treatise, and independently Poynting [9] (who incidentally had surveyed the Cavendish experiment [10]), realized that the new electromagnetic theory applied to radiation, and that it provided a quantitative estimate of the pressure exerted by light.

The following decade witnessed the observation of the elusive radiation pressure. Lebedev [11], and independently, Nichols and Hull [8], performed a series of experiments in 1901–1910 that managed to isolate the effect of radiation pressure due to light from a carbon-arc source, and demonstrated that the force was indeed as predicted by Maxwell's theory. Figure 1.1 depicts the apparatus used in the experiment.

With the triumph of Maxwell's theory on all fronts (telegraphy, telephony and radio communication being everyday examples), radiation pressure studies on table-top experiments faded into history, partly owing to a lack of intense sources of radiation to amplify its effect. However, its implications for astrophysical phenomena, where such sources are aplenty, continued to be investigated [12–15] (including its non-negligible effect for artificial satellites and interplanetary missions [16]).

Historically, a circuitous route had to be traversed to return to the question of how one may realize an intense and highly-directional source of light. But this development had its roots in two pillars of physics erected by Maxwell. Maxwell, together with Boltzmann and others, had arrived at a microscopic description of particles (for example, gaseous atoms in a box) in thermal equilibrium. Suffice it to say that attempts to apply these ideas to the description of light in thermal equilibrium led

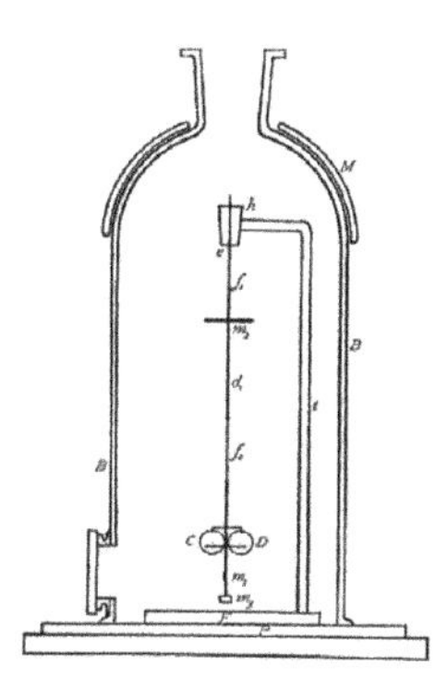

Fig. 1.1 **Nichols' radiometer.** [8] Two silvered mirrors (C and D), hang from a thin quartz wire. An approximately 150 mW light beam impinged upon one of the mirror imparting a force of 10^{-9} N. A much weaker light beam, reflected off the other mirror, formed an optical lever to measure the torque on the pendulum. The observations agreed to within 6% of the theoretical prediction. The glass bell jar was evacuated to eliminate the influence of thermal air currents that plagued Crookes' experiment [3]

to conceptual difficulties. Without going into the details [17], the resolution of these issues warranted a corpuscular description of light! Thus was born the *light quantum*, and with it, the theory of quantum mechanics [18]. Not only did quantum mechanics demand that light waves have a particulate character, but it also implied the reverse, namely, that what had been thought of as being intrinsically particulate (electrons, atoms etc.), have a complementary wave character. Another counter-intuitive prediction is the absence of a state of true rest—an incessant restless residual motion that cannot be quenched. These *vacuum fluctuations*, needless to say, are tiny compared to the typical size of objects—hence very difficult to observe. The qualitative idea of vacuum fluctuations is captured in a basic tenet of quantum theory—Heisenberg's uncertainty principle—roughly stating that both the position and velocity of objects cannot be simultaneously known. The present thesis studies an incarnation of this prediction, and experimentally observes it.

The weird predictions of quantum mechanics aren't confined to the realm of pure thought. The laser, invented in the 1960s, relies crucially on the principles of quantum theory. The invention of the laser, finally made it possible to access intense light fields in the laboratory. Primarily intended as a diagnostic tool to study the absorption and emission characteristics of atoms (i.e. spectroscopy [19, 20]), the laser soon became a tool to actively manipulate and control atomic-scale matter. The work of Ashkin and colleagues in the 1970s [21, 22] showed that the radiation pressure from a focused laser beam could be used to trap and move small (wavelength-scale) electrically neutral particles.[2] Closely related techniques to cool and eventually stop atoms were proposed [25, 26], and demonstrated [27] in the same decade. Ultimately, spectacular progress along this path led to the use of lasers to trap and control individual atoms [28–31]. These experiments have managed to reveal the

[2] On the other hand, electrically charged particles (electrons and ions) were being manipulated using radio-frequency electromagnetic fields, which naturally forces a charged particle [23, 24].

vacuum fluctuations of photons [32, 33], individual atoms [34], and collections of ($\approx 10^4 - 10^6$) atoms [35, 36]. In other words, it is now known for a fact that the fantastic predictions of quantum theory are at least valid for atomic-scale objects.

The confluence of two parallel threads of scientific inquiry led to a renewed interest in radiation pressure forces in the mid-1990s. On the one hand, progress in atomic physics led to the question of whether it is possible to witness quantum mechanical effects on larger assemblies of atoms—macroscopic objects—where the effect of quantum mechanical fluctuations would be even smaller. On the other hand, the question arose as to how measurements precise enough to see the already tiny effects could be devised. In some sense, these two quests are conceptually intertwined—a part of this thesis exposes this connection in an experiment.

1.1 Precise Position Measurements

Before the question of how a sufficiently precise measurement of the position of an object can be made, it is pertinent to address a more important question of principle. If the object is nominally "at rest", would a sufficiently precise measurement of its position reveal quantum fluctuations?

As it turns out there is a purely classical (i.e. non-quantum-mechanical) effect that defies the notion of "at rest" for most objects. Most ubiquitously, objects have a temperature, which means that the atoms that constitute them are in an agitated state—temperature being a measure of this agitation. This random movement of objects, called *thermal motion*, is small for large objects, however, typically much bigger than the quantum vacuum fluctuations. To give a sense of scale, the thermal motion of the mirror hanging off the torsion pendulum used by Nichols (Fig. 1.1), weighing a few grams, is of the order of 10^{-9} m, while its vacuum fluctuations are of the order of 10^{-16} m. In comparison, the motion induced by radiation pressure that Nichols and Hull managed to measure was of the order of 10^{-4} m; about 5 order of magnitude lacking in sensitivity to observe the thermal motion, and 8 orders of magnitude away from observing vacuum fluctuations.

Thermal motion was however observed in the early 1800s, and played a key role in the development of the atomic hypothesis. Botanist Robert Brown was famously puzzled by the random motion of pollen grains in water which he observed using a microscope.[3] The origin and nature of the apparent spontaneous motion soon became a fountain-head of scientific speculation that continued through the century.

Einstein, in 1905 (his annus mirabilis), conjectured that the *Brownian motion* (the thermal motion observed by Brown) was essentially due to the atomic constituents of water randomly hitting the pollen grain [37]—a bold prediction that, if confirmed, could provide proof for Dalton's theory that all matter is made of atoms. In fact, Maxwell and Boltzmann had used Dalton's suggestion as a metaphor to construct

[3]Remarkably, the Roman poet-philosopher Titus Lucretius, in his poem *De Rerum Natura* (ca. 60 BC), described the spontaneous motion of dust particles suspended in a sun beam falling across a dark room, and conjectured the presence of an invisible agency responsible for the movement.

the kinetic theory of gases, wherein the properties of gases could be derived from simple assumptions regarding a hypothetical atomic constitution. Einstein's theory was tantamount to the statement that if one placed a large object in the gas, the atoms that made up the gas, would knock the object; in fact, the more frequent the knocks, if the pressure or temperature of the gas were higher. Experimental verification of these predictions was immediate—Jean Perrin not only verified the theory, but also managed to extract the Avogadro number from his measurements [38]—putting the atomic hypothesis on firm experimental foundation. This idea, of measuring what is essentially noise, to discern something useful, continues to be a strong tradition in physics, this thesis being no exception.

Two questions immediately arise: firstly, if the temperature is zero, would there be no Brownian motion? If the object were in contact with no "gas" (i.e. if the pressure were zero), would the motion subside? The answers to these questions bring us back to quantum mechanics. When the temperature is zero, indeed there is no Brownian motion—however, vacuum fluctuations remain. When the object is perfectly isolated, and one tries to verify this fact by performing a measurement, the rules of quantum mechanics dictate that the act of measurement disturbs the object so as to impart an additional motion precisely equal to its vacuum fluctuation, resulting in two units worth of vacuum fluctuations in the observation. Part of this thesis investigates the latter phenomena.

The answers to the above questions also indicate what is necessary in order to observe vacuum fluctuations: a near-zero-temperature environment, a supremely well-isolated system, and a measuring device that is so exquisite that it is only limited by the laws of quantum mechanics.

In the early 1920s, a decade after Perrin's conclusive measurements of the Brownian motion of microscopic particulate matter suspended in liquids, the first observation of similar thermal motion of a macroscopic object were made. Willem Einthoven, in his Nobel prize (in medicine, for the invention of the electrocardiogram) lecture mentions the curious movement of the extremely light galvanometer needle used in his apparatus, conjecturing it to be the thermal Brownian motion of the needle. In subsequent investigations he found qualitative agreement between the motion and the predictions of the by-then more mature theory of thermal motion [39]. These findings were quickly replicated in a series of experiments by Moll and Burger [40] on a galvanometer suspended on a spring—drastically different from the free movement allowed for the particles in Perrin's experiments and the free needle in Einthoven's galvanometer. The effect of the spring is to render the needle a harmonic oscillator, that responds to a limited range of frequencies with a large amplitude; consequently, these experiments witness the thermal motion of the needle with unprecedented signal-to-noise. The theory of the thermal motion of a harmonic oscillator is quickly furnished by Ising [41] and Ornstein [42]. In fact, Ornstein's work suggests a cause for the thermal motion—the thermal motion of electrons in the galvanometer circuit play the role of gas particles kicking the needle. This pre-empts similar conclusions by Johnson and Nyquist. Throughout the mid-20th century, incarnations of the galvanometer needle were measured with ever increas-

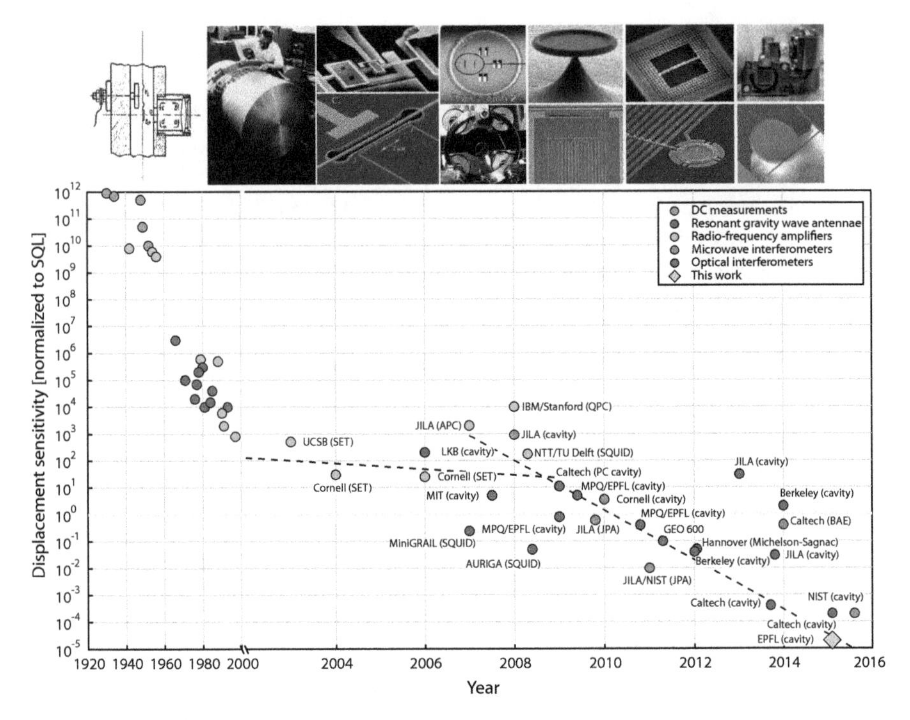

Fig. 1.2 Displacement sensitivity over the last century. Plot shows the improvement over 17 orders of magnitude improvement in the sensitivity in the measurement of position fluctuations over the past 100 years. The past decade has seen spectacular progress owing to the use of low-noise electromagnetic cavities that lead to a build-up of intense fields, interrogating low-mass high-quality-factor nano-mechanical oscillators. Here, the sensitivity is expressed in a natural unit, the sensitivity at the so-called *standard quantum limit* (SQL), which essentially amounts to the sensitivity required to resolve vacuum fluctuations. Work reported in this thesis is the orange diamond on the lower right corne—the most precise measurement of the position of a mechanical oscillator relative to its zero-point motion, at the time of writing

ing precision [43, 44], largely without paying heed to the fundamental limitations of the measuring device(s) being employed.

These early observations (see Fig. 1.2) were limited by the large thermal noise inherent in the measuring device itself—mainly in the form of thermal motion of electrons in electronic circuits [45, 46]. All this changes in the 1960s, with the invention of two low-noise sensing platforms—the laser, and the SQUID (superconducting quantum interference device) [47]. In the decades leading up to the 2000s, both these devices, essentially based on interferometric techniques, pushed the frontier of precision displacement measurements.

The ability to fabricate low-mass (hence large vacuum fluctuations) mechanical oscillators of small dimensions, i.e. nano-mechanical oscillators, featuring very high mechanical quality, brought the dream of witnessing the vacuum fluctuations of a macroscopic object ever closer. Integrating such oscillators with sensitive radio-frequency amplifiers, like single-electron transistors [48–50], quantum-point contacts [51, 52], or SQUIDs [53], proved to be fruitful in terms of approaching the sensitivity required to see vacuum fluctuations (i.e. the *standard quantum limit* (SQL)). These efforts were curtailed by the lack of a strong enough coupling between the oscillator and the sensor, or the presence of excess noise in the sensor, or both.

Ultimately, the integration of radio-frequency nano-mechanical oscillators, with high quality optical micro-cavity based interferometer [54], or microwave cavity with a Josephson parametric amplifier [55], proved to be successful in achieving the long-standing goal of measurements of position with the sensitivity at the SQL. The former was demonstrated in the group of Prof. Tobias Kippenberg, few years prior to the commencement of this thesis in the same laboratory.

Both these approaches typify the burgeoning field of *cavity optomechanics* [56], wherein the emphasis has been to tightly integrate high quality mechanical oscillators with a high quality optical (or microwave) cavity. Figure 1.3 shows the principle of cavity optomechanical measurement of the position of mechanical oscillators. In the generic scheme, shown in Fig. 1.3a, light is injected into a space formed by two mirrors facing each other. The multiple reflections from either mirror traps the light in the enclosed space, forming an optical cavity. One of the mirrors, mounted on a spring, forms the mechanical oscillator. Its position $x(t)$ is recorded as a phase shift proportional to $\frac{x(t)}{\lambda}$, where λ is the wavelength of light used. The cavity amplifies this tiny phase shift by recycling the light a large number of times, given by the *finesse* $\mathcal{F}$. Light leaking out of the cavity thus features a phase change given by $\phi(t) \propto \frac{\mathcal{F}}{\lambda} x(t)$. Therefore, a good optical cavity, having a small operating wavelength λ, and a large finesse $\mathcal{F}$, can transduce the small motion $x(t)$ into a larger phase shift $\phi(t)$. A type of optical cavity (see Fig. 1.3b) formed by continually bending light around a curved path—a *whispering-gallery* mode cavity [57]—is used in this thesis. Such cavities, operating at a wavelength of $\lambda \approx 8 \cdot 10^{-7}$ m, can have $\mathcal{F} \approx 10^6$, meaning that a typical vacuum fluctuation amplitude (of the string shown in Fig. 1.3c) $x \approx 10^{-15}$ m, would lead to a phase shift of $\phi \approx 0.1$ rad for the light emanating from the cavity. Although small, this phase shift can be measured precisely by comparing

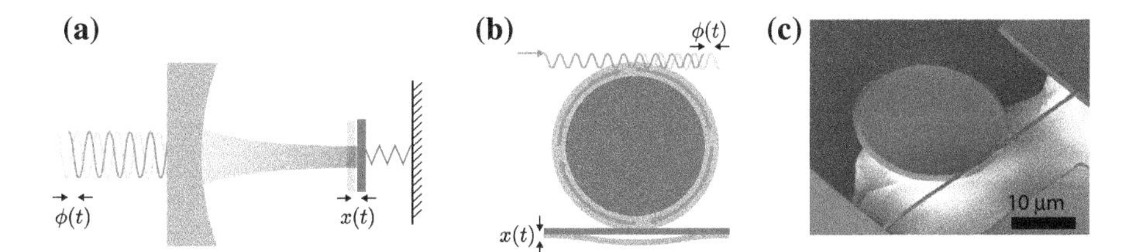

Fig. 1.3 Principle of cavity optomechanics. a Generic cavity optomechanical system, consisting of two mirrors facing each other. The mirrors form an optical cavity, trapping light between them (i.e. a Fabry-Perot cavity); one of the mirrors, mounted on a spring, forms the mechanical oscillator. **b** A cavity optomechanical system where the mechanical oscillator is a string, and the optical cavity is formed by continually bending light around a circle (i.e. a whispering gallery mode cavity). **c** False-coloured scanning electron micrograph of the system studied in this thesis; scale bar shows reference for the size of the device

the cavity transmission with a reference light beam, in a technique called *homodyne interferometry*.

However, a fundamental impediment presents itself in trying to make this measurement. Since light is itself a quantum mechanical entity, the ability to measure its phase is impeded by the particle nature of photons: the random arrival times of the photons constitute an uncertainty in the timing information in the measurement, essentially scrambling the phase. This uncertainty can be minimized by employing a larger flux of photons, so that (in a naive sense) the inter-arrival intervals of the photon stream is reduced. The price paid for this choice is that the larger photon flux exerts a larger radiation pressure force on the oscillator being measured—in trying to make a precise measurement, the thing being measured gets disturbed—a manifestation of Heisenberg's uncertainty principle.

The disturbance, called *measurement back-action*, plays a prominent role as the precision of the measurement approaches what is required to resolve the vacuum fluctuation of the oscillator—the SQL. The ability to measure position with a sensitivity at the SQL is sufficient to see the vacuum fluctuations of the oscillator if it were free of thermal motion. For typical macroscopic mechanical oscillators, this is never true, unless the environment is maintained at a temperature of 10^{-4} K—an enormous technical challenge. Nevertheless, the preparation and stabilization of the vacuum state of the oscillator is crucial.

1.2 Outline of this Thesis

In the work reported in this thesis (in Chap. 6), the problem of stabilizing the vacuum state of the oscillator is solved by turning it on its head. Firstly, a measurement of the position fluctuations of an oscillator is made, which is sufficiently sensitive to resolve the vacuum fluctuations of the oscillator before it is overwhelmed by the environment. For the system employed, this corresponds to operating a factor 10^4 below the SQL. The measurement back-action associated with such a measurement is observed in our experiment. As the next step, this information is used to actively stabilize the vacuum state of the oscillator using feedback [58]. At the time of writing, this remains the sole example of active feedback of a mechanical oscillator to near its ground state.

In a second experiment (in Chap. 7), the capability to feedback control the oscillator near its ground state is employed to study the subtle nature of the act of measurement itself [59]. In particular, information regarding the "system" (the mechanical oscillator) is transferred to the "meter" (light) via correlations created between either system. These tiny correlations—whose magnitude is comparable to the vacuum fluctuation of the oscillator—are measured using the feedback technique. Feedback is used to suppress the back-action due to the measurement, so that the underlying correlations can be revealed.

Naively, it appears that feedback control can be used to completely circumvent measurement back-action; however, we show that this is not the case. In order to understand why this is, one needs to realize that the state of the "meter" (light) is not directly accessible—a final measurement step, involving a *classical* "detector" (a photo-absorptive detector) is necessary. The output of such a detector invariably contains traces of the vacuum fluctuations of the meter. In our experiment, when this output is used to perform feedback control, the vacuum fluctuations of the meter therefore act back on the system, leading to what we call *feedback back-action*. Therefore, although feedback suppresses measurement back-action, feedback back-action cannot be suppressed. We experimentally observe this to be the case, and interpret the result as a fundamental limit of measurement-based feedback control.

1.2.1 Organization of Thesis

The experimental results reported in this thesis—in Chaps. 6 and 7—rely on a wide range of theoretical and experimental techniques. Chapters 2 to 5 provide a review of these preliminaries. The excursion begins in Chap. 2 with a concise (and therefore necessarily abstract) description of linear quantum systems, including continuous linear measurements. The theoretical formalism and vernacular developed therein, forms the conceptual backbone of much of the thesis. Chapter 3 delves into a description of the actors that take part in the experiment—phonons and photons. Following a brief description of low-energy phonons and their quantisation in Sects. 3.1 and 3.2 gives an account of the electromagnetic field. In particular, Sect. 3.2 develops a formalism to describe the fluctuations of the electromagnetic field, and their measurements. Chapter 4 details how photons trapped in a cavity couple to phonons describing the effective elastic deformations of the cavity; a simplified (single-mode) description of this phenomena gives the standard theory of cavity optomechanics [56]. This chapter then describes the phenomenology of cavity optomechanics and how such a coupling can be used to realize a continuous linear measurement of the position of an oscillator. Chapter 5 details the concrete realisation in which phonons and photons interact linearly—a near-field cavity optomechanics system deployed in a ^{3}He cryostat. Chapters 6 and 7 present the central experimental results of this thesis.

References

1. J. Kepler, De cometis libelli tres (1619)
2. I. Newton, *Opera quae Exstant Omnia*, vol. III (Londinium, 1782)
3. W. Crookes, Phil. Trans. Roy. Soc. **164**, 501 (1874)
4. J.H. Poynting, Phil. Mag. **9**, 393 (1905)
5. H. Cavendish, Phil. Trans. Roy. Soc. **88**, 469 (1798)
6. A.E. Woodruff, Isis **57**, 188 (1966)
7. J.C. Maxwell, *A Treatise on Electricity and Magnetism* (Clarendon Press, UK, 1873)
8. E.F. Nichols, G.F. Hull, Phys. Rev. (Series I) **17**, 26 (1903)

9. J.H. Poynting, Phil. Trans. Roy. Soc. **175**, 343 (1884)
10. J.H. Poynting, *The Mean Density of the Earth* (C. Griffin & Co., London, 1894)
11. P. Lebedew, Ann. der Phys. **311**, 433 (1901)
12. J.H. Poynting, Phil. Trans. Roy. Soc. A **202**, 525 (1904)
13. M.N. Saha, Astrophys. J. **50**, 220 (1919)
14. H.P. Robertson, Month. Not. R. Astron. Soc. **97**, 423 (1937)
15. J.A. Burns, P.L. Lamy, S. Soter, Icarus **40**, 1 (1979)
16. P. Musen, J. Geophys. Res **65**, 1391 (1960)
17. M. Jammer, *The Conceptual Development of Quantum Mechanics* (McGraw Hill, New York, 1966)
18. P.A.M. Dirac, *The Principles of Quantum Mechanics*, 4th edn. (Clarendon Press, Oxford, 1982)
19. J.L. Hall, Science **202**, 147 (1978)
20. T.W. Hänsch, H. Walther, Rev. Mod. Phys **71**, S242 (1999)
21. A. Ashkin, Phys. Rev. Lett **24**, 156 (1970)
22. A. Ashkin, Science **210**, 1081 (1980)
23. W. Paul, Rev. Mod. Phys **62**, 531 (1990)
24. H. Dehmelt, Science **247**, 539 (1990)
25. D.J. Wineland, H. Dehmelt, Bull. Amer. Phys. Soc. **20**, 637 (1975)
26. T.W. Hänsch, A.L. Schawlow, Opt. Commun. **13**, 68 (1975)
27. D.J. Wineland, R.E. Drullinger, F.L. Walls, Phys. Rev. Lett **40**, 1639 (1978)
28. S. Chu, Rev. Mod. Phys. **70**, 685 (1998)
29. C. Cohen-Tannoudji, Rev. Mod. Phys. **70**, 707 (1998)
30. C.E. Wieman, D.E. Pritchard, D.J. Wineland, Rev. Mod. Phys. **71**, S253 (1999)
31. D. Leibfried, R. Blatt, C. Monroe, D.J. Wineland, Rev. Mod. Phys. **75**, 281 (2003)
32. M. Brune, F. Schmidt-Kaler, A. Maali, J. Dreyer, E. Hagley, J.M. Raimond, S. Haroche, Phys. Rev. Lett. **76**, 1800 (1996)
33. S. Haroche, J.-M. Raimond, *Exploring the Quantum: Atoms, Cavities and Photons* (Oxford University Press, Oxford, 2006)
34. F. Diedrich, J.C. Bergquist, W.M. Itano, D.J. Wineland, Phys. Rev. Lett. **62**, 403 (1989)
35. W. Ketterle, Rev. Mod. Phys. **74**, 1131 (2002)
36. E.A. Cornell, C.E. Wieman, Rev. Mod. Phys. **74**, 875 (2002)
37. A. Einstein, *Investigations on the Theory of the Brownian Movement* (Dover, NY, USA, 1926)
38. J. Perrin, *Brownian Movement and Molecular Reality* (Taylor & Francis, London, 1910) (translated by F. Soddy of the original in Ann. Chimi. Phys. **18**, 5 (1909))
39. W. Einthoven, F.W. Einthoven, W. van der Holst, H. Hirschfeld, Physica **5**, 358 (1925)
40. W.J.H. Moll, H.C. Burger, Phil. Mag. **50**, 626 (1925)
41. G. Ising, Phil. Mag. **1**, 827 (1926)
42. L.S. Ornstein, H.C. Burger, J. Taylor, W. Clarkson, Proc. Roy. Soc. A **115**, 391 (1927)
43. R.V. Jones, C.W. McCombie, Phil. Trans. Roy. Soc. A **244**, 205 (1952)
44. J.M.W. Milatz, J.J. Van Zolingen, Physica **19**, 181 (1953)
45. R.B. Barnes, S. Silverman, Rev. Mod. Phys. **6**, 162 (1934)
46. C.W. McCombie, Rep. Prog. Phys. **16**, 266 (1953)
47. J. Clarke, Proc. IEEE **77**, 1208 (1989)
48. R.G. Knobel, A.N. Cleland, Nature **424**, 291 (2003)
49. M.D. LaHaye, O. Buu, B. Camarota, K.C. Schwab, Science **304**, 74 (2004)
50. A. Naik, O. Buu, M.D. Lahaye, A.D. Armour, A.A. Clerk, M.P. Blencowe, K.C. Schwab, Nature **443**, 193 (2006)
51. N.E. Flowers-Jacobs, D.R. Schmidt, K.W. Lehnert, Phys. Rev. Lett. **98**, 096804 (2007)
52. M. Poggio, M.P. Jura, C.L. Degen, M.A. Topinka, H.J. Mamin, D. Goldhaber-Gordon, D. Rugar, Nat. Phys. **4**, 635 (2008)
53. S. Etaki, M. Poot, I. Mahboob, K. Onomitsu, H. Yamaguchi, H.S.J. van der Zant, Nat. Phys. **4**, 785 (2008)
54. G. Anetsberger, O. Arcizet, Q.P. Unterreithmeier, R. Rivière, A. Schliesser, E.M. Weig, J.P. Kotthaus, T.J. Kippenberg, Nat. Phys. **5**, 909 (2009)

55. J.D. Teufel, T. Donner, M.A. Castellanos-Beltran, J.W. Harlow, K.W. Lehnert, Nat. Nanotechnol. **4**, 820 (2009)
56. M. Aspelmeyer, T.J. Kippenberg, F. Marquardt, Rev. Mod. Phys. **86**, 1391 (2014)
57. K.J. Vahala, Nature **424**, 839 (2003)
58. D.J. Wilson, V. Sudhir, N. Piro, R. Schilling, A. Ghadimi, T.J. Kippenberg, Nature **524**, 325 (2015)
59. V. Sudhir, D.J. Wilson, R. Schilling, H. Schütz, A. Ghadimi, A. Nunnenkamp, T.J. Kippenberg, Phys. Rev. X **7**, 011001 (2017)

Chapter 2
Quantum Fluctuations in Linear Systems

Here are some words which have no place in a formulation with any pretension to physical precision: system, apparatus, environment, microscopic, macroscopic, reversible, irreversible, observable, information, measurement.

John Bell

Despite John Bell's eloquent tirade against the arbitrary division of the universe into a *system* surrounded by an *environment*, the experimental physicist, due to his limited means of enquiry, is forced to subscribe to a well-defined notion of what is considered the *system* under study. In the example relevant to this thesis, it is a macroscopic mechanical oscillator. Everything beyond, is the *environment*. In this sense, the measuring device—the *meter*—is a form of environment, the crucial difference being that the experimenter has the ability to prepare it in well-characterised (quantum) states. In this thesis, the meter is an electromagnetic field that interacts with the oscillator.

The purpose of this chapter is to provide a reasonably self-contained presentation of a few basic results pertaining to a certain class of interactions—linear interactions—between a system and its environment. When the environment in question is a thermal bath at finite temperature, classical (thermal) fluctuations drive the system; at zero temperature, quantum (vacuum) fluctuations remain. When the environment is a meter, fluctuations from the meter excite the system, an effect called *measurement back-action*; when the meter is prepared in a pure quantum state—measurement back-action is due to quantum fluctuations in its degrees of freedom, i.e. *quantum back-action*. Ultimately, for an ideal measurement chain—a system in contact with a zero-temperature thermal environment measured by a meter limited by quantum fluctuations—the output of the meter will feature an additional source of

V. Sudhir, *Quantum Limits on Measurement and Control of a Mechanical Oscillator*, Springer Theses, https://doi.org/10.1007/978-3-319-69431-3_2

quantum fluctuations, called *measurement imprecision*, which, together with quantum back-action, constraints the precision with which the system can be measured. The rest of the chapter systematically unravels this tale.

2.1 Kinematics of Fluctuations in Quantum Mechanics

In the following we adopt the standard mathematical formalism of quantum mechanics [1, 2]: to every system (not necessarily *the* system) is associated a Hilbert space[1]; the states of the system are the positive, unit-trace operators. The relation between this abstract construct and the outcomes of experiments is through a set of distinguished operators called *observables*, defined as follows.

Definition 2.1 (*Observable*) The observables are the self-adjoint operators in the Hilbert space of the system.

If the system is repeatedly prepared in a definite state, say $\hat{\rho}$, and one of the observables, say $\hat{X}$, is measured per preparation, the outcomes will be random real numbers drawn from the eigenspectrum of the observable (self-adjointedness guarantees that the eigenspectrum is real [1, 2, 5]). This random variable is drawn according to a probability distribution.[2] The fluctuations of the random variable can be associated with the operator,

$$\delta\hat{X} := \hat{X} - \left\langle \hat{X} \right\rangle, \quad \text{where,} \quad \left\langle \hat{X} \right\rangle = \mathrm{Tr}[\hat{X}\hat{\rho}].$$

In a large variety of cases, the statistical dispersion in the random variable may be quantified by the variance,

$$\mathrm{Var}\left[\hat{X}\right] := \left\langle \delta\hat{X}^{\dagger}\delta\hat{X} \right\rangle = \left\langle \delta\hat{X}^{2} \right\rangle. \tag{2.1.1}$$

In light of the following fact, the variance is positive.

Lemma 2.1 (Operator positivity) *For any operator* $\hat{A}$*, not necessarily self-adjoint, it is true that* $\left\langle \hat{A}^{\dagger}\hat{A} \right\rangle \geq 0$*; i.e.* $\hat{A}^{\dagger}\hat{A}$ *is a positive operator.*

Proof Consider that the state $\hat{\rho}$ over which the expectation is taken, is represented as,

---

[1] It turns out that, on technical grounds, the framework of Hilbert space is too restrictive to realize the flexibility of Dirac's formulation of quantum mechanics [3, 4]; here, we will however be satisfied with using the Dirac formalism rather than justifying each step of the usage rigorously.

[2] A peculiarity of quantum mechanics is that although the value taken by each observable, for a fixed state, can be assumed to be drawn from a classical probability distribution (exhibited in Appendix A), there is generally no joint probability distribution for the values of a set of operators [6].

$$\hat{\rho} = \sum_i p_i |\psi_i\rangle\langle\psi_i|,$$

where, $p_i \geq 0$, and, $\langle\psi_i|\psi_j\rangle = \delta_{ij}$; such a representation is always possible [2]. Evaluating the expectation value gives,

$$\left\langle \hat{A}^\dagger \hat{A} \right\rangle = \mathrm{Tr}[\hat{A}^\dagger \hat{A}\hat{\rho}] = \mathrm{Tr}\left[\hat{A}^\dagger \hat{A} \sum_i p_i |\psi_i\rangle\langle\psi_i|\right] = \sum_i p_i \langle\psi_i|\hat{A}^\dagger \hat{A}|\psi_i\rangle = \sum_i p_i \| \hat{A}|\psi_i\rangle \|^2,$$

which is always positive by the property of norms on Hilbert space. □

The mathematical structure of quantum mechanics dictates that the variances of a pair of observables, say $\hat{X}, \hat{Y}$, satisfies the inequality (see Appendix A for a proof and further discussion),

$$\mathrm{Var}\left[\hat{X}\right]\mathrm{Var}\left[\hat{Y}\right] \geq \frac{1}{4}\left|\left\langle\left\{\delta\hat{X}, \delta\hat{Y}\right\}\right\rangle\right|^2 + \frac{1}{4}\left|\left\langle\left[\delta\hat{X}, \delta\hat{Y}\right]\right\rangle\right|^2 \geq \frac{1}{4}\left|\left\langle\left[\delta\hat{X}, \delta\hat{Y}\right]\right\rangle\right|^2, \tag{2.1.2}$$

conventionally called the *uncertainty principle*. The first inequality (due to Robertson [7] and Schrödinger [8] for pure states) is saturated for pure states defined as eigenstates of the operator $\alpha_X \delta\hat{X} + \alpha_Y \delta\hat{Y}$, with the constants $\alpha_{X,Y}$ chosen to maximize the correlation term $\langle\{\delta\hat{X}, \delta\hat{Y}\}\rangle$ [9]. In contrast, the second (looser) inequality (due to Heisenberg [10], Kennard [11], and Weyl [12]), obtained by omitting the correlation term, is saturated by the same eigenstates for any value of the constants $\alpha_{X,Y}$.

The physical content of the uncertainty principle (Eq. 2.1.2) is that the measurement outcomes of the pair of observables $\hat{X}, \hat{Y}$, on *identical and independent* preparations of the system, have a fundamental statistical dispersion. It is thus a purely kinematic statement devoid of any a priori relevance to the notion of "simultaneous", or "sequential" measurements.[3] It is best interpreted to mean that there exists no state which is jointly dispersion-free for certain pairs of observables—a distinctly quantum mechanical feature [6, 17].

Before describing an approach to treating outcomes of continuous measurements, and in an act of foresight, we generalize the definition of the variance of an observable, given in Eq. (2.1.1), to the case of a general operator. Following [18], the variance of an operator $\hat{A}$, not necessarily self-adjoint, is defined by,

$$\mathrm{Var}\left[\hat{A}\right] := \frac{1}{2}\left\langle\left\{\delta\hat{A}^\dagger, \delta\hat{A}\right\}\right\rangle. \tag{2.1.3}$$

When $\hat{A}$ is self-adjoint, this reduces to the standard definition in Eq. (2.1.1); but when it isn't, Lemma 2.1 still ensures that,

---

[3] Attempts to formulate inequalities applicable to *sequential* measurements [13–16] give results very different from Eq. (2.1.2).

$$\text{Var}\left[\hat{A}\right] \geq 0.$$

Since any non-self-adjoint operator has a Cartesian decomposition in terms of two self-adjoint operators, i.e. $\hat{A} = \hat{X} + i\hat{Y}$, for $\hat{X}, \hat{Y}$ self-adjoint, the uncertainty inequality in Eq. (2.1.2) satisfied by the Cartesian components implies a bound for the variance of the corresponding non-self-adjoint operator. The following lemma codifies the resulting inequality.

Lemma 2.2 *For any (not necessarily self-adjoint) operator $\hat{A}$, the following inequality holds [18],*

$$\text{Var}[\hat{A}] \geq \frac{1}{2} \left|\left\langle\left[\hat{A}^{\dagger}, \hat{A}\right]\right\rangle\right|. \tag{2.1.4}$$

Proof Denoting the Cartesian decomposition, $\hat{A} = \hat{X} + i\hat{Y}$, direct computation shows that the variance, defined by Eq. (2.1.3), takes the form,

$$\text{Var}\left[\hat{A}\right] = \text{Var}\left[\hat{X}\right] + \text{Var}\left[\hat{Y}\right].$$

The sum on the right-hand side can be bounded by the arithmetic-geometric mean inequality,[4] and subsequently the Heisenberg form of the inequality in Eq. (2.1.2), leading to,

$$\text{Var}\left[\hat{A}\right] \geq 2\sqrt{\text{Var}\left[\hat{X}\right]\text{Var}\left[\hat{X}\right]} \geq \left|\left\langle\left[\delta\hat{X}, \delta\hat{Y}\right]\right\rangle\right| = \left|\left\langle\frac{1}{2i}\left[\delta\hat{A}^{\dagger}, \delta\hat{A}\right]\right\rangle\right|.$$

□

2.1.1 Operational Description of Fluctuations in Time

In order to treat system observables varying in time, the Heisenberg picture is most convenient: the system is in some time-independent state $\hat{\rho}$, while its observables undergo fluctuations due to the pervasive *environment* that the system is in contact with. These fluctuations are reflected in the observables as deviations from their mean values, viz.

$$\delta\hat{X}(t) = \hat{X}(t) - \text{Tr}\left[\hat{\rho}\,\hat{X}(t)\right].$$

The fluctuating part, $\delta\hat{X}(t)$, represents a continuous random variable—stochastic process—taking values in the set of observables.

[4] For positive real numbers x, y, it is true that $x + y \geq 2\sqrt{xy}$; this follows from the identity, $(\sqrt{x} - \sqrt{y})^2 \geq 0$.

In the following, we employ an operational description of the statistical properties of the operator-valued stochastic process.[5] In order to resolve the variance of the process over the different time scales over which the fluctuations happen, we consider the windowed Fourier transform,[6]

$$\delta\hat{X}^{(T)}[\Omega] := \frac{1}{\sqrt{T}}\int_{-T/2}^{T/2} \delta\hat{X}(t)e^{i\Omega t}\,\mathrm{d}t, \tag{2.1.5}$$

which is in general non-hermitian. The definition of the variance of a non-hermitian operator in Eq. (2.1.3) then implies,

$$\begin{aligned}\mathrm{Var}\left[\delta\hat{X}^{(T)}[\Omega]\right] &= \frac{1}{2}\left\langle\left\{\delta\hat{X}^{(T)}[\Omega]^\dagger, \delta\hat{X}^{(T)}[\Omega]\right\}\right\rangle \\ &= \frac{1}{T}\int_{-T/2}^{T/2}\frac{1}{2}\left\langle\left\{\delta\hat{X}(t), \delta\hat{X}_i(t')\right\}\right\rangle e^{i\Omega(t-t')}\,\mathrm{d}t\,\mathrm{d}t' \\ &= \frac{1}{T}\int_{-T/2}^{T/2}\frac{1}{2}\left\langle\left\{\delta\hat{X}(t-t'), \delta\hat{X}(0)\right\}\right\rangle e^{i\Omega(t-t')}\,\mathrm{d}t\,\mathrm{d}t' \\ &= \int_{-T/2}^{T/2}\frac{1}{2}\left\langle\left\{\delta\hat{X}(\tau), \delta\hat{X}(0)\right\}\right\rangle e^{i\Omega\tau}\left(1-\frac{|\tau|}{T}\right)\mathrm{d}\tau.\end{aligned}$$

Note that here and henceforth we assume processes are weak-stationary, i.e. that their first and second moments are time-translation invariant. In the limit $T \to \infty$ (i.e. the limit of infinite resolution in frequency), this variance defines the function,

$$\bar{S}_{XX}[\Omega] := \lim_{T\to\infty}\mathrm{Var}\left[\delta\hat{X}^{(T)}[\Omega]\right] = \int_{-\infty}^{\infty}\left\langle\frac{1}{2}\left\{\delta\hat{X}(t), \delta\hat{X}(0)\right\}\right\rangle e^{i\Omega t}\,\mathrm{d}t, \tag{2.1.6}$$

characterising the distribution of the variance of the process about each frequency. The second equality, giving the value of the limit, is the analogue of the Wiener-Khinchine theorem [25, 26].

---

[5]There exists two, progressively finer, levels of description of the evolution of a quantum system in contact with an environment. The coarse description concerns itself with the time evolution of observables, and some of its statistics. The finer description addresses the question of how the quantum state itself changes. The former is subsumed by the latter in a variety of equivalent ways [19–23].

[6]The normalisation warrants clarification: if the integrand were a classical Brownian process, its root-mean-square diverges as the square root of the observation window, i.e. as $T^{1/2}$, which is checked by the normalisation. For a wide class of classical stochastic processes, a theorem due to Donsker [24] guarantees that the integral limits to a Brownian process (a "functional central limit theorem")—the $T^{-1/2}$ normalisation is necessary. This result from classical probability theory suffices to justify the normalisation.

The function is reminiscent of the classical notion of a power spectral density. Firstly, being a variance, $\bar{S}_{XX}[\Omega] \geq 0$, at all frequencies and for any operator-valued process. Secondly, being a distribution (obtained by applying the Fourier inversion theorem [25] to Eq. (2.1.6)),

$$\mathrm{Var}\left[\delta\hat{X}(t)\right] = \left\langle \delta\hat{X}(0)^2 \right\rangle = \int_{-\infty}^{\infty} \bar{S}_{XX}[\Omega] \frac{\mathrm{d}\Omega}{2\pi}, \tag{2.1.7}$$

which exhibits the complementary aspect that the integral of the power spectral density is the variance of the process $\delta\hat{X}(t)$. Equations (2.1.6) and (2.1.7) are fundamental properties of the *symmetrised spectrum*[7] so defined, that render it useful (irrespective of whether it is generically measured in an experiment [26–28]).

2.1.2 Spectral Densities and Uncertainty Relations

A formal hierarchy of spectral distributions generalise the above concept of the symmetrized spectrum of an observable. For a general (i.e. not necessarily hermitian) operator $\hat{A}$, define its Fourier transform,

$$\hat{A}[\Omega] = \int_{-\infty}^{+\infty} \hat{A}(t) e^{i\Omega t}\, \mathrm{d}t, \tag{2.1.8}$$

and its inverse,

$$\hat{A}(t) = \int_{-\infty}^{+\infty} \hat{A}[\Omega] e^{-i\Omega t} \frac{\mathrm{d}\Omega}{2\pi}. \tag{2.1.9}$$

We shall denote by, $\hat{A}^\dagger[\Omega]$, the Fourier transform of $\hat{A}^\dagger(t)$; and by, $\hat{A}[\Omega]^\dagger$, the hermitian conjugate of $\hat{A}[\Omega]$. With this convention, $\hat{A}^\dagger[\Omega] = \hat{A}[-\Omega]^\dagger$. For an observable, say $\hat{X}(t)$, it is further true that, $\hat{X}[\Omega]^\dagger = \hat{X}[-\Omega]$, and so $\hat{X}^\dagger[\Omega] = \hat{X}[\Omega]$.

The *unsymmetrized (cross-)spectrum* of two operators $\hat{A}, \hat{B}$ (not necessarily equal) is defined as the Fourier transform of their unsymmetrized two-time correlation function, i.e.,

$$S_{AB}[\Omega] := \int_{-\infty}^{+\infty} \left\langle \delta\hat{A}^\dagger(t)\delta\hat{B}(0) \right\rangle e^{i\Omega t}\, \mathrm{d}t = \int_{-\infty}^{+\infty} \left\langle \delta\hat{A}^\dagger[\Omega]\delta\hat{B}[\Omega'] \right\rangle \frac{\mathrm{d}\Omega'}{2\pi}, \tag{2.1.10}$$

which is in general a complex number at each Fourier frequency Ω; here, the second equality follows from replacing the operators with their Fourier transforms (i.e. Eq. (2.1.9)). When the operators involved are weak-stationary, i.e.,

$$\left\langle \hat{A}^\dagger(t)\hat{B}(t') \right\rangle = \left\langle \hat{A}^\dagger(t-t')\hat{B}(0) \right\rangle,$$

[7] Short for "symmetrised power spectral density", by abuse of terminology.

their unsymmetrized spectrum is directly related to their two-point correlation in the frequency domain; specifically,

$$S_{AB}[\Omega] \cdot 2\pi\,\delta[\Omega + \Omega'] = \left\langle \delta\hat{A}^{\dagger}[\Omega]\delta\hat{B}[\Omega'] \right\rangle$$

$$\text{i.e.,} \quad S_{AB}[\Omega] \cdot 2\pi\,\delta[0] = \left\langle \delta\hat{A}[-\Omega]^{\dagger}\delta\hat{B}[-\Omega] \right\rangle.$$

The last form makes explicit the symmetry,

$$S_{AB}[\Omega]^* = S_{BA}[\Omega]. \tag{2.1.11}$$

The other algebraic property that is practically useful is bilinearity, which can be expressed as follows: consider an operator which is a linear superposition of another pair, i.e. $\hat{A}[\Omega] = \alpha_1[\Omega]\hat{B}_1[\Omega] + \alpha_2[\Omega]\hat{B}_2[\Omega]$, then,

$$S_{AA}[\Omega] = \begin{cases} \alpha_1[-\Omega]S_{AB_1}[\Omega] + \alpha_2[-\Omega]S_{AB_2}[\Omega] \\ \alpha_1^*[-\Omega]S_{B_1A}[\Omega] + \alpha_2^*[-\Omega]S_{B_2A}[\Omega]. \end{cases} \tag{2.1.12}$$

These two properties allow for the practical computation of the spectra of operators defined as linear superpositions of other operators. Concretely, if a set of operators (arranged into a column vector) $\hat{\mathbf{A}} = [\hat{A}_i]^T$ are related to another set, $\hat{\mathbf{B}} = [\hat{B}_i]^T$, as,

$$\hat{A}_i[\Omega] = \sum_k \alpha_{ik}[\Omega]\hat{B}_k[\Omega],$$

$$\text{equivalently,} \quad \hat{\mathbf{A}}[\Omega] = \boldsymbol{\alpha}[\Omega]\hat{\mathbf{B}}[\Omega],$$

for some (matrix $\boldsymbol{\alpha}$ of) coefficients α_{ik}, then,

$$S_{A_iA_j}[\Omega] = \sum_{k,l} \alpha_{ik}^*[-\Omega]S_{B_kB_l}[\Omega]\alpha_{jl}[-\Omega]$$

$$\text{equivalently,} \quad S_{\mathbf{AA}}[\Omega] = \boldsymbol{\alpha}[-\Omega]^* S_{\mathbf{BB}}[\Omega]\boldsymbol{\alpha}[-\Omega]^T.$$

The second form expresses the first as a matrix equation, where $S_{\mathbf{AA}}$ denotes the matrix whose elements are $S_{A_iA_j}$, i.e. it is the unsymmetrised covariance matrix of $\hat{\mathbf{A}}$ in the frequency domain.

The physical motivation for the definition of the unsymmetrized spectrum becomes obvious when considering the properties of the spectrum of a single operator, viz.,

$$S_{AA}[\Omega] \cdot 2\pi\,\delta[0] = \left\langle \delta\hat{A}[-\Omega]^{\dagger}\delta\hat{A}[-\Omega] \right\rangle \geq 0; \tag{2.1.13}$$

specifically, S_{AA} is real, and positive. Mathematically, the positivity follows from lemma 2.1; its physical content is that S_{AA} can be interpreted as (being proportional to) the transition probability of a process mediated by an interaction that couples the system to its environment via the operator $\hat{A}[-\Omega]$ [29–31]. In classic examples in quantum optics [30, 31], the operator may be the destruction operator of a photon in which case the spectrum is the output spectrum of a photodetector [32], or, it may be the raising/lowering operator for an atomic level in which case the spectrum is the absorption/emission spectrum of that level [33].

Pairs of spectra of operators, in analogy with the variances of pairs of observables, satisfy an inequality reminiscent of the (Robertson-Schrodinger) uncertainty principle (Eq. (2.1.2)).

Proposition 2.1 (Spectral uncertainty relation I) *The spectra of any pair of operator-valued stochastic processes, $\hat{A}(t)$, $\hat{B}(t)$, that are weak-stationary, satisfies the inequality,*

$$S_{AA}[\Omega]S_{BB}[\Omega] - |S_{AB}[\Omega]|^2 \geq 0. \tag{2.1.14}$$

Proof A slick proof follows by a direct adaptation of the one used by Roberston to originally establish the uncertainty relation in Eq. (2.1.2) (see Appendix A), as done for example in [20]. For a pair of observables, a simpler method is as follows: define, $\hat{M}_\lambda(t) = \hat{A}(t) + \lambda \hat{B}(t)$, for some complex λ. From Eq. (2.1.13), it must be that, $S_{M_\lambda M_\lambda}[\Omega] \geq 0$ for all λ. Writing this out explicitly using the bilinearity (Eq. (2.1.12)) and symmetry (Eq. (2.1.11)):

$$S_{M_\lambda M_\lambda} = S_{AA} + |\lambda|^2 \, S_{BB} + 2\mathrm{Re}\; \lambda S_{AB} \geq 0.$$

This trivial inequality can be tightened by replacing $S_{M_\lambda M_\lambda}$ with $\min_\lambda S_{M_\lambda M_\lambda}$. A straightforward exercise shows that the minimum is achieved for,

$$\lambda = \lambda_{\min} := \frac{|S_{AB}|}{S_{BB}} \exp\left(i \; \arg S^*_{AB}\right),$$

for which the inequality reduces to the required result. □

Returning back to physics, it may happen that in some situations, distinguishing between an emission and an absorption event may not be possible. To model the outcomes of such cases, we introduce the *symmetrised spectrum*,

$$\bar{S}_{AB}[\Omega] := \int_{-\infty}^{+\infty} \left\langle \frac{1}{2} \left\{ \delta \hat{A}^\dagger(t), \delta \hat{B}(0) \right\} \right\rangle e^{i\Omega t} \, \mathrm{d}t = \frac{1}{2} \left(S_{AB}[\Omega] + S_{B^\dagger A^\dagger}[-\Omega] \right),$$

which is a complex quantity in general. For the case of an observable, say $\hat{X}$, with $\hat{X}^\dagger = \hat{X}$, we have,

$$\bar{S}_{XX}[\Omega] = \frac{1}{2}(S_{XX}[\Omega] + S_{XX}[-\Omega]), \tag{2.1.15}$$

i.e. symmetrisation in ordering is equivalent to symmetrisation in frequency. Note that this formally-motivated definition is equivalent to the physically-motivated one given in Eq. (2.1.6), allowing the symmetrized spectra of observables to be interpreted as the variance of the observable process.

The frequency symmetry,

$$\bar{S}_{XX}[\Omega] = \bar{S}_{XX}[-\Omega], \tag{2.1.16}$$

suggests that the *single-sided* spectrum defined by,

$$\bar{S}_X[\Omega] := 2\ \bar{S}_{XX}[\Omega], \quad \text{for} \quad \Omega \geq 0,$$

encodes the full information contained in the double-sided symmetrised spectrum $\bar{S}_{XX}[\Omega]$ of an observable $\hat{X}$. In terms of the single-sided spectrum, the variance of the process is,

$$\mathrm{Var}\left[\delta\hat{X}(t)\right] = \int_0^\infty \bar{S}_X[\Omega]\,\frac{\mathrm{d}\Omega}{2\pi}.$$

The single-sided spectrum thus defined is apparently equivalent to the conventional definition of the spectral density of a real-valued classical stochastic process [34].

Despite similarities to classical spectral densities at the level of definition, the lack of commutativity of time-dependent observables amongst each other, and even amongst the same observable at different times, leads to certain basic quantum mechanical conditions on the symmetrized spectra. Firstly, any observable will feature a fundamental level of statistical dispersion, preventing it from saturating the naive lower bound in $\bar{S}_{XX}[\Omega] \geq 0$; secondly, two (or more) observables will never be jointly dispersion-free at all frequencies. These constraints, expressed respectively in Propostitions 2.2 and 2.3 that follow, may be viewed as the irreducible content of quantum mechanics expressed at the level of spectra.

Proposition 2.2 (Spectral minimum) *Any observable $\hat{X}$ of a quantum mechanical system satisfies the inequality,*

$$\bar{S}_{XX}[\Omega] \geq \frac{1}{2}\left|\int_{-\infty}^{\infty}\left\langle\left[\delta\hat{X}(t), \delta\hat{X}(0)\right]\right\rangle e^{i\Omega t}\,\mathrm{d}t\right|. \tag{2.1.17}$$

Proof Using the definition of the spectral density Eq. (2.1.6), together with the uncertainty relation Eq. (2.1.4) gives the crux of the inequality, viz.

$$\bar{S}_{XX}[\Omega] = \lim_{T\to\infty} \mathrm{Var}\left[\delta\hat{X}^{(T)}[\Omega]\right] \geq \lim_{T\to\infty}\frac{1}{2}\left|\left\langle\left[\delta\hat{X}^{(T)}[\Omega], \delta\hat{X}^{(T)}[\Omega]^\dagger\right]\right\rangle\right|.$$

Expressing the windowed Fourier transform in terms of the time domain operator, and employing the weak-stationary property results in,

$$\bar{S}_{XX}[\Omega] \geq \frac{1}{2}\left|\lim_{T\to\infty}\frac{1}{T}\int_{-T/2}^{T/2}\left\langle\left[\delta\hat{X}(t-t'),\delta\hat{X}(0)\right]\right\rangle e^{i\Omega(t-t')}\,\mathrm{d}t\,\mathrm{d}t'\right|$$
$$= \frac{1}{2}\left|\lim_{T\to\infty}\int_{-\infty}^{\infty}\left\langle\left[\delta\hat{X}(\tau),\delta\hat{X}(0)\right]\right\rangle\left(1-\frac{|\tau|}{T}\right)e^{i\Omega\tau}\,\mathrm{d}\tau\right|$$
$$= \frac{1}{2}\left|\int_{-\infty}^{\infty}\left\langle\left[\delta\hat{X}(\tau),\delta\hat{X}(0)\right]\right\rangle e^{i\Omega\tau}\,\mathrm{d}\tau\right|;$$

the second equality is obtained from a change of variables, while the third follows from evaluating the limit inside the integral. □

If $\hat{X}(t)$ were classical, its spectrum $\bar{S}_{XX}$ could in principle exhibit no statistical dispersion—when X is deterministic—in which case, $\bar{S}_{XX}[\Omega] = 0$. However, as per Propositition 2.2, such a dispersion-free situation is untenable for a quantum mechanical process, unless $[\delta\hat{X}(t), \delta\hat{X}(0)] = 0$. This motivates the following definition of a *continuous observable*.[8]

Definition 2.2 (*Continuous observable*) An observable $\hat{X}(t)$ is said to be a continuous observable iff.

$$[\hat{X}(t), \hat{X}(t')] = 0. \tag{2.1.18}$$

They are also called "quantum non-demolition" observables [37, 38], to emphasize the fact that generic observables do not satisfy this constraint, and therefore cannot be measured without causing disturbance.

Given a pair of continuous observables—i.e. which individually feature no statistical dispersion—they may still exhibit a joint statistical dispersion; a feature that is classically impossible. The following proposition encodes this idea and it may be viewed as a generalization of the Robertson-Schrodinger inequality given in Eq. (2.1.2) to the case of continuous observables.

Proposition 2.3 (Spectral uncertainty relation II) *A pair of continuous observables $\hat{X}$, $\hat{Y}$ of a quantum mechanical system satisfying the (cross-)commutation relation,*

$$\left[\hat{X}(t), \hat{Y}(t')\right] = i\hat{C}_{XY}(t-t'), \quad \textit{or,} \quad \left[\hat{X}[\Omega], \hat{Y}[\Omega']\right] = i\hat{C}_{XY}[\Omega]\cdot 2\pi\ \delta[\Omega+\Omega']$$

satisfy the following inequality for their symmetrised spectra:

$$\bar{S}_{XX}[\Omega]\bar{S}_{YY}[\Omega] \geq \left|\bar{S}_{XY}[\Omega]\right|^2 + \frac{1}{4}\left|\left\langle\hat{C}_{XY}[\Omega]\right\rangle\right|^2 \tag{2.1.19}$$

[8] Note that the notion of a *continuous observable*, as introduced here, is very different from that of a *continuous variable* used in the context of quantum information [35, 36]. The latter refers to hermitian operators (i.e. observables) whose eigenspectrum is continuous. The former, as used here, refers to time-dependent observables (with a continuous, or discrete, eigenspectrum) which can (in principle) be continuously monitored in time.

Proof The strategy is to specialise the uncertainty relation for unsymmetrized spectra, given in Proposition 2.1, to the case of continuous observables. The uncertainty relation gives,

$$S_{XX}[\Omega]S_{YY}[\Omega] - |S_{XY}[\Omega]|^2 \geq 0.$$

To translate it into symmetrized spectra, note the following identity,

$$\begin{aligned}S_{XY}[\Omega] &= \int \left\langle \delta\hat{X}(t)\delta\hat{Y}(0)\right\rangle e^{i\Omega t}\,\mathrm{d}t \\ &= \int \left\langle \tfrac{1}{2}\left\{\delta\hat{X}(t), \delta\hat{Y}(0)\right\} + \tfrac{1}{2}\left[\delta\hat{X}(t), \delta\hat{Y}(0)\right]\right\rangle e^{i\Omega t}\,\mathrm{d}t \\ &= \bar{S}_{XY}[\Omega] + \frac{i}{2}\left\langle \hat{C}_{XY}[\Omega]\right\rangle.\end{aligned}$$

Inserting this expression for S_{XY} into the inequality and simplifying gives the result. □

Having developed the theoretical apparatus to deal with statistical properties of operator-valued stochastic processes, the next two sections will apply them to the case where a system is coupled to a thermal environment, and a meter, respectively.

2.2 Dynamics Due to a Thermal Environment

Consider a system, with a prescribed average energy, in equilibrium with an environment. The state of the system—described by a single parameter, the temperature T—is the one with the maximal entropy compatible with the average energy [2]. This unique state is the canonical thermal state,

$$\hat{\rho}_\beta = \frac{e^{-\beta\hat{H}_0}}{Z}, \tag{2.2.1}$$

where $\beta = (k_B T)^{-1}$ is the inverse temperature, $\hat{H}_0$ is the free hamiltonian of the system, and $Z = \mathrm{Tr}\, e^{-\beta\hat{H}_0}$ is the partition function that ensures the normalisation of the state, i.e. $\mathrm{Tr}\,\hat{\rho}_\beta = 1$. At zero temperature ($\beta \to \infty$) the canonical thermal state picks out the ground state of the hamiltonian, i.e., $\hat{\rho}_{\beta\to\infty} = |0\rangle\langle 0|$.

In a thermal state, the observables of the system, $\{\hat{X}_i\}$, are weak-stationary; i.e., their mean values are constant, while their second moments are time-translation invariant:

$$\begin{aligned}\left\langle \hat{X}_i(t)\right\rangle &= \left\langle \hat{X}_i(0)\right\rangle \\ \left\langle \hat{X}_i(t)\hat{X}_j(t')\right\rangle &= \left\langle \hat{X}_i(t-t')\hat{X}_j(0)\right\rangle.\end{aligned} \tag{2.2.2}$$

These identities follow from the observation that the hamiltonian $\hat{H}_0$, and hence the propagator $\hat{U}_t = \exp(-\frac{i}{\hbar}\hat{H}_0 t)$, commutes with the thermal state. Such states—*stationary states*—feature weak-stationarity of observables. However, the thermal states are distinguished among the stationary states by the following property, which is a generalisation of the principle of detailed balance.

Lemma 2.3 (Kubo-Martin-Schwinger [39, 40]) *The observables, $\{\hat{X}_i\}$, of a system in a thermal state at inverse temperature β obey the identity,*

$$\left\langle \hat{X}_i(t)\hat{X}_j(0)\right\rangle = \left\langle \hat{X}_j(0)\hat{X}_i(t+i\hbar\beta)\right\rangle \tag{2.2.3}$$

or equivalently,

$$S_{X_iX_j}[\Omega] = e^{\beta\hbar\Omega}\, S_{X_jX_i}[-\Omega]. \tag{2.2.4}$$

Proof Ignoring questions of rigour (see [41] for a remedy), the proof follows through a straightforward algebraic manipulation viz.,

$$\begin{aligned}
\left\langle \hat{X}_i(t)\hat{X}_j(0)\right\rangle &= \mathrm{Tr}\left[\hat{\rho}_\beta\, \hat{X}_i(t)\hat{X}_j(0)\right] \\
&= \mathrm{Tr}\left[e^{-\beta\hat{H}_0}\cdot \hat{U}_t^\dagger \hat{X}_i(0)\hat{U}_t \cdot e^{\beta\hat{H}_0}e^{-\beta\hat{H}_0}\cdot \hat{X}_j(0)\right] Z^{-1} \\
&= \mathrm{Tr}\left[e^{-\beta\hat{H}_0}\hat{X}_j(0)\cdot (e^{-\beta\hat{H}_0}\hat{U}_t^\dagger)\hat{X}_i(0)(\hat{U}_t e^{\beta\hat{H}_0})\right] Z^{-1} \\
&= \mathrm{Tr}\left[\hat{\rho}_\beta\, \hat{X}_j(0)\cdot \hat{U}_{t+i\hbar\beta}^\dagger \hat{X}_i(0)\hat{U}_{t+i\hbar\beta}\right] \\
&= \left\langle \hat{X}_j(0)\hat{X}_i(t+i\hbar\beta)\right\rangle.
\end{aligned}$$

The frequency domain form, in Eq. (2.2.4), can be proven starting by taking the Fourier transform of both sides and using the stationarity property, viz.,

$$\begin{aligned}
S_{X_iX_j}[\Omega] &:= \int \left\langle \hat{X}_i(t)\hat{X}_j(0)\right\rangle e^{i\Omega t}\,\mathrm{d}t = \int \left\langle \hat{X}_j(0)\hat{X}_i(t+i\hbar\beta)\right\rangle e^{i\Omega t}\,\mathrm{d}t \\
&= \int \left\langle \hat{X}_j(-t-i\hbar\beta)\hat{X}_i(0)\right\rangle e^{i\Omega t}\,\mathrm{d}t = \int \left\langle \hat{X}_j(t')\hat{X}_i(0)\right\rangle e^{i\Omega(-t'-i\hbar\beta)}\,\mathrm{d}t' \\
&= e^{\beta\hbar\Omega}\int \left\langle \hat{X}_j(t')\hat{X}_i(0)\right\rangle e^{i(-\Omega)t'}\,\mathrm{d}t' = e^{\beta\hbar\Omega} S_{X_jX_i}[-\Omega].
\end{aligned}$$

□

On the one hand, the KMS identity (Eq. (2.2.3)) may be seen as controlling the commutativity of observables in a thermal state: in the high temperature limit ($\beta \to 0$), it implies that all observables commute—evocative of classical behaviour. On the other hand, its frequency domain form (Eq. (2.2.4)) may be interpreted as a detailed balance principle: the ratio of the forward and reverse transition probabilities, represented by the ratio of the unsymmetrised spectra, is given by the thermal exponent. Perhaps more profoundly, it can be shown that[9] for a time-translation invariant system, if *all* operators satisfy the KMS identity pairwise, then the system is in the canonical thermal state $\hat{\rho}_\beta$. Thus the properties expressed in Eqs. (2.2.3) and (2.2.2) (almost completely) characterize the kinematics of thermal equilibrium.

Having thus described the essential structure of a thermal state, in the following, fluctuations due to a system being in such a state will be analysed. Essentially, this requires a model that describes the interaction between the system and its thermal environment, and then a procedure to calculate the dynamics of the system variables. We suppose that the interaction is mediated by a *linear* coupling between the system and the environment, modelled by an interaction hamiltonian,

$$\hat{H}_{\text{int}} = \sum_i \hat{X}_i \hat{F}_i, \tag{2.2.5}$$

that couples the observable $\hat{X}_i$ to a *generalized force* $\hat{F}_i$ which is a self-adjoint environment operator. Once the interaction is fixed, multiple approaches exist to treat the dynamics of the system [23, 30, 42]; we adopt the *linear response* formalism [39, 43, 44], enshrined in the following celebrated result.

[9] An outline of a proof is as follows (see [41] for the setup required to justify some of the steps). Assume then that there is some state $\hat{\rho}$ for which all operators (not just observables) of the system satisfy Eq. (2.2.3); i.e. $\langle \hat{A}(t)\hat{B}(0)\rangle = \langle \hat{B}(0)\hat{A}(t+i\hbar\beta)\rangle$, for all operators $\hat{A}$, $\hat{B}$. Time-translation invariance means that only the case $t=0$ need to be considered, i.e., $\langle \hat{A}(0)\hat{B}(0)\rangle = \langle \hat{B}(0)\hat{A}(i\hbar\beta)\rangle \;\forall\; \hat{A}, \hat{B}$. Dropping the time argument and writing this out with the unknown state $\hat{\rho}$ explicitly,

$$\text{Tr}[\hat{\rho}\hat{A}\hat{B}] = \text{Tr}[\hat{\rho}\hat{B}e^{-\beta\hat{H}_0}\hat{A}e^{\beta\hat{H}_0}] \quad \forall\, \hat{A}, \hat{B}.$$

This can be expressed in two different ways. Firstly, since it applies for any $\hat{A}$, it must also apply for $\hat{A} = e^{\beta\hat{H}_0}$; in this case, $\text{Tr}[\hat{\rho}e^{\beta\hat{H}_0}\hat{B}] = \text{Tr}[e^{\beta\hat{H}_0}\hat{\rho}\hat{B}] \;\forall\; \hat{B}$, implying that,

$$\hat{\rho}e^{\beta\hat{H}_0} = e^{\beta\hat{H}_0}\hat{\rho}.$$

Secondly, permuting within the trace gives the alternate form, $\text{Tr}[\hat{B}\hat{\rho}\hat{A}] = \text{Tr}[e^{\beta\hat{H}_0}\hat{\rho}\hat{B}e^{-\beta\hat{H}_0}\hat{A}] \;\forall\; \hat{A}, \hat{B}$, implying that,

$$\hat{B}\hat{\rho} = e^{\beta\hat{H}_0}\hat{\rho}\hat{B}e^{-\beta\hat{H}_0}$$
$$\text{i.e.,}\quad \hat{B}\hat{\rho}e^{\beta\hat{H}_0} = e^{\beta\hat{H}_0}\hat{\rho}\hat{B} \quad \forall\, \hat{B}.$$

Combining the results from the two forms gives,

$$\hat{B}e^{\beta\hat{H}_0}\hat{\rho} = e^{\beta\hat{H}_0}\hat{\rho}\hat{B} \quad \forall\, \hat{B},$$

i.e. the operator $e^{\beta\hat{H}_0}\hat{\rho}$ commutes with every operator in the Hilbert space. This means that it must be proportional to the identity operator, i.e. $e^{\beta\hat{H}_0}\hat{\rho} \propto 1$, or, $\hat{\rho} \propto e^{-\beta\hat{H}_0}$. The normalization of the state fixes the proportionality constant.

Lemma 2.4 (Kubo) *If a system is presumed to be maintained in a thermal state by a linear coupling to the environment, i.e. by a hamiltonian of the form,*

$$\hat{H}_F(t) = \hat{H}_0 + \sum_i \hat{X}_i \hat{F}_i(t), \tag{2.2.6}$$

where $\hat{F}_i$ is the generalised force corresponding to $\hat{X}_i$, then fluctuations in the system observables are given by,

$$\begin{aligned} \delta\hat{X}_j(t) &= \int_{-\infty}^{\infty} \sum_k \chi_{jk}(t-t')\, \delta\hat{F}_k(t')\, \mathrm{d}t', \\ \textit{or,} \quad \delta\hat{X}_j[\Omega] &= \sum_k \chi_{jk}[\Omega]\, \delta\hat{F}_k[\Omega], \end{aligned} \tag{2.2.7}$$

where, the "susceptibilities" χ_{jk} are (here $\Theta(t)$ is the Heaviside step function),

$$\chi_{jk}(t) = -\frac{i}{\hbar}\Theta(t)\left\langle [\hat{X}_j(t), \hat{X}_k(0)] \right\rangle. \tag{2.2.8}$$

Proof Standard time-dependent perturbation theory as for example in [45]. □

Formally, the power of the Kubo formula in Eq. (2.2.8), is that it relates the response of the system to an external influence in terms of expectation values of the system operators taken on the equilibrium (thermal) state of the system. Practically, the great advantage of the linear response formalism is that by relating the fluctuations in the system's observables to the fluctuations of a generalised force, it suggests an avenue to probe the system: coherent response measurements—harmonically driving $\hat{F}_k$ and observing its effect in $\hat{X}_j$—give access to $\chi_{jk}[\Omega]$, which then predict the incoherent behaviour of the system in the absence of an explicit stimulus. Within the regime of its validity, the linear response formalism is pervasive in physics [26, 46–49].

For the set of observables that are assumed to directly couple to the environment—those in the interaction hamiltonian in Eq. (2.2.5)—the spectral uncertainty relation (Eq. (2.1.4)), and the Kubo formula, imply a couple of general properties.

2.2.1 Effect of Fluctuations from a Thermal Environment

Proposition 2.4 (Fundamental fluctuations) *Observables of the system that directly couple to the environment exhibit fluctuations, whose spectra $\bar{S}_{X_i X_i}[\Omega]$ have a minimum positive value,*

$$\bar{S}_{X_i X_i}[\Omega] \geq \hbar\, |\mathrm{Im}\ \chi_{ii}[\Omega]|\, . \tag{2.2.9}$$

Proof Proposition 2.2 already states that spectra of operator-valued stochastic processes have a minimum value dictated by the expectation value of its commutator (in any state), i.e.,

$$\bar{S}_{X_i X_i}[\Omega] \geq \frac{1}{2}\left|\int_{-\infty}^{\infty}\left\langle\left[\delta\hat{X}_i(t),\delta\hat{X}_i(0)\right]\right\rangle e^{i\Omega t}\,\mathrm{d}t\right|.$$

When the state is the thermal state, the Kubo formula Eq. (2.2.7) relates the expectation value of the commutator to the susceptibility. This can be incorporated by splitting the integral, and using the symmetries of the susceptibility, viz.,

$$\begin{aligned}\bar{S}_{X_i X_i}[\Omega] &\geq \frac{1}{2}\left|\int_{-\infty}^{0}\left\langle\left[\delta\hat{X}_i(t),\delta\hat{X}_i(0)\right]\right\rangle e^{i\Omega t}\,\mathrm{d}t + \int_{0}^{\infty}\left\langle\left[\delta\hat{X}_i(t),\delta\hat{X}_i(0)\right]\right\rangle e^{i\Omega t}\,\mathrm{d}t\right| \\ &= \frac{1}{2}\left|\int_{0}^{\infty}\left\langle\left[\delta\hat{X}_i(-t),\delta\hat{X}_i(0)\right]\right\rangle e^{-i\Omega t}\,\mathrm{d}t + \int_{0}^{\infty}\left\langle\left[\delta\hat{X}_i(t),\delta\hat{X}_i(0)\right]\right\rangle e^{i\Omega t}\,\mathrm{d}t\right| \\ &= \frac{1}{2}\left|\int_{0}^{\infty}\left\langle\left[\delta\hat{X}_i(t),\delta\hat{X}(0)\right]\right\rangle\left(e^{i\Omega t}-e^{-i\Omega t}\right)\,\mathrm{d}t\right| \\ &= \hbar\left|\int_{0}^{\infty}\chi_{ii}(t)\sin(\Omega t)\,\mathrm{d}t\right|\end{aligned}$$

The third equality follows from the odd property of the average of the commutator of a weak-stationary operator, i.e. $\langle[\delta\hat{X}_i(-t),\delta\hat{X}_i(0)]\rangle = -\langle[\delta\hat{X}_i(t),\delta\hat{X}_i(0)]\rangle$, while the fourth employs Eq. (2.2.8). Since the sine-transform is the imaginary part of the Fourier transform, the right-hand side becomes $\hbar\,|\mathrm{Im}\ \chi_{ii}|$. □

Proposition 2.4 signifies that once a system is coupled to a thermal environment via a linear coupling through its observable $\hat{X}_i$, then that observable exhibits a fundamental fluctuation that depends on the details of the coupling (i.e. the susceptibility), but not the temperature. Tentatively, and with foresight, the minimum value of the spectrum may be identified as the spectrum of *vacuum fluctuations* of that observable.

Clearly the imaginary part of the susceptibility plays a prominent role in determining the fluctuations in system observables in a thermal state. From the expression for the imaginary part of the susceptibility,

$$\mathrm{Im}\ \chi_{ij}[\Omega] = -\frac{i}{2}\int\left(\chi_{ij}(t)-\chi_{ij}(-t)\right)e^{i\Omega t}\,\mathrm{d}t.$$

is is clear that it characterises the lack of invariance to time-reversal $t \to -t$, and thus captures the irreversible character of the system once it is coupled to the environment. On the other hand, the coupling to the environment leads to fluctuations in the system's observables, characterised by the bound Eq. (2.2.9). It is therefore natural to enquire whether a precise equality exists between the imaginary part of the susceptibility and the spectrum of observables that codifies the shared origin of fluctuations and dissipation.

Proposition 2.5 (Fluctuation-Dissipation) *For a system maintained in a thermal state through its contact with an environment, the fluctuations in the observables that couple to the environment are characterised by the relation,*

$$\bar{S}_{X_iX_i}[\Omega] = \hbar(2n_\beta(\Omega) + 1)\text{Im } \chi_{ii}[\Omega], \tag{2.2.10}$$

where $n_\beta(\Omega)$ is the Bose occupation at frequency Ω and inverse temperature β,

$$n_\beta(\Omega) := (e^{\beta\hbar\Omega} - 1)^{-1}. \tag{2.2.11}$$

Proof First we prove a slightly general result and then specify to the case at hand. Starting from the left-hand side of Eq. (2.2.10) in the time domain:

$$\text{Im } \chi_{ij}(t) = -\frac{i}{2}\left(\chi_{ij}(t) - \chi_{ij}(t)^*\right) = -\frac{i}{2}\left(\chi_{ij}(t) - \chi_{ji}(-t)\right).$$

Using the Kubo formula (Eq. (2.2.7)), and employing time-translation invariance, the susceptibilities can be expressed in terms of correlators (which are the inverse Fourier transforms of the unsymmetrised spectra),

$$\chi_{ij}(t) = -\frac{i}{\hbar}\Theta(t)\left(S_{X_iX_j}(t) - S_{X_jX_i}(-t)\right)$$
$$\chi_{ji}(-t) = -\frac{i}{\hbar}\Theta(-t)\left(S_{X_jX_i}(-t) - S_{X_iX_j}(t)\right),$$

which gives,

$$\text{Im } \chi_{ij}(t) = \frac{-1}{2\hbar}\left(S_{X_iX_j}(t) - S_{X_jX_i}(-t)\right).$$

Now using the KMS condition (Eq. (2.2.3)), the order of observables in the second correlator can be reversed, i.e. $S_{X_jX_i}(-t) = S_{X_iX_j}(t - i\hbar\beta)$. Inserting this back gives,

$$\text{Im } \chi_{ij}(t) = \frac{-1}{2\hbar}\left(S_{X_iX_j}(t) - S_{X_iX_j}(t - i\hbar\beta)\right).$$

Fourier transforming each side and re-arranging results in

$$S_{X_iX_j}[\Omega] = \frac{2\hbar}{1 - e^{-\beta\hbar\Omega}}\text{Im } \chi_{ij}[\Omega], \tag{2.2.12}$$

which relates the unsymmetrised cross-spectra with the susceptibility. For the required result, we consider the case $\hat{X}_j = \hat{X}_i$, and the symmetrised spectral density,

$$\begin{aligned}\bar{S}_{X_iX_i}[\Omega] &= \frac{1}{2}\left(S_{X_iX_i}[\Omega] + S_{X_iX_i}[-\Omega]\right)\\ &= \frac{1}{2}(1+e^{-\beta\hbar\Omega})S_{X_iX_i}[\Omega]\\ &= \hbar\left(\frac{e^{\beta\hbar\Omega}+1}{e^{\beta\hbar\Omega}-1}\right)\text{Im }\chi_{ii}[\Omega];\end{aligned}$$

here, the first equality is the symmetric property of the spectral density (Eq. (2.1.15)), the second follows from the detailed balance condition (Eq. (2.2.4)), and the third from Eq. (2.2.12). Replacing the exponentials in terms of the Bose occupation (Eq. (2.2.11)) gives the result. □

The fluctuation-dissipation theorem (Eq. (2.2.10)) relates the fluctuations in the system to the system-environment coupling, and the environment state (determined by the single parameter, temperature). The bound in Proposition 2.4 (Eq. (2.2.9)), on the other hand, follows from the non-commutativity of the observable and not on the properties of the environment, and is therefore a more general statement. Notably, the zero-temperature limit ($\beta \to \infty$, for which $n_\beta[\Omega] \to 0$) of Eq. (2.2.10) gives Eq. (2.2.9), motivating the interpretation that the lower-bound in the latter is due to intrinsic—vacuum—fluctuations in the system.

An important corollary of the fluctuation-dissipation theorem is that the spectrum of fluctuations of the system observable can be referred to an effective spectrum of the generalised force. In the case where only one observable, $\hat{X}$, is coupled to its generalised force $\hat{F}$, the respective spectra are given by,

$$\begin{aligned}&\bar{S}_{XX}[\Omega] = \hbar\left(2n_\beta(\Omega)+1\right)\text{Im }\chi[\Omega]\\ \Rightarrow\quad &\bar{S}_{FF}[\Omega] = |\chi[\Omega]|^{-2}\,\bar{S}_{XX}[\Omega] = \hbar\left(2n_\beta(\Omega)+1\right)\text{Im }\chi[\Omega]^{-1},\end{aligned}$$

where $\chi[\Omega]$ is the sole susceptibility involved.

2.3 Dynamics Due to a Meter

Quantum mechanically, a *meter*—a measuring device—is a specific form of environment from the perspective of the system. During an act of measurement, the system is coupled to a meter. The meter, being a quantum mechanical system itself, has intrinsic fluctuations in its variables. Via the measurement interaction, this leads to additional fluctuations in the system variables. Such fluctuations are called *measurement back-action*. Unlike a thermal environment however, the meter needs to be ideally prepared in some non-equilibrium state,[10] meaning that the fluctuations

[10]This is because the meter is expected to output a classical record of the system observable being measured; this can only be arranged for if the states of the meter corresponding to the various values taken by the system observable are macroscopically distinguishable [50].

imparted by it are not determined by the fluctuation-dissipation theorem. However, bounds for the imparted fluctuations can be derived under the minimal assumption of the system-meter coupling being linear and weak.

Very generally, a continuous linear measurement of the observable $\hat{X}$ may be described by an operator $\hat{Y}$ corresponding to the output of a detector. Linearity means that[11] $\hat{Y}(t) \propto \hat{X}(t)$. Since we assume that $\hat{Y}(t)$ is the output of a detector—i.e. the *measurement record*—it must certainly be a continuous observable, i.e.,

$$\left[\hat{Y}(t), \hat{Y}(t')\right] = 0. \tag{2.3.1}$$

However, in general, the system observable, $\hat{X}$, is not a continuous observable. For $\hat{Y}$ to commute with itself, while $\hat{X}$ does not, it is necessary that the record be contaminated by some additional process $\hat{X}_n(t)$, arising from the meter,[12] so that the combination,

$$\hat{Y}(t) = \hat{X}(t) + \hat{X}_n(t) \tag{2.3.2}$$

is a continuous observable. The two equations above operationally characterise the class of so-called *continuous linear measurements* [20, 26].

2.3.1 Effect of Fluctuations from a Meter

Proposition 2.6 (Standard Quantum Limit) *When a meter provides a continuous linear record $\hat{Y}(t)$, of the observable $\hat{X}(t)$ of a system, the spectrum of the output is,*

$$\bar{S}_{YY}[\Omega] \geq 2 \cdot min\, \bar{S}_{XX}[\Omega] = 2\hbar \left|\text{Im}\ \chi_{XX}[\Omega]\right|, \tag{2.3.3}$$

when no correlations exist between the system and meter. In other words, the measurement record contains at least twice the minimum noise in the observable being measured.

Proof The spectrum of the output $\hat{Y}$ (in Eq. (2.3.2)) is,

$$\bar{S}_{YY}[\Omega] = \bar{S}_{XX}[\Omega] + \bar{S}_{X_n X_n}[\Omega] + 2\text{Re}\ \bar{S}_{X X_n}[\Omega].$$

Assuming no correlations between the system and meter, the last term can be neglected, and so,

---

[11]The most general linear relationship is of the form $\hat{Y}(t) = \int f(t)\hat{X}(t-t')\,\mathrm{d}t'$, corresponding to a filtered version of the observable. However, without loss of generality, the filtering may be considered as happening on the classical measurement record, after the detector.

[12]On the other hand, if it can be arranged that the observable $\hat{X}$ already satisfies $[\hat{X}(t), \hat{X}(t')] = 0$, i.e. it is a continuous observable in the sense defined in Eq. (2.1.18), then there is in principle no additional contamination.

$$\bar{S}_{YY}[\Omega] = \bar{S}_{XX}[\Omega] + \bar{S}_{X_n X_n}[\Omega].$$

The bound set by vacuum fluctuations (Eq. (2.2.9)) implies a lower bound for the spectrum of the system observable $\bar{S}_{XX}$: i.e. $\bar{S}_{XX} \geq \hbar \, |\mathrm{Im} \, \chi_{XX}|$. The remaining task is therefore to lower bound $\bar{S}_{X_n X_n}$. The continuous observability condition (Eq. (2.3.1)) implies that the commutators of $\hat{X}_n$ and $\hat{X}$ are related, viz.,

$$[\hat{X}_n(t), \hat{X}_n(t')] = -[\hat{X}(t), \hat{X}(t')].$$

Now applying the minimum noise bound, in Proposition 2.2, to $\hat{X}_n$ gives,

$$\begin{aligned} \bar{S}_{X_n X_n}[\Omega] &\geq \frac{1}{2} \left| \int_{-\infty}^{\infty} \left\langle \left[\delta\hat{X}_n(t), \delta\hat{X}_n(0) \right] \right\rangle e^{i\Omega t} \, \mathrm{d}t \right| \\ &= \frac{1}{2} \left| \int_{-\infty}^{\infty} \left\langle \left[\delta\hat{X}(t), \delta\hat{X}(0) \right] \right\rangle e^{i\Omega t} \, \mathrm{d}t \right| \\ &\geq \hbar \, |\mathrm{Im} \, \chi_{XX}[\Omega]| \, . \end{aligned}$$

Here, the last inequality follows from arguments given in the proof of Eq. (2.2.9). Ultimately,

$$\bar{S}_{YY}[\Omega] \geq 2 \cdot \hbar \, |\mathrm{Im} \, \chi_{XX}[\Omega]| = 2 \cdot \min \bar{S}_{XX}[\Omega].$$

□

Conceptually, the standard quantum limit (Eq. (2.3.3)) states that quantum mechanics extorts a penalty twice: once in the form of the vacuum fluctuations of the observable (as in Eq. (2.2.9)), and once more, the same price, in the form of unavoidable fluctuations in the linear measurement process. This factor of two may also be understood if the action of the meter is considered to be that of an abstract linear amplifier [18, 51] whose role is to amplify the values taken by the system observable into a classically recordable signal. This perspective sheds light on the relationship between the standard quantum limit derived here for a general scenario, and the specific example of the vacuum-equivalent noise that is added when simultaneously measuring the canonically conjugate variables of a harmonic oscillator [52–56].

The standard quantum limit rests on the validity of the assumptions basic to its existence being fulfilled in a given situation: (1) the system-meter coupling is linear, (2) continuous, (3) stationary, and, (4) the system and meter states are uncorrelated. (Presumably, the adjective "standard" refers to this *standard* configuration.) A violation of one or more of these assumptions can beat the bound in Eq. (2.3.3). In the context of interferometric position measurement [57]—a prototypical example of a continuous linear measurement—all these loop holes have been exploited as a means to improve measurement sensitivity beyond the standard quantum limit.

For example, quantum non-demolition techniques to measure position rely on a time-dependent coupling between the system and meter [37, 38], violating the continuity and/or stationarity assumptions. Injection of squeezed light into the interferometer [58–60], or the use of squeezing generated within the interferometer [60, 61], relies on harnessing system-meter correlations.

References

1. P.A.M. Dirac, *The Principles of Quantum Mechanics*, 4th edn. (Clarendon Press, 1982)
2. J. von Neumann, *Mathematical Foundations of Quantum Mechanics* (Princeton University Press, 1955)
3. J.E. Roberts, J. Math. Phys. **7**, 1097 (1966)
4. J.-P. Antoine, J. Math. Phys. **10**, 53 (1969)
5. E. Prugovečki, *Quantum Mechanics in Hilbert Space* (Academic Press, 1981)
6. A. Fine, J. Math. Phys. **23**, 1306 (1982)
7. H.P. Robertson, Phys. Rev. **46**, 794 (1934)
8. E. Schrödinger, Sitzungsberichte Preus. Akad. Wiss. **19**, 296 (1930)
9. V.V. Dodonov, E.V. Kurmyshev, V.I. Man'ko, Phys. Lett. A **79**, 150 (1980)
10. W. Heisenberg, Z. Phys. **43**, 172 (1927)
11. E.H. Kennard, Z. Phys. **44**, 326 (1927)
12. H. Weyl, *Theory of Groups and Quantum Mechanics* (Dover, 1932)
13. M. Ozawa, Phys. Rev. A **67**, 042105 (2003)
14. P. Busch, T. Heinonen, P. Lahti, Phys. Rep. **452**, 155 (2007)
15. T. Heinosaari, M.M. Wolf, J. Math. Phys. **51**, 092201 (2010)
16. J. Distler, S. Paban, Phys. Rev. A **87**, 062112 (2013)
17. V.S. Varadarajan, Comm. Pure Appl. Math. **15**, 189 (1962)
18. C.M. Caves, Phys. Rev. D **26**, 1817 (1982)
19. C. Caves, G. Milburn, Phys. Rev. A **36**, 5543 (1987)
20. V.B. Braginsky, F.Y. Khalili, *Quantum Measurement* (Cambridge University Press, 1992)
21. A. Barchielli, Phys. Rev. A **34**, 1642 (1986)
22. A. Barchielli, M. Gregoratti, *Quantum Trajectories and Measurements in Continuous Time* (Springer, 2009)
23. H.M. Wiseman, G.J. Milburn, *Quantum Measurement and Control* (Cambridge University Press, 2010)
24. M.D. Donsker, Mem. Am. Math. Soc. **6**, 12 (1951)
25. N. Wiener, *Fourier Integrals and Certain of Its Applications* (Cambridge University Press, 1933)
26. A.A. Clerk, M.H. Devoret, S.M. Girvin, F. Marquardt, R.J. Schoelkopf, Rev. Mod. Phys. **82**, 1155 (2010)
27. G.B. Lesovik, R. Loosen, J. JETP Lett. **65** (1997)
28. U. Gavish, Y. Levinson, Y. Imry, Phys. Rev. B. **62**, R10637 (2000)
29. H. Eksteinand, N. Rostoker, Phys. Rev. **100**, 1023 (1955)
30. H.J. Carmichael, *An Open Systems Approach to Quantum Optics* (Springer, 1993)
31. C. Cohen-Tannoudji, J. Dupont-Roc, G. Grynberg, *Atom-Photon Interactions* (Wiley, 1998)
32. J. Eberly, K. Wodkiewicz, J. Opt. Soc. Am. **67**, 1252 (1977)
33. B. Mollow, Phys. Rev. **188**, 1969 (1969)
34. W.B. Davenport, W.L. Root, *An Introduction to the Theory of Random Signals and Noise* (Wiley-IEEE, 1987)
35. S.L. Braunstein, P. van Loock, Rev. Mod. Phys. **77**, 513 (2005)
36. C. Weedbrook, S. Pirandola, R. García-Patrón, N.J. Cerf, T.C. Ralph, J.H. Shapiro, S. Lloyd, Rev. Mod. Phys. **84**, 621 (2012)

37. C.M. Caves, M. Zimmermann, K.S. Thorne, R.W. Drever, Rev. Mod. Phys. **52**, 341 (1980)
38. V.B. Braginsky, Y.I. Vorontsov, K.S. Thorne, Science **209**, 547 (1980)
39. R. Kubo, J. Phys. Soc. Japan **12**, 570 (1957)
40. P. Martin, J. Schwinger, Phys. Rev. **115**, 1342 (1959)
41. R. Haag, N.M. Hugenholtz, M. Winnink, Comm. Math. Phys. **5**, 215 (1967)
42. E.B. Davies, *Quantum Theory of Open Systems* (Academic Press, 1976)
43. W. Bernard, H. Callen, Rev. Mod. Phys. **31**, 1017 (1959)
44. N. Pottier, *Nonequilibrium Statistical Physics* (Oxford University Press, 2014)
45. R. Kubo, Rep. Prog. Phys. **29**, 255 (1966)
46. L.P. Kadanoff, P.C. Martin, Ann. Phys. **24**, 419 (1963)
47. D. Zubarev, Sov. Phys. Usp. **3**, 320 (1960)
48. T. Dittrich, P. Hanggi, G.-L. Ingold, B. Kramer, G. Schön, W. Zwerger, *Quantum Transport and Dissipation* (Wiley-VCH, 1998)
49. Y. Nazarov (ed.), *Quantum Noise in Mesoscopic Physics* (Springer, 2002)
50. W.H. Zurek, Rev. Mod. Phys. **75**, 715 (2003)
51. R.J. Glauber, Ann. NY Acad. Sci. **480**, 336 (1986)
52. E. Arthurs, J. Kelly, Bell. Sys. Tech. J. **44**, 725 (1965)
53. C.Y. She, H. Heffner, Phys. Rev. **152**, 1103 (1966)
54. H.P. Yuen, Phys. Lett. A **91**, 101 (1982)
55. J. Shapiro, S. Wagner, IEEE. J. Quant. Elec. **20**, 803 (1984)
56. U. Leonhardt, H. Paul, J. Mod. Opt. **40**, 1745 (1993)
57. A. Abramovici, W.E. Althouse, R.W.P. Drever, Y. Gürsel, S. Kawamura, F.J. Raab, D. Shoemaker, L. Sievers, R.E. Spero, K.S. Thorne, R.E. Vogt, R. Weiss, S.E. Whitcomb, M.E. Zucker, Science **256**, 325 (1992)
58. C.M. Caves, Phys. Rev. D **23**, 1693 (1981)
59. M. Jaekel, S. Reynaud, Europhys. Lett. **13**, 301 (1990)
60. H.J. Kimble, Y. Levin, A.B. Matsko, K.S. Thorne, S.P. Vyatchanin, Phys. Rev. D **65**, 022002 (2001)
61. A. Buonanno, Y. Chen, Phys. Rev. D **64**, 042006 (2001)

Chapter 3
Phonons and Photons

The career of a young theoretical physicist consists of treating the harmonic oscillator in ever-increasing levels of abstraction.

Sidney Coleman

The objective of this chapter is to introduce the dramatis personae of the quantum measurement problem studied in this thesis. Nominally, the *system* is a mechanical oscillator formed by a well-defined mode of a solid-state elastic resonator, excited by its thermal environment under ambient conditions. The *meter* is an electromagnetic mode of an optical (micro-)cavity, excited by a laser source. In Sect. 3.1, a formal description of the quantum mechanics of an elastic resonator is given, followed by a brief treatment of a single mode of such a resonator in thermal equilibrium. Section 3.2 tackles the analogous development for the electromagnetic field, first describing the travelling wave field that excites the optical cavity, and then the coupling of the cavity to the travelling wave field. True to Sidney Coleman's observation, both the mechanical mode and the optical mode are formally harmonic oscillators. The concrete implementation of either oscillator will be introduced later in Chap. 5.

3.1 Phonons: Quantised Linear Elastodynamics

Bulk matter, existing in a state where its constituent atoms are bound to each other, maybe assumed to form a continuum. When this system is in mechanical equilibrium, its state at each instant of time maybe specified by the set of positions $\mathbf{r} \in V \subset \mathbb{R}^3$

V. Sudhir, *Quantum Limits on Measurement and Control of a Mechanical Oscillator*, Springer Theses, https://doi.org/10.1007/978-3-319-69431-3_3

of the material constituents within the domain V that forms the continuum.[1] This domain is the elastic body. Changes in the state of the body are described by the transformation,

$$\mathbf{r} \mapsto \mathbf{r} + \mathbf{u}(\mathbf{r}), \tag{3.1.1}$$

wherein the constituents at position $\mathbf{r}$ get displaced to their new position $\mathbf{u}(\mathbf{r})$. $\mathbf{u}$ is the *displacement field*, and the above map represents the deformation of the body. We shall be concerned with elastodynamic phenomena that can be described by a smooth displacement field, thus excluding phenomena like dislocations and fracture.

Note that Eq. (3.1.1) is essentially a geometric transformation of the body; we shall therefore strive to describe and analyse its consequences in suitable language.

The strain tensor, defined by[2]

$$u_{ij}^{(1)}(\mathbf{r}) := \frac{\partial u_i}{\partial r_j}$$

is essential to the description of how distances between points in the body change due to the deformation $\mathbf{r} \mapsto \mathbf{r} + \mathbf{u}$ in Eq. (3.1.1). In fact, the infinitesimal length element[3] $\mathrm{d}s^2 := \mathrm{d}r_i \mathrm{d}r_i$ changes to [7],

$$\begin{aligned}
\mathrm{d}s^2 &\to (\mathrm{d}r_i + \mathrm{d}u_i)(\mathrm{d}r_i + \mathrm{d}u_i) \\
&= \left(\mathrm{d}r_i + \frac{\partial u_i}{\partial r_j}\mathrm{d}r_j\right)\left(\mathrm{d}r_i + \frac{\partial u_i}{\partial r_k}\mathrm{d}r_k\right) \\
&\approx \mathrm{d}r_i \mathrm{d}r_i + \left(u_{ij}^{(1)} + u_{ji}^{(1)}\right)\mathrm{d}r_i \mathrm{d}r_j \\
&= \left(\delta_{ij} + \left(u_{ij}^{(1)} + u_{ji}^{(1)}\right)\right)\mathrm{d}r_i \mathrm{d}r_j .
\end{aligned}$$

Clearly, the symmetric part of $u^{(1)}$ plays the role of a metric tensor within the body. The tensor $u^{(1)}$ maybe decomposed into three components [8], each describing a possible motion of the body: (a) volume deformation—distances between constituents changing in the same sense throughout the body—characterised by the scalar $\operatorname{Tr} u^{(1)} = \nabla \cdot \mathbf{u}$; (b) shear motion—infinitesimal parallel planes sliding along each other—characterised by the traceless symmetric tensor, $\frac{1}{2}(u_{ij}^{(1)} + u_{ji}^{(1)}) - \frac{\delta_{ij}}{3}\operatorname{Tr} u^{(1)}$; and, (c) rigid rotation characterised by the anti-symmetric tensor, $\frac{1}{2}(u_{ij}^{(1)} - u_{ji}^{(1)})$. The case of rigid translational motion is described by a uniform-in-space displacement

---

[1]We implicitly assume a non-relativistic setting; in the contrary setting, the specification of the preferred state $\mathbf{r}$ is untenable [1, 2]. However formulations which extend to the relativistic case exist [2–5]; see [6, Chap. 15] for a historical review of the conceptual subtleties in a relativistic theory of elastodynamics.

[2]Note the difference from the standard definition, as considered for example in [7]; we follow [8].

[3]For the rest of Sect. 3.1, we adopt the summation convention that whenever two indices are repeated, they are implicitly summed over. For instance, $\mathrm{d}r_i \mathrm{d}r_i = \sum_i \mathrm{d}r_i \mathrm{d}r_i$ etc.

field, and therefore has a null strain tensor. Elastodynamics does not concern with rigid motions, henceforth, we may consider $u^{(1)}$ to be symmetric.

In addition to the strain tensor $u^{(1)}$, we consider the object,

$$u^{(2)}_{ijk} := \frac{\partial^2 u_i}{\partial r_j \partial r_k}, \tag{3.1.2}$$

which describes the curvature of lines and surfaces due to the elastic deformation. For example, the (flat) coordinate plane $r_i = 0$ at $t = 0$, is mapped to the (curved) surface $r_i = u_i(\mathbf{r}, t)$ at later times $t > 0$. The curvature of the deformed surface is quantified by the eigenvalues of the matrix (indexed by j, k) $u^{(2)}_{ijk}$ [9].

Fortunately, still higher derivatives of the displacement field need not be considered. For, it is a theorem [9] that all the local geometric properties of surfaces in three-dimensions are captured by combinations of the two tensors $u^{(1)}, u^{(2)}$. This concludes the essential aspects of the kinematics of the displacement field.

3.1.1 *Classical Description of Navier-Euler-Bernoulli Elastic Field*

In order to derive the dynamics (i.e., equations of motion) of $\mathbf{u}$, we appeal to the principle of least action [10]. In Hamilton's form, it dictates that the continuous sequence of deformations $t \mapsto \mathbf{u}(\mathbf{r}, t)$, realised at each point in time, is the one that renders the action,

$$\mathscr{S}[\![u_i]\!] := \int \mathrm{d}t\, L\left(t, r_i, u_i, \frac{\partial u_i}{\partial t}, \frac{\partial u_i}{\partial r_j}, \frac{\partial^2 u_i}{\partial r_j \partial r_k}\right) \tag{3.1.3}$$

stationary. Here L is the Lagrangian which is in general a function of time, spatial coordinates, displacement field, and its derivatives. Note that the action is a *functional* of the deformation field $\mathbf{u}$, associating a real number with a given configuration $\mathbf{u}(\mathbf{r}, t)$ defined over the spatial domain V. Simple principles maybe invoked to fix the form of L, and thence to derive the equations of motion.[4]

In order to clearly identify and delineate the physical symmetry principles involved, and the inference of L therein, we go through them in a sequence of steps:

1. The elastic body is assumed to conform to some loose notion of locality, so that the Lagrangian L is a sum over a Lagrangian density[5] $\mathscr{L}$ defined for each infinitesimal sub-domain of V; in other words,

$$\mathscr{S}[\![u_i]\!] = \int \mathrm{d}t \int_V \mathrm{d}^3 r\, \mathscr{L}\left(t, r_i, u_i, \frac{\partial u_i}{\partial t}, \frac{\partial u_i}{\partial r_j}, \frac{\partial^2 u_i}{\partial r_j \partial r_k}\right). \tag{3.1.4}$$

[4]See [11, 12] for a lucid articulation of this general idea, and [13, 14] for examples.

[5]By abuse of terminology, $\mathscr{L}$ will also be referred to as a Lagrangian.

2. The assumption of the principle of Galilean relativity [15, 16]: the description of elastic phenomena is assumed independent of translations in time (by t_0) and space (by $\mathbf{r}_0$), uniform motion (by velocity $\mathbf{v}$), and rotation (by the infinitesimal rotation matrix R_{ij}), i.e. the transformations

$$\begin{aligned} t &\mapsto t + t_0 \\ r_i &\mapsto r_i + r_{0i} + v_i t + R_{ij} r_j . \end{aligned} \tag{3.1.5}$$

Invariance under translations in time and uniform motion are Newtonian precepts, which elasticity is expected to obey. Translations in space also fall in this category for systems which evolve freely (i.e. not under the influence of an external force)—here we consider elasticity in this form, where the deformation field evolves under self-consistent forces imposed by deformations in the material. Rotational invariance, on the other hand, requires additional assumptions about the nature of the material forming the body. The description of crystalline material are not invariant to arbitrary rotations, but only to a discrete set which describe its symmetry [17]. We limit ourselves to amorphous material, which is the case relevant to this thesis. For such materials, the various elements of the Galilean transformation Eq. (3.1.5) maybe analysed as follows:

   2.1 Translation invariance in time and space necessitates that the lagrangian be independent of t, r_i and u_i, i.e.,

$$\mathscr{L} = \mathscr{L}(\dot{u}_i, u^{(1)}_{ij}, u^{(2)}_{ijk}). \tag{3.1.6}$$

   2.2 Invariance under uniform motion only affects rigid translational motion, resulting in the deformation field $u_i(\mathbf{x}, t) = v_i t$. In this case, the only term in $\mathscr{L}$ is the one that depends on $\dot{u}_i$. This being a vector, rotational invariance further limits possible terms to ones that are functions of the invariant $\dot{u}_i \dot{u}_i$. The simplest non-trivial function provides the first term of the lagrangian, viz.,

$$\mathscr{L}(\dot{u}_i, u^{(1)}_{ij}, u^{(2)}_{ijk}) = \frac{\rho}{2} \dot{u}_i \dot{u}_i + \dots . \tag{3.1.7}$$

   Here ρ is a positive real number—dimensional analysis shows that it is in fact the mass density of the amorphous material.

   2.3 Rotational invariance further constrains additional terms that depend on the second and third rank tensor $u^{(1)}$, $u^{(2)}$. These could contribute terms in the Lagrangian so that $\mathscr{L}$ takes the form,

$$\mathscr{L} = \frac{\rho}{2} \dot{u}_i \dot{u}_i - \frac{1}{2} U^{(1)} [\![u^{(1)}]\!] - \frac{1}{2} U^{(2)} [\![u^{(2)}]\!], \tag{3.1.8}$$

where $U^{(i)}$ are functionals of $u^{(i)}$. The factors of $\frac{1}{2}$ are conventional, while the negative sign allows for the loose interpretation that $U^{(1)}$ ($U^{(2)}$) is the potential energy due to elastic stress (local curvature). In order to determine the form of $U^{(i)}$, we seek refuge in the theory of tensor invariants [18].

2.4 Firstly we deal with the second rank tensor $u^{(1)}$. Since $U^{(1)}$ a scalar formed from a second rank tensor, we choose the simplest such term,

$$U^{(1)} = \alpha_{ijkl} u^{(1)}_{ij} u^{(1)}_{kl}. \tag{3.1.9}$$

Recognising that $u^{(1)}$ commutes under multiplication gives the basic symmetry $\alpha_{ijkl} = \alpha_{klij}$. The assumed translation invariance in space and time implies that α_{ijkl} is a constant. Finally the choice $u^{(1)}_{ij} = u^{(1)}_{ji}$, implies $\alpha_{ijkl} = \alpha_{jikl} = \alpha_{ijlk}$. Thus,

$$\alpha_{ijkl} = \alpha_{jikl} = \alpha_{ijlk} = \alpha_{klij}. \tag{3.1.10}$$

However α_{ijkl} cannot be any tensor that satisfies this symmetry relation—the term $U^{(1)}$ must be invariant under rotations. For second rank tensors in three dimensions, such as $u^{(1)}$, there are three such invariants [18]: $\mathrm{Tr}\, u^{(1)}$, $\mathrm{Tr}\, [u^{(1)}]^2$, and $\det u^{(1)}$. Since $U^{(1)}$ is quadratic in $u^{(1)}$, it must be that $U^{(1)}$ is a linear superposition of only the first two, viz.,

$$\begin{aligned} U^{(1)} &= \mu_1 \left[\mathrm{Tr}\, u^{(1)}\right]^2 + \mu_2 \,\mathrm{Tr}\, \left[u^{(1)}\right]^2 \\ \text{i.e.,}\quad \alpha_{ijkl} u^{(1)}_{ij} u^{(1)}_{kl} &= \mu_1 [u^{(1)}_{ii}]^2 + \mu_2 u^{(1)}_{ij} u^{(1)}_{ji}. \end{aligned} \tag{3.1.11}$$

Together with the symmetry constraints in Eq. (3.1.10), this fixes the form of α_{ijkl}, viz.,

$$\alpha_{ijkl} = \mu_1\, \delta_{ij} \delta_{kl} + \frac{\mu_2}{2} (\delta_{ik} \delta_{jl} + \delta_{jk} \delta_{il}), \tag{3.1.12}$$

in terms of two constants μ_1, μ_2 carrying the dimension of elastic modulus; μ_1 and $\frac{\mu_2}{2}$ are the conventional Lamé constants [7]. Thus, we have an amendment to Eq. (3.1.8):

$$\mathscr{L} = \frac{\rho}{2} \dot{u}_i \dot{u}_i - \frac{1}{2} \alpha_{ijkl} u^{(1)}_{ij} u^{(1)}_{kl} + \ldots. \tag{3.1.13}$$

2.5 The case of the third rank tensor $u^{(2)}$ is much more complicated; its invariants in three dimensions are known [19, 20], and in principle can be used used to determine the form of $U^{(2)}$ (up to a few constants). However, brevity compels restricting attention to the two dimensional—a membrane with negligible thickness—or the one dimensional—a beam with negligible thickness and width—cases. In the former case, only the single transverse field—displacement of the membrane surface orthogonal to itself–is relevant, while in the latter case, there are two independent transverse directions. Choosing Cartesian coordinates where the relevant displacement is u_i ($i = 1$ for a membrane, $i = 1, 2$ for a beam), the lowest order invariant composed of the elements, $u^{(2)}_{i11}, u^{(2)}_{i12} = u^{(2)}_{i21}, u^{(2)}_{i22}$ are the invariants of the 2×2 matrix (indexed by j, k for each i) $u^{(2)}_{ijk}$ [21]. In particular, the two invariants are conveniently expressed as $\mathrm{Tr}[u^{(2)}_{ijk}]^2$ and $\det [u^{(2)}_{ijk}]$, so that,

$$U^{(2)}_i \propto (u^{(2)}_{i11} + u^{(2)}_{i22})^2 + (u^{(2)}_{i11} u^{(2)}_{i22} - (u^{(2)}_{i12})^2),$$

for each independent transverse motion u_i. Since $u^{(2)}$ carries a dimension of inverse length (unlike $u^{(1)}$ which is dimensionless), the proportionality factor depends on a length scale set by the dimensions of the body. The conventional choice is [21],

$$U^{(2)}_i = K_i M_i \left[(u^{(2)}_{i11})^2 + (u^{(2)}_{i22})^2 - 2(1 - \varsigma)(u^{(2)}_{i11} u^{(2)}_{i22} - (u^{(2)}_{i12})^2, \right],$$

where K_i is the elastic modulus, M_i is the moment of inertia about the axis orthogonal to i, and ς is Poisson's ratio. $U^{(2)}_i$ represents the potential energy due to curvature; for beams and membranes, a significant proportion of energy in higher order elastic modes is due to curvature. Indeed, the equations of motion including this term gives the conventional Euler-Bernoulli theory of beams [21]. We shall return to this later when treating the case of a nanobeam, in Sect. 5.1.

To recap, the action Eq. (3.1.3) is given by,

$$\mathscr{S}[\![u_i]\!] = \int \mathrm{d}t \int_V \mathrm{d}^3 r \, \mathscr{L}, \tag{3.1.14}$$

where the Lagrangian is (neglecting terms due to local curvature, $u^{(2)}$),

$$\mathscr{L} = \frac{\rho}{2} \dot{u}_i \dot{u}_i - \frac{1}{2} \alpha_{ijkl} (\partial_i u_j)(\partial_k u_l) = \frac{\rho}{2} \dot{u}_i \dot{u}_i - \frac{1}{2} t_{ij} s_{ij}. \tag{3.1.15}$$

In going to the second equality, we have defined the stress tensor,

$$t_{ij} := \alpha_{ijkl} u^{(1)}_{kl}. \tag{3.1.16}$$

This definition is essentially Hooke's law for a linear elastic medium, with the Hooke tensor α_{ijkl} given by (Eq. 3.1.10),

$$\alpha_{ijkl} = \mu_1 \, \delta_{ij}\delta_{kl} + \mu_2(\delta_{ik}\delta_{jl} + \delta_{jk}\delta_{il}), \tag{3.1.17}$$

where we have re-defined $\frac{\mu_2}{2} \to \mu_2$ for notational simplicity, and to conform with the definition of the Lamé constants currently in vogue [7].

The principle of least action asserts that the motion $\mathbf{u}(\mathbf{r}, t)$ is the one that minimises the action (Eq. 3.1.14), consistent with the appropriate spatial and temporal boundary conditions. Since the three displacement fields u_i are independent, such optimisation maybe performed in a fairly standard manner by setting the functional derivative $\frac{\mathrm{D}\mathscr{S}}{\mathrm{D}u_i} = 0$. This gives the equations of motion of elastodynamics [7, 10] (see Appendix B.1.1 for details),

$$\begin{aligned} \rho\, \ddot{u}_i &= \alpha_{ijkl}\, \partial_j \partial_l u_k = (\mu_1 + \mu_2)\partial_i \partial_j u_j + \mu_2\, \partial_j \partial_j u_i \\ \text{or,}\quad \rho\, \ddot{\mathbf{u}} &= (\mu_1 + 2\mu_2)\nabla(\nabla \cdot \mathbf{u}) - \mu_2\, \nabla \times (\nabla \times \mathbf{u}), \end{aligned} \tag{3.1.18}$$

and a set of natural conditions to be satisfied at the boundary ∂V (see Appendix B.1.2):

$$\begin{aligned} \text{free boundary:}\quad & t_{ij} A_j|_{\partial V} = 0 \\ \text{fixed-surface:}\quad & u_i|_{\partial V} = 0. \end{aligned} \tag{3.1.19}$$

The vectorial form of the equations of motion (Eq. 3.1.18) suggests the existence of two types of elastic excitations (see Appendix B.2): transverse waves (corresponding to $\nabla \cdot \mathbf{u} = 0$) propagating with the velocity $c_T = \sqrt{\mu_2/\rho}$; and, longitudinal waves (corresponding to $\nabla \times \mathbf{u} = 0$) propagating with the velocity $c_L = \sqrt{(\mu_1 + 2\mu_2)/\rho}$. These waves are the long-wavelength excitations of the underlying microscopic medium, described by the effective theory of elastodynamics.

3.1.2 Quantised Modes of the Elastic Field

These excitations maybe quantised on equal footing[6] via the canonical method [22, 23]. An alternate, less formal, route will be pursued here. Firstly, we note that the differential operator,

$$\hat{L}_{ik} := \frac{\alpha_{ijkl}}{\rho}\, \partial_j \partial_l \quad \text{or,} \quad \hat{\mathbf{L}} = c_L^2 \nabla(\nabla \cdot) - c_T^2 \nabla \times (\nabla \times), \tag{3.1.20}$$

is self-adjoint with respect to the inner product defined by (here $\mathrm{Vol}(V) = \int_V \mathrm{d}^3 r$),

[6] Unlike electrodynamics for example, where the transverse excitations are constrained [22].

$$\langle \mathbf{v}, \mathbf{u} \rangle := \frac{1}{\mathrm{Vol}(V)} \int_V v_i^*(\mathbf{r}) u_i(\mathbf{r})\, \mathrm{d}^3 r,$$

as long as one of the boundary conditions in Eq. (3.1.19) is satisfied, and the Hooke tensor satisfies the symmetry constraints in Eq. (3.1.10) (see [24], and Appendix B.3). Therefore, the eigenvectors $\tilde{\mathbf{u}}_n(\mathbf{x})$ defined by solutions of,

$$\hat{\mathbf{L}} \tilde{\mathbf{u}}_n = w_n^2 \tilde{\mathbf{u}}_n,$$

form an orthonormal set [25], i.e.

$$\langle \tilde{\mathbf{u}}_n, \tilde{\mathbf{u}}_{n'} \rangle = \delta_{nn'} \langle \tilde{\mathbf{u}}_n, \tilde{\mathbf{u}}_n \rangle,$$

and is complete [25], i.e.

$$\mathbf{u}(\mathbf{r}, t) = \sum_n \tilde{\mathbf{u}}_n(\mathbf{r}) x_n(t), \tag{3.1.21}$$

for any valid displacement field configuration[7] $\mathbf{u}(\mathbf{r}, t)$. Here, the dimensions of w_n are such that x_n carries the dimension of a length. The Lagrangian,

$$L = \int_V \mathrm{d}^3 r\, \mathscr{L} = \int_V \left(\frac{\rho}{2} \langle \dot{\mathbf{u}}, \dot{\mathbf{u}} \rangle - \frac{1}{2} \langle \mathbf{u}, \hat{\mathbf{L}} \mathbf{u} \rangle \right) \mathrm{d}^3 r,$$

expressed in terms of the mode expansion (in Eq. 3.1.21) takes the form:

$$L = \sum_n \frac{m_n}{2} \dot{x}_n^2(t) - \frac{k_n}{2} x_n^2(t) =: \sum_n L_n; \tag{3.1.22}$$

i.e., it describes a sum of simple harmonic oscillators, one for each elastic mode with generalised coordinates $x_n(t)$. The nth oscillator is characterised by a "mass" m_n [26], and "spring constant" k_n, respectively,[8]

$$m_n = \int_V \rho\, |\tilde{\mathbf{u}}_n(\mathbf{r})|^2\, \mathrm{d}^3 r, \quad \text{and,} \quad k_n = \int_V w_n^2\, |\tilde{\mathbf{u}}_n(\mathbf{r})|^2\, \mathrm{d}^3 r, \tag{3.1.23}$$

and it oscillates at the frequency, $\Omega_n = \sqrt{k_n/m_n}$.

[7] Since we limit attention to a domain V which is finite, i.e. $\mathrm{Vol}(V) < \infty$, the operator $\hat{\mathbf{L}}$ has a discrete eigenspectrum [25], and so the expansion is necessarily a sum.

[8] Note however that there exists a freedom in the definition of the the generalized coordinates $x_n(t)$: as per Eq. (3.1.21), scale transformations of the mode function, $\tilde{\mathbf{u}}_n(\mathbf{r}) \to \lambda \tilde{\mathbf{u}}_n(\mathbf{r})$, requires that the generalized coordinate transform as, $x_n(t) \to \lambda^{-1} x_n(t)$, so as to ensure that the physical elastic deformation $\mathbf{u}(\mathbf{r}, t)$ remains invariant. Such a scale transformation changes both the mass and the spring constant associated with the mode n, by the factor λ^2, as per Eq. (3.1.23); thus, neither of these quantities as defined is physical. However, the mode frequency Ω_n remains invariant to the scale transformation, and is therefore physical.

The lagrangian L, in Eq. (3.1.22), describes a discrete set of independent harmonic oscillators, each of which can be quantised independently. Following standard procedure [23], the commutation relations between the position operator $\hat{x}_n$ and its conjugate momentum $m_n \dot{\hat{x}}_n$, can be implemented in terms of non-hermitian operators $\hat{b}_n, \hat{b}_n^\dagger$, defined by

$$\hat{b}_n(t) = \frac{1}{2}\left(\frac{\hat{x}_n(t)}{\sqrt{\hbar/2m_n\Omega_n}} + i\frac{m_n \dot{\hat{x}}_n(t)}{\sqrt{\hbar m_n \Omega_n/2}}\right), \tag{3.1.24}$$

satisfying the equal-time commutation relations,

$$\left[\hat{b}_n(t), \hat{b}_{n'}^\dagger(t)\right] = \delta_{nn'}.$$

These quantised excitations, due to the above commutator, are bosons – phonons corresponding to the elastic deformation of the medium.[9] The dynamics of each quantised mode is most conveniently described by its hamiltonian [10], $\hat{H}_n = m_n \dot{\hat{x}}_n^2 - \hat{L}_n$, given by,

$$\hat{H}_n = \frac{m_n}{2}\dot{x}_n^2 + \frac{m_n\Omega_n^2}{2}x_n^2 = \hbar\Omega_n\left(\hat{b}_n^\dagger\hat{b}_n + \frac{1}{2}\right), \tag{3.1.25}$$

where the $\frac{1}{2}$ arises from the intrinsic vacuum fluctuation of the mode. Note that the mass of the oscillator only appears through the definition of $\hat{b}_n$ in Eq. (3.1.24).

3.1.3 Mechanical Oscillator in Thermal Equilibrium

Focusing on one of the harmonic modes (and dropping the mode index henceforth), with generalised position $x(t)$ and frequency Ω_m, we are interested in its description when it is in equilibrium with a thermal environment[10] at temperature T. Using the free hamiltonian $\hat{H}$ (in Eq. 3.1.25),

$$\hat{H} = \hbar\Omega_\mathrm{m}\left(\hat{b}^\dagger\hat{b} + \frac{1}{2}\right),$$

the equilibrium is described by the thermal state (see Eq. 2.2.1),

[9]These phonons are the quantised excitations of the long-distance effective theory of the bulk medium, i.e. elastodynamics; this has to be contrasted with the conventional phonon [27] that results from a quantisation of the short-distance theory of atoms in crystalline order. Consequently, the former—the one of interest here—falls at low energies (typically less than GHz), while the latter falls at much higher frequencies (typically THz).

[10]That this can be done, i.e. that the equipartition principle applied to each mode gives the same result as the thermal equilibrium of the full field $\mathbf{u}(\mathbf{r})$, requires explicit proof [28, 29].

$$\hat{\rho}_{\mathrm{m}} = \frac{\exp\left(-\frac{\hat{H}}{k_B T}\right)}{\mathrm{Tr}\,\exp\left(-\frac{\hat{H}}{k_B T}\right)} = \left(\frac{1}{n_{\mathrm{m,th}}+1}\right)\left(\frac{n_{\mathrm{m,th}}}{n_{\mathrm{m,th}}+1}\right)^{\hat{b}^\dagger \hat{b}}. \tag{3.1.26}$$

In the second equality, the state is parametrised in terms of the mean phonon occupation,

$$n_{\mathrm{m,th}} := \mathrm{Tr}\left[\hat{b}^\dagger \hat{b}\, \hat{\rho}_m\right] = \frac{1}{e^{\hbar\Omega_{\mathrm{m}}/k_B T} - 1} \xrightarrow[k_B T \gg \hbar\Omega_{\mathrm{m}}]{} \frac{k_B T}{\hbar\Omega_{\mathrm{m}}}.$$

The variance in the oscillator position, due to thermal fluctuations is given by,

$$\mathrm{Var}\left[\hat{x}\right] = \mathrm{Tr}[\hat{x}^2\hat{\rho}_{\mathrm{m}}] = (2n_{\mathrm{m,th}}+1)x_{\mathrm{zp}}^2, \quad \text{where,} \quad x_{\mathrm{zp}}^2 := \frac{\hbar}{2m\Omega_{\mathrm{m}}}. \tag{3.1.27}$$

It exhibits a contribution from vacuum fluctuations, i.e. the variance in the position when the mean occupation is zero ($n_{\mathrm{m,th}} = 0$), which defines the *zero-point motion* x_{zp}.

The development of Chap. 2 allows for a finer understanding of the total variance $\mathrm{Var}\left[\hat{x}\right]$; in particular, its distribution in time, or frequency. Specifically, the equilibrium thermal state in Eq. (3.1.26) may be modelled as being enforced by the coupling of the oscillator to a generalised force, $\delta\hat{F}_{\mathrm{th}}(t)$, that describes the fluctuations of the environment degrees of freedom [30–32]. A quantum Langevin equation can be used to describe the resulting dynamics of the oscillator position, viz. [32, 33],

$$\frac{\mathrm{d}^2\hat{x}}{\mathrm{d}t^2} + \Gamma_{\mathrm{m}}\frac{\mathrm{d}\hat{x}}{\mathrm{d}t} + \Omega_{\mathrm{m}}^2\hat{x} = \frac{\delta\hat{F}_{\mathrm{th}}}{m}. \tag{3.1.28}$$

The damping rate, Γ_{m}, introduced here characterises the coupling between the oscillator and its thermal environment. The Fourier transform of the equation,

$$\begin{aligned} \hat{x}[\Omega] &= \chi_x[\Omega]\delta\hat{F}_{\mathrm{th}}[\Omega], \\ \text{where,} \quad \chi_x[\Omega] &:= \left[m\left(-\Omega^2 + \Omega_{\mathrm{m}}^2 - i\Omega\Gamma_{\mathrm{m}}\right)\right]^{-1}, \end{aligned}$$

determines the susceptibility, χ_x, that relates the thermal force to the position. The role of the thermal force $\delta\hat{F}_{\mathrm{th}}$, is to maintain the oscillator in the thermal state $\hat{\rho}_{\mathrm{m}}$.[11] The fluctuation-dissipation theorem (Eq. 2.2.10), essentially codifying this constraint, implies that (in the case, $\Gamma_{\mathrm{m}} \ll \Omega_{\mathrm{m}}$),

[11] It is worthwhile to point out that in fact, the equation of motion in Eq. (3.1.28), is inconsistent with *any* legitimate quantum state when $k_B T \ll \hbar\Gamma_{\mathrm{m}}$, or, $\hbar\Gamma_{\mathrm{m}} \gtrsim \hbar\Omega_{\mathrm{m}}$ [34–36]. Either regimes are irrelevant to this thesis.

$$\left\langle \delta \hat{F}_{\text{th}}(t)\delta \hat{F}_{\text{th}}(0)\right\rangle \approx 2\hbar m\Omega_{\text{m}}\Gamma_{\text{m}}\left(n_{\text{m,th}}+\frac{1}{2}\right)\delta(t)$$

$$\text{and,}\quad \bar{S}^{\text{th}}_{FF}[\Omega] \approx 2\hbar m\Omega_{\text{m}}\Gamma_{\text{m}}\left(n_{\text{m,th}}+\frac{1}{2}\right) \underset{k_B T\gg\hbar\Omega_{\text{m}}}{\longrightarrow} 2m\Gamma_{\text{m}}k_B T.$$

The resulting (symmetrised) spectrum of the oscillator position, is given by,

$$\bar{S}_{xx}[\Omega] = |\chi_x[\Omega]|^2\,\bar{S}^{\text{th}}_{FF}[\Omega] \approx \frac{4x_{\text{zp}}^2}{\Gamma_{\text{m}}}\frac{(\Omega_{\text{m}}\Gamma_{\text{m}})^2}{(\Omega^2-\Omega_{\text{m}}^2)^2+(\Omega\Gamma_{\text{m}})^2}\left(n_{\text{m,th}}+\frac{1}{2}\right). \tag{3.1.29}$$

It is straightforward to verify that,

$$\int_{-\infty}^{\infty}\bar{S}_{xx}[\Omega]\,\frac{\mathrm{d}\Omega}{2\pi} = \text{Var}\left[\hat{x}\right],$$

confirming that the thermal force maintains the oscillator in a thermal equilibrium. Finally note that the spectrum of the zero-point motion of the oscillator,

$$\bar{S}^{\text{zp}}_{xx}[\Omega] := \bar{S}_{xx}[\Omega]|_{n_{\text{m,th}}=0},$$

achieved at zero temperature, exhibits a non-zero peak,

$$\bar{S}^{\text{zp}}_{xx}[\Omega_{\text{m}}] = \frac{2x_{\text{zp}}^2}{\Gamma_{\text{m}}}. \tag{3.1.30}$$

An alternate description sheds light on the origin of the vacuum fluctuation component exhibited in the position fluctuation spectrum $\bar{S}_{xx}[\Omega]$. Applying the alternate form of the fluctuation-dissipation theorem (given in Eq. 2.2.12) for the unsymmetrised spectrum, gives,

$$\begin{aligned} S_{xx}[\Omega>0] &= 2\hbar(n_{\text{m,th}}+1)\ \text{Im}\ \chi_x[\Omega] \\ S_{xx}[\Omega<0] &= 2\hbar\, n_{\text{m,th}}\ \text{Im}\ \chi_x[\Omega], \end{aligned}$$

where the second relation follows from using the detailed balanced condition (Eq. 2.2.4). These equations suggest an alternate, equivalent, description of the mechanical oscillator. The behaviour of $S_{xx}[\Omega]$ is determined by the poles of the imaginary part of the mechanical susceptibility Im χ_x. The four poles,

$$\Omega_* = \pm\Omega_{\text{m}}\left[1-\frac{\Gamma_{\text{m}}^2}{2\Omega_{\text{m}}^2}\mp i\frac{\Gamma_{\text{m}}}{\Omega_{\text{m}}}\left(1-\frac{\Gamma_{\text{m}}^2}{4\Omega_{\text{m}}^2}\right)^{1/2}\right]^{1/2}$$

coalesce to the two poles,

$$\Omega_* \approx \pm\Omega_{\text{m}} - \frac{i\Gamma_{\text{m}}}{2},$$

one each on the left/right half planes, in the regime where $\Gamma_m \ll \Omega_m$. This high-Q approximation essentially amounts to assuming that the resonance at positive (negative) frequency is due to processes that are independent of those at the negative (positive) frequency. It must therefore be possible to introduce degrees of freedom that describe these processes, and which, owing to the first-order nature of the pole at either frequency, obeys a first order differential equation. The creation/annihilation operators $\hat{b}, \hat{b}^\dagger$ are precisely the required degrees of freedom. The equation of motion [37],

$$\frac{d\hat{b}}{dt} = -\left(i\Omega_m + \frac{\Gamma_m}{2}\right) + \sqrt{\Gamma_m}\,\delta\hat{b}_{in}(t), \tag{3.1.31}$$

and its hermitian conjugate, model the two poles Ω_*. The noise operator $\delta\hat{b}_{in}$, satisfying

$$\begin{aligned} \left\langle \delta\hat{b}^\dagger_{in}(t)\delta\hat{b}_{in}(0)\right\rangle &= n_{m,th}\delta(t) \\ \left\langle \delta\hat{b}_{in}(t)\delta\hat{b}^\dagger_{in}(0)\right\rangle &= (n_{m,th}+1)\delta(t), \end{aligned} \tag{3.1.32}$$

models the thermal force due to the environment. Further, since the two processes at positive and negative frequencies are independent, the double-sided spectrum of the position fluctuations, $\hat{x} = x_{zp}(\hat{b} + \hat{b}^\dagger)$, may be expressed as,

$$\begin{aligned} S_{xx}[\Omega] &= x_{zp}^2\left(S_{b^\dagger b^\dagger}[\Omega] + S_{bb}[\Omega]\right) \\ &= \frac{4x_{zp}^2}{\Gamma_m}\left(\frac{(n_{m,th}+1)(\Gamma_m/2)^2}{(\Omega_m+\Omega)^2+(\Gamma_m/2)^2} + \frac{n_{m,th}(\Gamma_m/2)^2}{(\Omega_m-\Omega)^2+(\Gamma_m/2)^2}\right), \end{aligned}$$

i.e. with no cross-correlations between the terms at positive and negative frequency. The formal definition of the spectra of the (non-hermitian) creation/annihilation operators may be used to interpret the term containing the vacuum contribution (proportional to $(n_{m,th}+1)$) as arising from environmental processes that excite the oscillator, followed by a de-excitation, whereas the term devoid of vacuum fluctuation (proportional to $n_{m,th}$) as arising from processes that happen in reverse. Clearly, the oscillator, being quantized, cannot sustain a process where its vacuum state is annihilated. However, the unsymmetrised spectrum $S_{xx}[\Omega]$ is not typically measured; the symmetrised spectrum, $\bar{S}_{xx}[\Omega]$, which can be measured, can no longer distinguish between the two processes—Chap. 7 deals with this subtlety.

3.2 Photons: Description and Detection

The experiments reported in this thesis use coherent electromagnetic fields as a measuring instrument. However, these fields themselves have to be measured in a final step to decipher the information they carry. The realisation that the latter step needs to be analysed carefully arises from a certain conceptual tension: on the one hand, light is a quantum mechanical entity, while on the other, the output of detectors that measure it, are classical. The purpose of this section is to cut through the tension using a formalism that adopts a quantum mechanical description of light, together with a ideas borrowed from quantum measurement theory (outlined in Chap. 2). In addition, it will become apparent, later in Chap. 4, that a quantum mechanical description of light is central to the subject of this thesis.

The classical electromagnetic field propagating along a specific direction—solutions of Maxwell's equations with radiative boundary conditions [38]—in a homogeneous isotropic linear medium is fully described by the Cartesian components of its transverse vector potential.[12] Fixing the propagation direction to be along the z−axis, the transverse vector potential, $\mathbf{A}_\perp(z,t) = (A_x(z,t), A_y(z,t))$, satisfies the property, $\mathbf{A}_\perp(z,t) = \mathbf{A}_\perp(0, t - z/c)$, where c is the propagation speed. It is thus sufficient to consider the field as a function of time alone, i.e. the quantity, $\mathbf{A}_\perp(t) := \mathbf{A}_\perp(0,t)$. Further simplification is possible if the polarisation is fixed; in this case, coordinates in the transverse plane can be chosen to coincide with the direction of polarisation, allowing the propagating field to be described by the single time-dependent function, $A(t) = \mathbf{A}_\perp(t)\cdot\mathbf{e}_\perp$, where $\mathbf{e}_\perp$ is the direction of polarisation.

Quantisation of the electromagnetic field within this setting amounts to promoting the field to Heisenberg picture operators, resulting in the expression [39],

$$\hat{A}(t) = \int_0^\infty \left(\frac{\hbar c}{\mathscr{A}_\perp \epsilon_0 \omega}\right)^{1/2} i\left(\hat{a}[\omega]e^{-i\omega t} - \hat{a}[\omega]^\dagger e^{i\omega t}\right)\frac{\mathrm{d}\omega}{2\pi}, \tag{3.2.1}$$

together with the canonical commutation relations,

$$\left[\hat{a}[\omega], \hat{a}[\omega']^\dagger\right] = 2\pi\,\delta[\omega - \omega'], \tag{3.2.2}$$

with all other commutators vanishing. The electric field corresponding to the vector potential in Eq. (3.2.1) is,

$$\hat{E}(t) = \int_0^\infty E_0(\omega)\left(\hat{a}[\omega]e^{-i\omega t} + \hat{a}[\omega]^\dagger e^{i\omega t}\right)\frac{\mathrm{d}\omega}{2\pi}, \quad \text{where,} \quad E_0(\omega) := \left(\frac{\hbar\omega}{\mathscr{A}_\perp \epsilon_0 c}\right)^{1/2}. \tag{3.2.3}$$

[12]The electric and magnetic fields are not independent degrees of freedom of the electromagnetic field—this has to do with the two Gauss laws that constrain them. The scalar and vector potentials, on the other hand, do provide the necessary degrees of freedom. The constraint imposed by Gauss laws are identically satisfied by the vector potential, and fixes the scalar potential. Of the remaining degrees of freedom—the components of the vector potential—choice of gauge leaves two components free. These are the transverse components of the vector potential. See [22, 39] for details.

Each propagating mode, at frequency ω, is thus described by the operator $\hat{a}[\omega]$. Since they obey the commutation relation (Eq. 3.2.2) appropriate to a bosonic field, the electromagnetic field at each frequency is due to a flux of bosonic excitations—photons—in a quantum description. The operator $\hat{a}[\omega]^\dagger\hat{a}[\omega]$ describes the photon flux through the transverse area $\mathscr{A}_\perp$; it is thus customary to refer to $a[\omega]$ as the amplitude flux operator.

In order to describe a monochromatic field, consisting of a carrier of amplitude $\bar{a}$ at frequency ω_ℓ, it is necessary that the field $\hat{a}$ consist of additional fluctuations, viz.

$$\hat{a}[\omega] = \bar{a}\cdot 2\pi\,\delta[\omega-\omega_\ell] + \delta\hat{a}[\omega-\omega_\ell], \tag{3.2.4}$$

where $\delta\hat{a}[\omega-\omega_\ell]$ represent fluctuations around the carrier. This is necessitated by the basic commutation relation in Eq. (3.2.2) which precludes the possibility that $\delta\hat{a}[\omega]=0$. Considering field fluctuations in a frequency bandwidth 2Λ around ω_ℓ, i.e. $\omega\in(\omega_\ell-\Lambda,\omega_\ell+\Lambda)$, the electric field operator in Eq. (3.2.3) may be expressed as,

$$\begin{aligned}\hat{E}(t) &= E_0(\omega_\ell)(\bar{a}e^{-i\omega_\ell t}+\bar{a}^*e^{i\omega_\ell t})\\ &\quad+\int_{\omega_\ell-\Lambda}^{\omega_\ell+\Lambda} E_0(\omega)\left(\delta\hat{a}[\omega-\omega_\ell]e^{-i\omega t}+\delta\hat{a}[\omega-\omega_\ell]^\dagger e^{i\omega t}\right)\frac{\mathrm{d}\omega}{2\pi}\\ &= E_0(\omega_\ell)(\bar{a}e^{-i\omega_\ell t}+\bar{a}^*e^{i\omega_\ell t})\\ &\quad+\int_{-\Lambda}^{+\Lambda} E_0(\omega_\ell+\Omega)\left(\delta\hat{a}[\Omega]e^{-i(\Omega+\omega_\ell)t}+\delta\hat{a}[\Omega]^\dagger e^{i(\Omega+\omega_\ell)t}\right)\frac{\mathrm{d}\Omega}{2\pi};\end{aligned}$$

here, $\Omega=\omega-\omega_\ell$, denotes the frequency shift from the carrier. Typical optical fluctuations that are detected are at frequency offsets, $\Omega\ll\omega_\ell$; in this case, firstly $E_0(\omega_\ell+\Omega) = E_0(\omega_\ell)\sqrt{1+\Omega/\omega_\ell}\approx E_0(\omega_\ell)$, and secondly, the formal limit $\Lambda\to\infty$ may be taken, so that,

$$\frac{\hat{E}(t)}{E_0(\omega_\ell)}\approx\left(\bar{a}+\int_{-\infty}^{+\infty}\delta\hat{a}[\Omega]e^{-i\Omega t}\frac{\mathrm{d}\Omega}{2\pi}\right)e^{-i\omega_\ell t}+\left(\bar{a}^*+\int_{-\infty}^{+\infty}\delta\hat{a}[\Omega]^\dagger e^{i\Omega t}\frac{\mathrm{d}\Omega}{2\pi}\right)e^{i\omega_\ell t}, \tag{3.2.5}$$

Thus, the fluctuations about the carrier, $\delta\hat{a}[\Omega]$, may be identified as the (double-sided) Fourier transform of a time-domain operator $\delta\hat{a}(t)$ that varies slowly compared to ω_ℓ. The assumptions that lead up to the above expression defines the situation where an elaborate multi-mode description of the electromagnetic field becomes equivalent to that of a single time-varying mode,

$$\hat{a}(t) = \left(\bar{a}+\delta\hat{a}(t)\right)e^{-i\omega_\ell t}, \tag{3.2.6}$$

with fluctuations at frequencies $\Omega \ll \omega_\ell$ around the carrier. The operator $\hat{a}(t)$ describes the amplitude of the photon flux per unit time. The commutation relations in Eq. (3.2.2) (for the field at optical frequencies), imply that $\delta\hat{a}[\Omega]$ (representing fluctuations at much lower frequencies) satisfies,[13]

$$\begin{aligned} \left[\delta\hat{a}[\Omega], \delta\hat{a}[\Omega]^\dagger\right] &= 2\pi\, \delta[\Omega - \Omega'], \\ \text{or,} \quad \left[\delta\hat{a}(t), \delta\hat{a}(t)^\dagger\right] &= \delta(t - t'). \end{aligned} \tag{3.2.7}$$

3.2.1 *Quadrature, Number, and Phase Operators*

If the expression for the amplitude operator in Eq. (3.2.6) were a classical one (i.e. not an operator), then it could be equivalently expressed in two different ways: the complex amplitude could be expressed as a Cartesian decomposition, or as a polar decomposition, of two real numbers. Kinematically, the former would furnish a description in terms of canonically conjugate variables of the single-mode, while the latter, in terms of action-angle variables [10, 15].

In a quantum mechanical description, the first program can be succesfully carried out via the introduction of *quadrature* operators,

$$\begin{aligned} \delta\hat{q}(t) &:= \frac{1}{\sqrt{2}} \left(\delta\hat{a}(t) + \delta\hat{a}(t)^\dagger\right), \\ \delta\hat{p}(t) &:= \frac{1}{i\sqrt{2}} \left(\delta\hat{a}(t) - \delta\hat{a}(t)^\dagger\right), \end{aligned} \tag{3.2.8}$$

that satisfy,

$$\begin{aligned} \left[\delta\hat{q}(t), \delta\hat{p}(t')\right] &= i\, \delta(t - t'), \\ \text{or,} \quad \left[\delta\hat{q}[\Omega], \delta\hat{p}[\Omega']\right] &= i \cdot 2\pi\, \delta[\Omega - \Omega'], \end{aligned} \tag{3.2.9}$$

in terms of which, the amplitude operator takes the form,

$$\hat{a}(t) = \left(\bar{a} + \frac{\delta\hat{q}(t)}{\sqrt{2}} + i\frac{\delta\hat{p}(t)}{\sqrt{2}}\right) e^{-i\omega_\ell t}. \tag{3.2.10}$$

The second program turns out be far less trivial in quantum mechanics. Ideally, one would require an expression of the form, $\hat{a} = \sqrt{\hat{N}} \exp(i\hat{\Phi})$, with $\hat{N}$, $\hat{\Phi}$ self-adjoint, respectively identical to the photon number flux operator and phase operator. However, this turns out to be impossible to achieve in general in an infinite dimensional Hilbert space with a definite ground state [41, 42]. Physically of course, a phase operator should not be sensible for a single-mode of the field, since the outcome of

[13] These commutation relations are approximate, with corrections proportional to $\left(\frac{E_0(\Omega+\omega_\ell)}{E_0(\omega_\ell)}\right)^2 = 1 + \frac{\Omega}{\omega_\ell}$ [40].

any measurement modeled by such an operator must necessarily involve comparison with an auxiliary mode that provides a phase reference [43]. Nevertheless, a classical correspondence can be invoked to get operators that describes small fluctuations in the phase and number relative to the carrier [44].

Classical fluctuations in the number, denoted δn, and phase, denoted $\delta\phi$, may be modeled as fluctuations of the mean amplitude $\bar{a}$; i.e. via the replacement,

$$\bar{a} \mapsto \sqrt{n_a + \delta n}\, e^{i\delta\phi},$$

where $n_a := \langle \hat{a}^\dagger \hat{a} \rangle$ is the mean photon flux in the carrier. The linearised expression for small fluctuations,

$$\bar{a} \approx \sqrt{n_a} + \frac{\delta n}{2\sqrt{n_a}} + i\sqrt{n_a}\,\delta\phi,$$

is consistent with the Cartesian decomposition in Eq. (3.2.8), if the identification,

$$\delta\hat{q} \mapsto \delta\hat{q} + \frac{\delta n}{\sqrt{2n_a}}, \qquad \delta\hat{p} \mapsto \delta\hat{p} + \sqrt{2n_a}\,\delta\phi,$$

is made. This is the classical correspondence between number and phase fluctuations on the one hand, and quadrature fluctuations on the other. Thus, operators representing number and phase fluctuations may be defined by reading the correspondence backwards, i.e. (here, $\langle \hat{n} \rangle := n_a$)

$$\delta\hat{n} := \sqrt{2\langle \hat{n} \rangle}\,\delta\hat{q}, \qquad \delta\hat{\phi} := \frac{\delta\hat{p}}{\sqrt{2\langle \hat{n} \rangle}}. \tag{3.2.11}$$

These relations suggest the nomenclature, "amplitude" (for $\delta\hat{q}$) and "phase" (for $\delta\hat{p}$) quadratures. Note finally that these operators are canonically conjugate [45], i.e.,

$$[\delta\hat{n}(t), \delta\hat{\phi}(t')] = [\delta\hat{q}(t), \delta\hat{p}(t')] = i\delta(t - t');$$

the caveat being that the definitions in Eq. (3.2.11) does not hold for all states of the field.

3.2.2 *Quantum and Classical Fluctuations in the Optical Field*

The behaviour of the electric field, in Eq. (3.2.3), is specified through the state of the field at each mode at frequency[14] ω. In fact, the ansatz in Eq. (3.2.4), describing the

[14]Since each mode is independent, as indicated by the commutator Eq. (3.2.2), the states of the field live in a Hilbert space that is formed by an infinite continuous tensor product space $\otimes_\omega \mathscr{H}_\omega$,

separation of the field into a monochromatic carrier with vacuum fluctuations around it, corresponds to a state with a coherent excitation with complex amplitude $\bar{a}$ at the carrier frequency ω_ℓ,

$$|\bar{a}, \omega_\ell\rangle = \hat{D}\left(\bar{a}, \omega_\ell\right)|0\rangle := \exp\left(\bar{a}\,\hat{a}[\omega_\ell]^\dagger - \bar{a}^*\,\hat{a}[\omega_\ell]\right)|0\rangle, \tag{3.2.12}$$

the so-called coherent state [47, 48]. Indeed, the action of the *displacement operator*, $\hat{D}(\bar{a}, \omega_\ell)$, in the Heisenberg picture, viz.,

$$\hat{a}[\omega] \mapsto \hat{D}(\bar{a}, \omega_\ell)^\dagger \hat{a}[\omega] \hat{D}(\bar{a}, \omega_\ell) = \bar{a} \cdot 2\pi\, \delta[\omega - \omega_\ell] + \underbrace{\hat{a}[\omega]}_{\delta\hat{a}[\omega-\omega_\ell]}$$

is to induce a separation of the carrier at ω_ℓ from the fluctuations at other freqencies. The ansatz in Eq. (3.2.4) results from the identification that the transformed operator is $\delta\hat{a}[\Omega]$. The central outcome of such an identification is that the coherent state of $\hat{a}[\omega]$ becomes the vacuum state of $\delta\hat{a}[\Omega]$. To see this, we start from an equivalent expression for the coherent state, viz.

$$\hat{a}[\omega]|\bar{a}, \omega_\ell\rangle = \bar{a} \cdot 2\pi\, \delta[\omega - \omega_\ell]|\bar{a}, \omega_\ell\rangle,$$

which states that $|\bar{a}, \omega_\ell\rangle$ is the eigenstate of $\hat{a}[\omega_\ell]$, with eigenvalue (proportional to) $\bar{a}$. This equation can be manipulated using the properties of the displacement operator, viz.

$$\begin{aligned}
& \hat{a}[\omega]\hat{D}(\bar{a}, \omega_\ell)|0\rangle = \bar{a} \cdot 2\pi\, \delta[\omega - \omega_\ell]\, \hat{D}(\bar{a}, \omega_\ell)|0\rangle \\
\Rightarrow\ & \hat{D}(\bar{a}, \omega_\ell)^\dagger \hat{a}[\omega]\hat{D}(\bar{a}, \omega_\ell)|0\rangle = \bar{a} \cdot 2\pi\, \delta[\omega - \omega_\ell]\, \hat{D}(\bar{a}, \omega_\ell)^\dagger \hat{D}(\bar{a}, \omega_\ell)|0\rangle \\
\Rightarrow\ & \left(\bar{a} \cdot 2\pi\, \delta[\omega - \omega_\ell] + \delta\hat{a}[\Omega]\right)|0\rangle = \bar{a} \cdot 2\pi\, \delta[\omega - \omega_\ell]\,|0\rangle \\
\Rightarrow\ & \delta\hat{a}[\Omega]|0\rangle = 0,
\end{aligned}$$

showing that the operator $\delta\hat{a}[\Omega]$ annihilates the vacuum at all sideband frequencies $\Omega = \omega - \omega_\ell$; complementarily, $\delta\hat{a}[\Omega]^\dagger$, creates excitations from the vacuum.[15] Thus, when the field is in the ideal coherent state, $\delta\hat{a}[\Omega]$ represents vacuum fluctuations at all frequencies about the carrier. This equivalence allows us to work exclusively with the fluctuation operator $\delta\hat{a}$, and its quadratures $\delta\hat{q}, \delta\hat{p}$, when describing the noise

---

one for each mode. Such objects may be dealt with using the normal rules of Hilbert spaces (i.e. by using the rules applicable to a denumerable tensor product), if there exists a state $|0\rangle$ such that $\int d\omega\, \hat{a}[\omega]^\dagger \delta\hat{a}[\omega]|0\rangle = 0$ [46]. This state, the *vacuum* state, will henceforth be assumed to exist.

[15]Instead of a coherent state, had we started from a state of the form, $\hat{D}|\psi\rangle$, then, a trivial extension of the proof above would show that the fluctuations in the sidebands are described by the state $|\psi\rangle$. A slightly less trivial extension shows that, if the state of the field were a mixed state of the form, $\hat{D}^\dagger \hat{\rho} \hat{D}$, then the fluctuations in the sidebands are described by the mixed state $\hat{\rho}$ (see [49, 50] for examples).

properties of the field in the coherent state. In particular, the annihilation property above, together with the commutation relations in Eq. (3.2.9) gives the two-time correlators of the quadratures,

$$\begin{pmatrix} S_{qq}(t) & S_{qp}(t) \\ S_{pq}(t) & S_{pp}(t) \end{pmatrix} := \begin{pmatrix} \langle \delta\hat{q}(t)\delta\hat{q}(0)\rangle & \langle \delta\hat{q}(t)\delta\hat{p}(0)\rangle \\ \langle \delta\hat{p}(t)\delta\hat{q}(0)\rangle & \langle \delta\hat{p}(t)\delta\hat{p}(0)\rangle \end{pmatrix} = \frac{1}{2}\begin{pmatrix} 1 & i \\ -i & 1 \end{pmatrix}\delta(t),$$

or equivalently, the two-time correlators of the creation/annihilation operators (following the notational convention introduced in Eq. 2.1.10),

$$\begin{pmatrix} S_{a^\dagger a}(t) & S_{a^\dagger a^\dagger}(t) \\ S_{aa}(t) & S_{aa^\dagger}(t) \end{pmatrix} := \begin{pmatrix} \langle \delta\hat{a}(t)\delta\hat{a}(0)\rangle & \langle \delta\hat{a}(t)\delta\hat{a}(0)^\dagger\rangle \\ \langle \delta\hat{a}^\dagger(t)\delta\hat{a}(0)\rangle & \langle \delta\hat{a}(t)^\dagger\delta\hat{a}(0)^\dagger\rangle \end{pmatrix} = \begin{pmatrix} 0 & 1 \\ 0 & 0 \end{pmatrix}\delta(t).$$

The resulting non-zero value of, $S_{aa}[\Omega] = 1\,(\text{photons/s})/\text{Hz}$, represents a flux of quantum noise in the optical field at all sideband frequencies due to vacuum fluctuations; this noise is equally distributed about the amplitude and phase quadratures, quantified by, $S_{qq}[\Omega] = S_{pp}[\Omega] = \frac{1}{2}\,(\text{photons/s})/\text{Hz}$.

A description of realistic laboratory fields demands a description of fluctuations in excess of the vacuum at sideband frequencies Ω. In this case, the operator $\delta\hat{a}(t)$ (or equivalently, $\delta\hat{q}(t), \delta\hat{p}(t)$) must decribe classical noise in addition to the intrinsic quantum noise—clearly the field in this situation is not in a coherent state. In principle, a detailed knowledge of the source should provide with the correct state to describe the emitted field, including any classical fluctuations (see footnote 15). Such detailed knowledge is however cumbersome to obtain.[16] Therefore, in the following, an operational description is provided that circumvents the issue of specifying the underlying state. A criterion to demarcate classical noise form quantum noise in such a pheonomenological description is also briefly mentioned.

The approach is to construct the most general form of the covariance matrix of the quadratures, allowed by quantum mechanics. Firstly, S_{qq}, S_{pp} are real since the quadratures are self-adjoint and commute among themselves. Secondly, the mutual commutation relation (Eq. 3.2.9) implies,

$$S_{qp}(t) - S_{pq}(t) = i\,\delta(t),$$

which is identically satisfied by the choice,

$$S_{qp}(t) = \frac{+i}{2}\delta(t) + n_{qp}(t), \quad S_{pq}(t) = \frac{-i}{2}\delta(t) + n_{qp}(t),$$

[16]For example, even when the source is an ideal laser, the coherent state is only an approximation of the emitted state [51, 52].

for some function n_{qp}; computing the symmetrised correlation shows that in fact $n_{qp}(t) = \bar{S}_{qp}(t)$. The ideal field state—the coherent state (Eq. 3.2.12)—consisting of vacuum fluctuations at all sideband frequencies has, $S_{qq}(t) = \frac{1}{2}\delta(t) = S_{pp}(t)$. The ansatz,

$$S_{qq}(t) = \frac{1}{2}\delta(t) + n_{qq}(t), \quad S_{pp}(t) = \frac{1}{2}\delta(t) + n_{pp}(t),$$

separates the component due to vacuum fluctuations, while introducing functions $n_{qq(pp)}$ that presumably represent excess classical noise. Thus, we have,

$$\begin{pmatrix} S_{qq}(t) & S_{qp}(t) \\ S_{pq}(t) & S_{pp}(t) \end{pmatrix} = \frac{1}{2}\begin{pmatrix} 1 & i \\ -i & 1 \end{pmatrix}\delta(t) + \begin{pmatrix} n_{qq}(t) & n_{qp}(t) \\ n_{qp}(t) & n_{pp}(t) \end{pmatrix}.$$

that provides an operational framework to describe fluctuations of general field states that is consistent with the commutation relations of the quadrature. The resulting symmetrized spectra of the quadratures,

$$\begin{pmatrix} \bar{S}_{qq}[\Omega] & \bar{S}_{qp}[\Omega] \\ \bar{S}_{pq}[\Omega] & \bar{S}_{pp}[\Omega] \end{pmatrix} = \begin{pmatrix} n_{qq}[\Omega] + \frac{1}{2} & n_{qp}[\Omega] \\ n_{qp}[\Omega] & n_{pp}[\Omega] + \frac{1}{2} \end{pmatrix}, \tag{3.2.13}$$

suggests the interpretation that $n_{qq(pp)}$ is the average photon flux representing classical noise in the amplitude (phase) quadrature. For example, for the field in a thermal coherent state [49], $n_{qq}[\Omega] = n_{pp}[\Omega] = \frac{1}{2}n_\beta(\Omega)$, i.e. half the thermal photon occupation per quadrature; here $n_\beta(\Omega)$ is the Bose occupation given in Eq. 2.2.11.

Consistency with the commutation relations is only a necessary condition for the ansatz chosen above to be physical. The sufficient condition is that the two-time correlators arise as expectation values over some quantum state. The following proposition addresses this requirement.

Proposition 7 ([53]) *The necessary and sufficient condition for the ansatz in Eq.* (3.2.13) *to be physical is that the quadratures satisfy the spectral uncertainty principle of proposition 3, i.e.*

$$\bar{S}_{qq}[\Omega]\bar{S}_{pp}[\Omega] - \bar{S}_{qp}^2[\Omega] \geq \frac{1}{4}. \tag{3.2.14}$$

In terms of the number and phase operators defined in Eq. (7.2.24), and assuming no number-phase correlations, the inequality above takes the form,

$$\bar{S}_{nn}[\Omega]\bar{S}_{\phi\phi}[\Omega] \geq \frac{1}{4}, \tag{3.2.15}$$

for a propagating electromagnetic field. A realistic field is said to be *quantum(-noise)-limited*, when the bound in Eq. (3.2.14) (or, Eq. 3.2.15) is saturated.

The inequality in Eq. (3.2.14) also serves a utilitarian purpose: it allows to bound the (hard-to-measure) classical cross-correlation n_{qp} in terms of the easily measured classical excess noise $n_{qq(pp)}$. Specifically, inserting the expressions for the spectra (in Eq. 3.2.13) into the inequality in Eq. (3.2.14) gives,

$$n_{qp} \leq \left(\tfrac{1}{2}(n_{qq} + n_{pp}) + n_{qq} n_{pp}\right)^{1/2}. \tag{3.2.16}$$

3.2.3 *Detection of Optical Fluctuations*

In this section, we describe the three most commonly employed strategies used to detect fluctuations in optical fields. The basic detecting element in all three strategies is a photoelectric detector, which ultimately emits an electric current corresponding to the incident optical flux. By this very definition, it does not respond to the variation of the electric field at optical frequencies ($\omega_\ell \approx 2\pi \cdot 400\,\mathrm{THz}$). In fact, we shall see that the formalism developed above, for the description of fluctuations about a carrier, anticipates the quantities of the optical field that are detected.

3.2.3.1 Detection of Amplitude Quadrature: Photodetection

A convenient and common way to detect a propagating optical field, described by the amplitude flux operator $\hat{a}(t)$ (as in Eq. 3.2.6), is to couple it to a detector which absorbs a photon and emits an electron via the photoelectric effect; the resulting electric current is called *photocurrent.*

Real photodetectors rarely produce an electron for every photon that is absorbed. The *quantum efficiency* of the detection process, η ($\eta \leq 1$), may be modelled as transmission through a lossy channel of transmissivity η, followed by an ideal photodetector, as shown in Fig. 3.1. The field that falls on the ideal detector, the transmission of the beam-splitter [54],

$$\hat{a}_\eta(t) = \sqrt{\eta}\,\hat{a}(t) + i\sqrt{1-\eta}\,\delta\hat{a}_0(t), \tag{3.2.17}$$

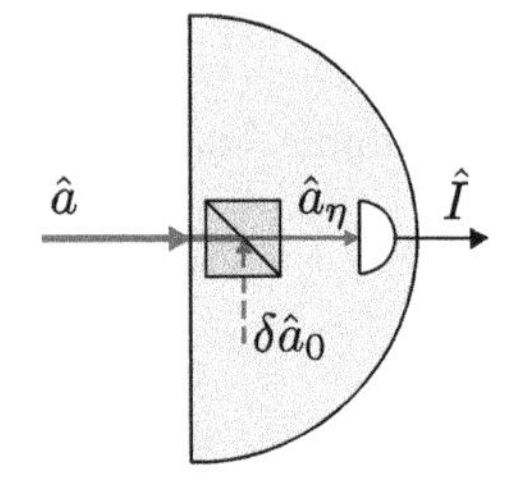

Fig. 3.1 Realistic photodetection. $\hat{a}$, the field to be detected, experiences a fractional loss of η, modelled as being mixed at a beam-splitter of transmissivity η, followed by ideal potodetection

consists of the field to be detected $\hat{a}(t) = (\bar{a} + \delta\hat{a}(t))e^{-i\omega_\ell t}$ as in Eq. (3.2.6), and the vacuum fluctuations $\delta\hat{a}_0$ entering through the other input port. On general grounds the photocurrent operator, $\hat{I}$, that models the detector output, is given by[17] the electron flux; in terms of the incident photon flux $\hat{n}_\eta = \hat{a}_\eta^\dagger \hat{a}_\eta$,

$$\hat{I}(t) = q_e\, \hat{n}_\eta(t) = \mathscr{R}\, \hat{P}_\eta(t) \tag{3.2.18}$$

where q_e is the electron charge, $\mathscr{R}$ is the *responsivity* of the detector given by,

$$\mathscr{R} = \frac{q_e}{\hbar\omega_\ell} = \frac{q_e\lambda_\ell}{hc} \approx (0.63\,\mathrm{A/W})\left(\frac{\lambda_\ell}{780\,\mathrm{nm}}\right), \tag{3.2.19}$$

and, $\hat{P}_\eta = \hbar\omega_\ell\hat{n}_\eta$ is the operator representing the incident optical power. Assuming a strong incident carrier ($|\bar{a}|^2 \gg 1$ photons/s), the photon flux can be linearised as,

$$\hat{n}_\eta = \hat{a}_\eta^\dagger(t)\hat{a}_\eta(t) \approx |\bar{a}|\, \eta\left(|\bar{a}| + \sqrt{2}\,\delta\hat{q}(t)\right) + |\bar{a}|\, \sqrt{2\eta(1-\eta)}\,\delta\hat{p}_0, \tag{3.2.20}$$

were, $\delta\hat{q}$ ($\delta\hat{p}_0$) is the signal (vacuum) amplitude (phase) quadrature fluctuation. The mean and the fluctuations in the incident photon flux, are therefore,

$$\begin{aligned}\langle\hat{n}_\eta\rangle &= \eta\, |\bar{a}|^2 \\ \delta\hat{n}_\eta(t) := \hat{n}_\eta(t) - \langle\hat{n}_\eta\rangle &= \sqrt{2}\, |\bar{a}| \left(\eta\,\delta\hat{q}(t) + \sqrt{\eta(1-\eta)}\,\delta\hat{p}_0(t)\right).\end{aligned}$$

Inserting these in Eq. (3.2.18) gives the mean and fluctuations in the photocurrent,

$$\begin{aligned}\left\langle\hat{I}\right\rangle &= \eta q_e\, |\bar{a}|^2 = \eta\mathscr{R}P \\ \delta\hat{I}(t) &= \sqrt{2}\, q_e\, |\bar{a}| \left(\eta\,\delta\hat{q}(t) + \sqrt{\eta(1-\eta)}\,\delta\hat{p}_0(t)\right),\end{aligned} \tag{3.2.21}$$

where $P = \hbar\omega_\ell\, |\bar{a}|^2$ is the power measured at the entrance of the realistic photodetector—the experimentally accessible optical power. Note that the photocurrent is a continuous observable, in the sense of Eq. (2.3.1), i.e. $\left[\delta\hat{I}(t), \delta\hat{I}(t')\right] = 0$; in fact, when illuminated by a field with a large coherent component, photodetectors perform a linear measurement of the amplitude quadrature fluctuations of the incident field (and not photon-counting [57]).

[17] An argument due to Glauber [55] goes as follows: the state of the optical field $|\psi\rangle$ that arrives at the detector, undergoes the transformation $|\psi\rangle \to \hat{a}|\psi\rangle$, corresponding to the absorption of a photon by the detector; the probability that this happens is proportional to the norm $\|\hat{a}|\psi\rangle\| = \langle\psi|\hat{a}^\dagger\hat{a}|\psi\rangle = \langle\hat{n}\rangle$; finally, the photocurrent operator describes the probability of this process, referred to an electron flux. This heuristic argument can be put on rigorous foundation by analysing the state transformation as a quantum jump process [56].

A photodetector does not distinguish between photons that lie at symmetric frequency offsets from the carrier [57, 58]; thus, it is the symmetrised photocurrent spectrum that is relevant. The (double-sided) spectral density of the photocurrent, computed using Eq. (3.2.21), is,

$$\bar{S}_{II}[\Omega] = 2q_e^2\,|\bar{a}|^2\left(\eta^2\,\bar{S}_{qq}[\Omega] + \eta(1-\eta)\,\bar{S}^0_{pp}[\Omega]\right) \tag{3.2.22}$$

A further approximation may be made at this point. The typical case for photodetection (relevant to this thesis) is where the incident field quadrature carries a signal atop its vacuum fluctuations, viz.

$$\delta\hat{q}(t) = \delta\hat{q}_{\rm sig}(t) + \delta\hat{q}_{\rm sig,0}(t),$$

with the additional assumption that the signal and vacuum are uncorrelated.[18] In this case, $\bar{S}_{qq} = \bar{S}^{\rm sig}_{qq} + \bar{S}^{\rm sig,0}_{qq}$, with, $\bar{S}^{\rm sig,0}_{qq} = \bar{S}^0_{pp} = \frac{1}{2}$; inserting this in Eq. (3.2.22) gives the (single-sided) photocurrent spectrum,

$$\bar{S}_I[\Omega] = 2\eta q_e^2\,|\bar{a}|^2\left(1 + \frac{\eta}{2}\,\bar{S}^{\rm sig}_q[\Omega]\right).$$

The spectral content of the signal rides on a background of photocurrent shot-noise,

$$\bar{S}^{\rm shot}_I[\Omega] = 2\eta q_e^2\,|\bar{a}|^2 = 2q_e\left\langle \hat{I}\right\rangle = 2q_e\cdot\eta\mathscr{R}P,$$

due to the amplified vacuum fluctuations of the incident optical field,[19] with the signal-to-noise determined by the overall detection efficiency.

Realistic photodetectors have an additional source of output noise–noise from the detector electronics—that determines the smallest optical power fluctuation that can be detected. This, the noise equivalent power (NEP) of the detector, $\bar{S}^{\rm NE}_P[\Omega]$, leads to a photocurrent-equivalent spectrum,

$$\bar{S}^{\rm det}_I[\Omega] = \mathscr{R}^2\cdot\bar{S}^{\rm NE}_P[\Omega].$$

Taking this into account gives the expression for the photocurrent spectrum of a realistic photodetector, viz.,

[18] This assumption fails when the incident field has amplitude squeezing—strong correlations between the signal in the amplitude quadrature and the amplitude vacuum fluctuations, in which case, the photocurrent spectrum would contain a term due to the correlation between the signal and vacuum.

[19] The latter expression, in terms of the average photocurrent, may be derived by assuming that the ejected photoelectrons are discrete [59]; this semi-classical interpretation dispenses with the need to attribute any quantum-mechanical character to photodetector shot noise, at least when illuminated by coherent states of the optical field.

$$\bar{S}_I[\Omega] = \underbrace{\mathscr{R}^2 \bar{S}_P^{\rm NE}[\Omega]}_{\bar{S}_I^{\rm det}} + \underbrace{2\eta \cdot q_e \mathscr{R} \cdot P}_{\bar{S}_I^{\rm shot}} + \underbrace{\eta^2 q_e \mathscr{R} P \, \bar{S}_q^{\rm sig}[\Omega]}_{\bar{S}_I^{\rm sig}} . \tag{3.2.23}$$

The signal-to-noise ratio in direct photodetection converges to its maximum possible value when the detector noise is overwhelmed by shot noise. Figure 3.2 shows a measurement of the detector noise and shot noise contributions for a photodetector (NewFocus 1801) widely employed in this thesis. The fits to the shot noise and detector noise model, Eq. (3.2.23), enables extraction of the total quantum efficiency, $\eta \approx 0.78$, consistent with typical quantum efficiencies of $\eta \approx 0.8$ for silicon detectors [60].

3.2.3.2 Detection of an Arbitrary Quadrature: Homodyne

Direct photodetection, having no reference for the phase of the incident field, measures the fluctuations in the amplitude quadrature $\delta\hat{q}$. Other, general quadratures of the form,

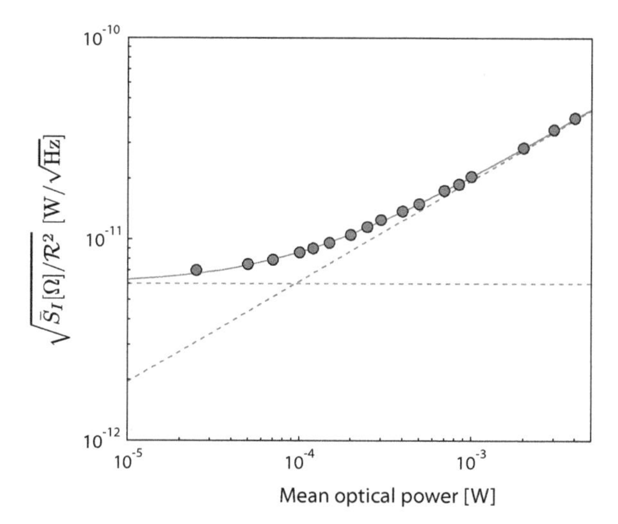

Fig. 3.2 Input-referred photocurrent noise. Measured photocurrent noise at a Fourier frequency $\Omega = 2\pi \cdot 5$ MHz from the optical carrier, referred back to optical power fluctuations. The detector used here is a NewFocus 1801. Below input powers of $\langle P \rangle < 100\,\mu$W, the detector NEP is the dominant source of noise, while above that power, optical shot noise begins to dominate. Solid line shows fit to the model in Eq. (3.2.23); dashed lines show detector and shot noise components of the model. Fit enables inference of $\eta \approx 0.78$

$$\delta\hat{q}^{\theta}(t) := \delta\hat{q}(t)\cos\theta + \delta\hat{p}(t)\sin\theta = \frac{1}{\sqrt{2}}\left(\delta\hat{a}(t)e^{-i\theta} + \delta\hat{a}^{\dagger}(t)e^{i\theta}\right), \tag{3.2.24}$$

furnish continuous observables of interest. From the commutation relation (implied by Eq. (3.2.24) and Eq. 3.2.9),

$$\left[\delta\hat{q}^{\theta}(t), \delta\hat{q}^{\theta'}(t')\right] = i\delta(t-t')\sin(\theta-\theta'), \tag{3.2.25}$$

it is clear that any quadrature $\delta\hat{q}^{\theta}(t)$ is a continuous observable (the case $\theta' = \theta$), while pairs of quadratures $\delta\hat{q}^{\theta}(t)$, $\delta\hat{q}^{\theta+\frac{\pi}{2}}(t)$ are canonically conjugate (the case $\theta' = \theta + \frac{\pi}{2}$). A homodyne detector measures the former; while a heterodyne detector attempts to measure the conjugate observables simultaneously.

Figure 3.3a shows a typical balanced homodyne detector. A local oscillator (LO) and signal field impinge on a balanced (i.e. transmissivity $\eta_t = 0.5$ ideally) beam-splitter such that their transverse mode profiles overlap in both output arms. The output fields [54],

$$\begin{pmatrix}\hat{a}_+\\ \hat{a}_-\end{pmatrix} = \begin{pmatrix}\sqrt{\eta_t} & i\sqrt{1-\eta_t}\\ i\sqrt{1-\eta_t} & \sqrt{\eta_t}\end{pmatrix}\begin{pmatrix}\hat{a}_{\mathrm{sig}}\\ \hat{a}_{\mathrm{LO}}\end{pmatrix}, \tag{3.2.26}$$

are directed onto identical independent photodetectors. The respective photocurrents, $\hat{I}_{\pm}(t) = q_e\hat{a}^{\dagger}_{\pm}(t)\hat{a}_{\pm}(t)$, are subtracted to obtain the homodyne signal,

$$\begin{aligned}\hat{I}_{\mathrm{hom}}(t) &= \hat{I}_+(t) - \hat{I}_-(t)\\ &= q_e(1-2\eta_t)\left(\hat{n}_{\mathrm{LO}}(t) - \hat{n}_{\mathrm{sig}}(t)\right) + 2q_e\sqrt{\eta_t(1-\eta_t)}\, i\left(\hat{a}^{\dagger}_{\mathrm{sig}}(t)\hat{a}_{\mathrm{LO}}(t) - \hat{a}^{\dagger}_{\mathrm{LO}}(t)\hat{a}_{\mathrm{sig}}(t)\right),\end{aligned} \tag{3.2.27}$$

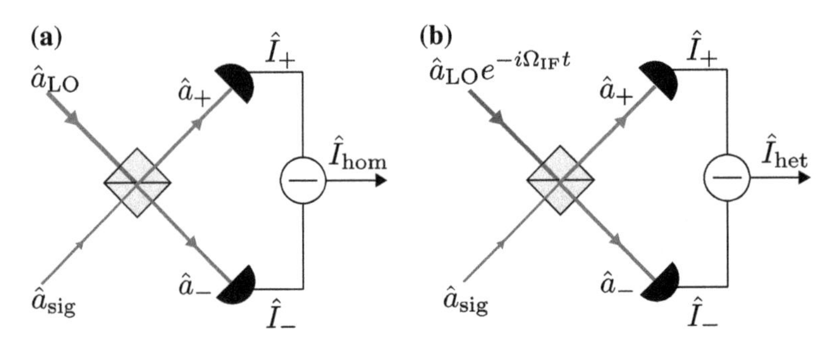

Fig. 3.3 Balanced homodyne and heterodyne detectors. a A strong local oscillator field (LO) overlaps with a (weaker) signal beam on a balanced beam-splitter. The resulting output fields are directed onto independent photodetectors. Their difference photocurrent is the homodyne signal. **b** Compared to a homodyne detector, the LO is frequency shifted with respect to the signal by Ω_{IF}

where $\hat{n}_{\mathrm{LO(sig)}}(t) = \hat{a}^{\dagger}_{\mathrm{LO(sig)}}(t)\hat{a}_{\mathrm{LO(sig)}}(t)$ is the LO (signal) photon flux. The expression for the homodyne photocurrent may be linearised under the assumption that both the LO and signal are coherent. Their amplitudes are then of the form (see Eq. 3.2.6),

$$\hat{a}_{\mathrm{LO(sig)}} = \left(\sqrt{\langle \hat{n}_{\mathrm{LO(sig)}} \rangle} + \delta\hat{a}_{\mathrm{LO(sig)}}(t)\right) e^{-i(\omega_\ell t + \theta_{\mathrm{LO(sig)}})}, \tag{3.2.28}$$

where the mean amplitude is expressed in terms of the mean photon flux, and $\theta_{\mathrm{LO(sig)}}$ are the mean phases of the LO and signal. The mean photocurrent takes the form,

$$\left\langle \hat{I}_{\mathrm{hom}} \right\rangle \approx q_e(1 - 2\eta_t)\left(\langle \hat{n}_{\mathrm{LO}} \rangle - \langle \hat{n}_{\mathrm{sig}} \rangle\right) - q_e\sqrt{2\eta_t(1-\eta_t)}\sqrt{4\langle \hat{n}_{\mathrm{LO}} \rangle \langle \hat{n}_{\mathrm{sig}} \rangle}\cos\theta_{\mathrm{hom}} \tag{3.2.29}$$

while the fluctuation part is,

$$\begin{aligned}\delta\hat{I}_{\mathrm{hom}}(t) \approx\, & q_e(1 - 2\eta_t)\left(\sqrt{2\langle \hat{n}_{\mathrm{LO}} \rangle}\delta\hat{q}^{0}_{\mathrm{LO}} - \sqrt{2\langle \hat{n}_{\mathrm{sig}} \rangle}\delta\hat{q}^{0}_{\mathrm{sig}}\right) \\ & + q_e\sqrt{2\eta_t(1-\eta_t)}\left(\sqrt{2\langle \hat{n}_{\mathrm{LO}} \rangle}\delta\hat{q}^{\theta_{\mathrm{hom}}}_{\mathrm{sig}} + \sqrt{2\langle \hat{n}_{\mathrm{sig}} \rangle}\delta\hat{q}^{-\theta_{\mathrm{hom}}}_{\mathrm{LO}}\right),\end{aligned} \tag{3.2.30}$$

where,

$$\theta_{\mathrm{hom}} := \theta_{\mathrm{sig}} - \theta_{\mathrm{LO}} + \frac{\pi}{2},$$

is the mean phase difference between the signal and LO fields after the combining beam-splitter (including a phase $\pi/2$ due to the beam-splitter).

The fluctuating part of the photocurrent (Eq. 3.2.30) suggests that the homodyne detector measures a combination of the LO and signal quadratures at various angles. The signal quadrature $\delta\hat{q}^{\theta_{\mathrm{hom}}}_{\mathrm{sig}}$ may be singled out by employing a configuration where: (1) the LO is much more powerful than the signal, i.e. $\langle \hat{n}_{\mathrm{LO}} \rangle \gg \langle \hat{n}_{\mathrm{sig}} \rangle$, and, (2) by balancing the combining beam-splitter, i.e. $\eta_t = \frac{1}{2}$. The latter offers the additional technical advantage that excess classical noise in the strong LO (first term in $\delta\hat{I}_{\mathrm{hom}}$ in Eq. 3.2.30) is cancelled [61] (see also Appendix C). In fact, under these two assumptions, the mean and the fluctuations of the photocurrent (Eqs. 3.2.29 and 3.2.30) simplify to,

$$\begin{aligned}\left\langle \hat{I}_{\mathrm{hom}} \right\rangle &\approx -2q_e\sqrt{\langle \hat{n}_{\mathrm{LO}} \rangle \langle \hat{n}_{\mathrm{sig}} \rangle}\cos\theta_{\mathrm{hom}} \\ \delta\hat{I}_{\mathrm{hom}}(t) &\approx q_e\sqrt{\langle 2\hat{n}_{\mathrm{LO}} \rangle}\,\delta\hat{q}^{\theta_{\mathrm{hom}}}_{\mathrm{sig}}(t),\end{aligned} \tag{3.2.31}$$

so that the homodyne photocurrent is a continuous observable providing a linear measurement of the signal quadrature fluctuations $\delta\hat{q}_{\mathrm{sig}}^{\theta_{\mathrm{hom}}}$. In this sense, homodyne detection may be identified as a realisation of a phase-sensitive linear amplifier of the signal quadrature (with gain provided by the LO).

Following manipulations similar to the ones followed for the analysis of photodetection (leading up to Eq. (3.2.23)), the spectrum of the homodyne photocurrent, taking into account non-ideal detection efficiency $\eta \leq 1$, and photodetector noise, is

$$\bar{S}_I^{\mathrm{hom}}[\Omega] = \underbrace{\mathscr{R}^2 \bar{S}_P^{\mathrm{NE}}[\Omega]}_{\bar{S}_I^{\mathrm{hom,det}}} + \underbrace{2\eta \cdot q_e \mathscr{R} \cdot P_{\mathrm{LO}}}_{\bar{S}_I^{\mathrm{hom,shot}}} + \underbrace{2\eta^2 q_e \mathscr{R} P_{\mathrm{LO}}\, \bar{S}_{q_{\theta_{\mathrm{hom}}}}^{\mathrm{sig}}[\Omega]}_{\bar{S}_I^{\mathrm{hom,sig}}}. \tag{3.2.32}$$

Alluding to the analogy between homodyne detection and phase-sensitive linear amplification, homodyne photocurrent shot noise may be interpreted as the vacuum fluctuations of the signal field amplified by the coherent LO, it is thus the intrinsic quantum noise of the signal quadrature, consistent with the quantum limit of a phase-preserving linear amplifier[20] [64].

Conventionally, homodyne detectors are employed to measure the phase quadrature $\delta\hat{p}$ (corresponding to $\theta_{\mathrm{hom}} = \pi/2$). In this case, using the expression for the signal phase operator in Eq. (3.2.11), the homodyne photocurrent spectrum may be expressed in terms of the signal phase noise spectrum, viz., (omitting the detector noise contribution for brevity),

$$\begin{aligned}
\bar{S}_I^{\mathrm{hom}}[\Omega]|_{\theta_{\mathrm{hom}}=\pi/2} &= 2\eta \cdot q_e \mathscr{R} \cdot P_{\mathrm{LO}} + 4\eta^2 \mathscr{R}^2 P_{\mathrm{LO}} P_{\mathrm{sig}} \bar{S}_{\phi\phi}^{\mathrm{sig}}[\Omega] \\
&= 4\eta^2 \mathscr{R}^2 P_{\mathrm{LO}} P_{\mathrm{sig}} \left(\bar{S}_{\phi\phi}^{\mathrm{sig}}[\Omega] + \frac{q_e}{2\eta \mathscr{R} P_{\mathrm{sig}}} \right).
\end{aligned}$$

The second term in the second line gives the phase imprecision in the homodyne signal due to photocurrent shot noise, viz.,

$$\bar{S}_{\phi\phi}^{\mathrm{hom,imp}}[\Omega] = \frac{1}{2\eta} \frac{\hbar\omega_\ell}{P_{\mathrm{sig}}}.$$

Since photocurrent shot noise arises from the strong LO amplifying the signal vacuum fluctuations, this represents the so-called "standard noise limit" for interferometric phase measurement, where the scaling is due to the quantum statistics of coherent states [65, 66].

---

[20] Indeed, for microwave signals, quantum-noise-limited homodyne detection is implemented by using a phase-preserving linear amplifier [62, 63].

3.2.3.3 Design and Operation of a Homodyne Detector

Irrespective of the quadrature being measured, the primary experimental challenge of operating a homodyne detector is to maintain a stable phase $\theta_{\rm hom}$, between the signal and LO fields.[21] The strategy employed in this thesis to attain this is detailed below.

Figure 3.4a depicts the essential layout of the homodyne interferometer employed in this thesis. At its heart is an optical interferometer in Mach-Zehnder configuration. Light from a laser source (at 780 nm, either an external cavity diode laser—NewFocus Velocity, or a Ti:Sa—Sirah Matisse) is appropriately attenuated, and intensity stabilised (IS, Thorlabs LCC3112/M). Polarisation is then cleaned and aligned, before passing through a broadband electro-optic modulator (EOM, NewFocus 4002, bandwidth DC$-$100 MHz). A subsequent half-wave plate divides the light at a polarising beam-splitter to derive the LO and signal fields; the half-wave plate orientation controls their respective powers. The signal field polarisation may be adjusted accordingly thereafter, before being coupled into an optical fiber that is $\approx$10 m long. The LO field is also coupled into a fiber. Both input fiber couplers rest on translation stages, the signal coupler on a manual micrometer stage, while the LO coupler on an electronically controlled one. Both fields subsequently exit into free space. The LO field is reflected off of a mirror mounted on a piezoelectric stack (PZT). The LO and signal are combined at a non-polarising beam-splitter, after their polarisations are aligned. The outputs of the combining beam-splitter are focused onto the two ports of a balanced photodetector (BPD, Femto HCA-S, bandwidth DC$-$125 MHz).

In order to enforce a stable homodyne phase $\theta_{\rm hom}$, both the LO and signal are derived from the same laser source. However, the path length difference between the LO and signal arms of the interferometer determines the homodyne phase,

$$\theta_{\rm hom} = \frac{2\pi}{\lambda}\left(\frac{L_{\rm sig}}{\nu_{\rm sig}} - \frac{L_{\rm LO}}{\nu_{\rm LO}}\right) \approx \frac{2\pi}{\lambda \nu_{\rm eff}}(L_{\rm sig} - L_{\rm LO}), \tag{3.2.33}$$

where λ is the wavelength of light used, $L_{\rm sig(LO)}$ is the physical length of the signal (LO) path, $\nu_{\rm sig(LO)}$ is the refractive index of the signal (LO) path, and $\nu_{\rm eff} \approx 1.5$ is the

[21] Another technical challenge is the ability to realise perfect balancing of powers at the interfering beam-splitter (i.e. $\eta_t = \frac{1}{2}$). Any deviation leads to imperfect cancellation of LO excess noise (see Eq. (3.2.30), and discussion below it). Assuming that $P_{\rm LO} \gg P_{\rm sig}$ (typically, $P_{\rm LO} \gtrsim 100 P_{\rm sig}$, in this thesis), Eq. (3.2.30) implies that the signal quadrature imprecision from imperfect LO noise cancellation is,

$$\bar{S}^{\rm sig,imp}_{q_{\theta_{\rm hom}}}[\Omega] = \frac{(1-2\eta_t)^2}{2\eta_t(1-\eta_t)}\bar{S}^{\rm LO}_q[\Omega] \approx 8\left(\eta_t - \tfrac{1}{2}\right)^2 \bar{S}^{\rm LO}_q[\Omega].$$

Our design enables $\eta_t - \frac{1}{2} \approx \pm 0.05$; combined with the large Fourier frequencies we work at (a few MHz), where the LO can be shot-noise limited up to $P_{\rm LO} = 1 - 2$ mW, this source of homodyne imprecision is negligible [67].

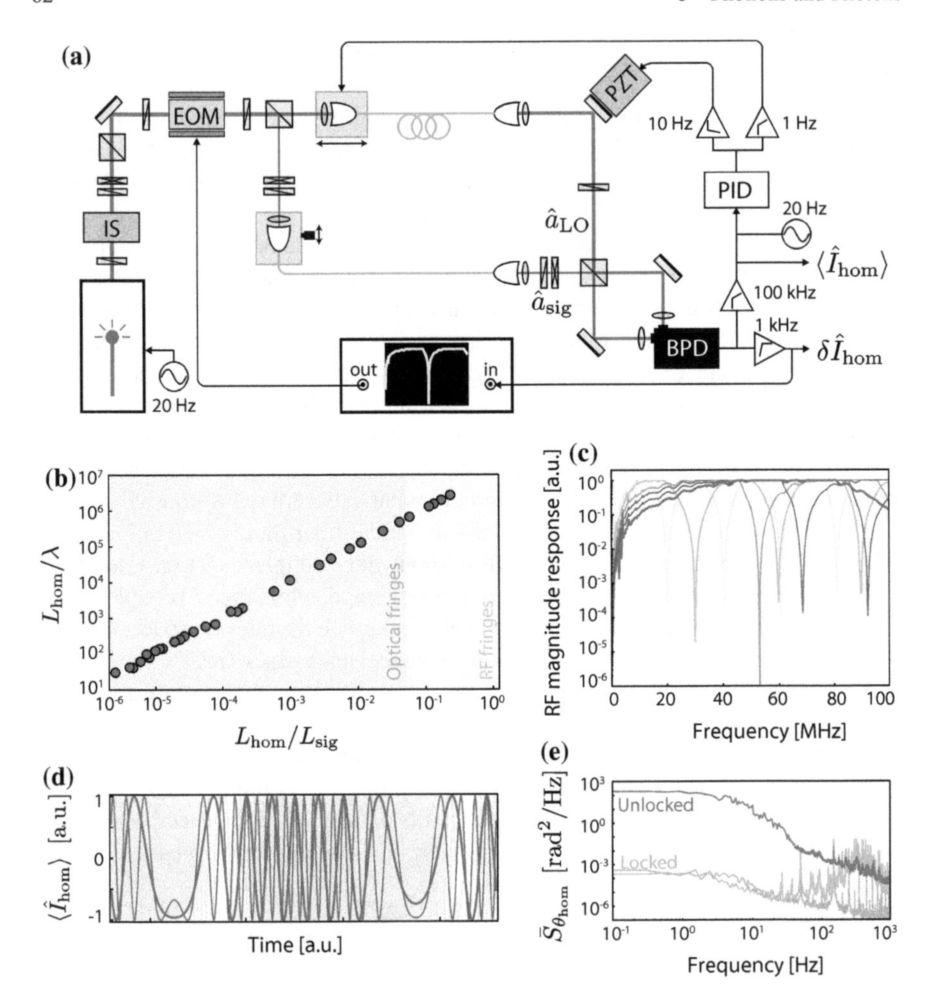

Fig. 3.4 Design and operation of homodyne interferometer. **a** Essential design of the homodyne interferometer, in Mach-Zehnder configuration, that is used in experiment reported in this thesis. Red lines denote free space optical beams, green lines are optical fibres and black lines are radio-frequency electric cables (BNC or SMA). See text for further details. **b** Balancing the interferometer by deterministically changing the physical path length difference L_{hom}. At each stage, counting radio-frequency or optical interference fringes, allows estimation of the imbalance L_{hom}/λ. **c** Magnitude response of the interferometer to an input optical phase modulation injected using the EOM. A radio-frequency network analyser is used to monitor the resulting RF interference. **d** Optical interference fringes in the final stages of balancing. Light and dark blue shows fringe count reduced by micron-scale changes in L_{hom}. Red shows the error signal used to perform active stabilisation of the interferometer. **e** Noise in the homodyne phase θ_{hom} compared for the case where the interferometer is locked (green) vs. free-running (red). Gray trace shows the limit set by electronic noise. (See text for details)

approximation assumed for the relevant case where the signal and LO predominantly propagate through an optical fiber. Since we typically need $0 \leq \theta_{\mathrm{hom}} \leq \frac{\pi}{2}$, the fractional path length difference,[22]

$$\frac{L_{\mathrm{hom}}}{\lambda} := \frac{L_{\mathrm{sig}} - L_{\mathrm{LO}}}{\lambda}, \tag{3.2.34}$$

needs to be stabilised to order unity for the Mach-Zehnder arms (see Fig. 3.4); this is done in three steps.

The first two steps of the procedures rely on counting the interference fringes in the photocurrent $\left\langle \hat{I}_{\mathrm{hom}} \right\rangle \propto \sin\theta_{\mathrm{hom}}$ (Eq. (3.2.29)) to estimate the length imbalance [70]. In the first step, a radio-frequency (RF) response measurement is performed on the optical interferometer. This is done by driving a phase modulator placed at the interferometer input using a network analyser (Agilent, E5061B), and demodulating the response of the interferometer. Figure 3.4c shows a series of such measurements, for varying L_{hom}. The phase modulation $\phi(t)$, effectively a frequency modulation $\omega_\ell(t) = \frac{\mathrm{d}\phi}{\mathrm{d}t}$, leads to a photocurrent,

$$\langle I_{\mathrm{hom}}(t)\rangle \propto \sin\left(\frac{2\pi L_{\mathrm{hom}}}{c v_{\mathrm{eff}}}\omega_\ell(t)\right) \approx \sin\left(\frac{2\pi L_{\mathrm{hom}}}{c v_{\mathrm{eff}}}\omega_\ell(0) + 2\pi t \cdot \frac{L_{\mathrm{hom}}}{c v_{\mathrm{eff}}}\frac{\mathrm{d}\omega_\ell}{\mathrm{d}t}\right),$$

exhibiting interference fringes with frequency, $f_{\mathrm{fringe}} = \frac{L_{\mathrm{hom}}}{c v_{\mathrm{eff}}}\frac{\mathrm{d}\omega_\ell}{\mathrm{d}t}$. Thus, the fringe frequency measured (shown in Fig. 3.4c) provides the length imbalance relative to the laser wavelength, and the RF frequency modulation amplitude. The length imbalance so inferred, is reduced by physically cutting and re-splicing the LO optical fiber. This technique however loses sensitivity once the fringe frequency surpasses the photodetector cutoff; typically this happens at $|L_{\mathrm{hom}}| \approx 10\,\mathrm{cm}$.

In a second step, sensitivity to input frequency changes is increased by working at optical frequencies. The diode laser driving the interferometer is wavelength-modulated[23] $\lambda(t) = \lambda(0) + d\lambda(t)$, and the interference fringes in the DC photocurrent (using Eqs. (3.2.29) and (3.2.33)),

[22] For the detection of signals at a few MHz from the carrier, such a stringent condition is not necessary. However, the ability to achieve broadband cancellation of excess phase noise injected at the input of the interferometer [68, 69], for example when using semiconductor diode lasers as in experiments reported in Chap. 6, demands an interferometer whose arms are length-balanced to within an optical wavelength. Indeed, the the contribution of input phase noise in the photocurrent of an imbalanced homodyne interferometer takes the form, $\bar{S}_I^{\mathrm{hom}}[\Omega] \propto \sin^2(\Omega\tau/2)\bar{S}_\phi^{\mathrm{in}}[\Omega]$, where τ is the time delay between the two arms (see Appendix C), and $\bar{S}_\phi^{\mathrm{in}}$ is the spectrum of excess input phase noise.

[23] Formally equivalent to a frequency modulation $d\omega_\ell = -2\pi\frac{c}{\lambda(0)} \cdot \frac{d\lambda}{\lambda(0)}$; however, for an ECDL, by construction, its frequency is modulated via the diode current, providing access to $d\omega_\ell \approx 2\pi \cdot 100\,\mathrm{GHz}$, whereas the wavelength is modulated by mechanically changing the laser cavity length, leading to $d\lambda \approx 10\,\mathrm{nm}$, equivalent to $d\omega_\ell \approx 2\pi \cdot 5\,\mathrm{THz}$.

$$\left\langle \hat{I}_{\rm hom}(t) \right\rangle \propto \sin\left(\frac{2\pi L_{\rm hom}}{\nu_{\rm eff}\lambda(t)}\right) \approx \sin\left(\frac{2\pi L_{\rm hom}}{\nu_{\rm eff}\lambda(0)} + 2\pi t \cdot \frac{L_{\rm hom}}{\nu_{\rm eff}\lambda(0)} \cdot \frac{1}{\lambda(0)}\frac{{\rm d}\lambda}{{\rm d}t}\right),$$

monitored on an oscilloscope (Tektronix DPO3034). The imbalance, reflected in the frequency of the interference fringes [70] $f_{\rm fringe} = \frac{L_{\rm hom}}{\nu_{\rm eff}\lambda(0)} \cdot \frac{1}{\lambda(0)}\frac{{\rm d}\lambda}{{\rm d}t}$, is reduced by cutting and re-splicing the fiber as before—below $|L_{\rm hom}| \approx 1$ cm, it becomes difficult to precisely cut the fiber. Subsequent adjustment is made by using micro-meter stages carrying the signal and LO input fiber couplers (see Fig. 3.4). Interference fringe count gradually reduces, as shown in Fig. 3.4d (blue traces), to a point where they become particularly sensitive to slight external disturbances—typically at $L_{\rm hom} \approx 10\lambda$. Figure 3.4b shows the length imbalance relative to the wavelength inferred from the fringe frequency as the physical length $L_{\rm hom}$ is reduced—the different sensitivities of RF and optical measurements is due to the much smaller wavelength of the latter [70, 71].

Beyond this point, the interferometer has to be actively stabilised. The error signal is the DC photocurrent $\left\langle \hat{I}_{\rm hom} \right\rangle \propto \sin\theta_{\rm hom}$, generated by modulating the length $L_{\rm hom}(t)$. The red trace in Fig. 3.4d shows a typical error signal, corresponding to a few cycles of the phase $\theta_{\rm hom}(t) = \frac{2\pi}{\lambda\nu_{\rm eff}} L_{\rm hom}(t)$ about zero. The length is changed by dithering a mirror placed on a a fast piezo-electric stack in the LO free space path (PZT, in Fig. 3.4a). The error signal is sent through a PID controller with a slow (1 Hz low-pass) and a fast (10-300 Hz bandpass) branch. (The filters are implemented using Stanford Research Systems SR560 pre-amplifiers, running off of its internal battery to reduce sensitivity to 50 Hz fluctuations from power lines. In practice, it is found that appropriate filtering at the PID input is also necessary.) The fast branch actuates the piezo-electric stack, and suppresses high frequency length fluctuations (mostly limited by the onset of piezo-electric resonances at a few kHz). The slow branch actuates a linear motor in the LO path, and is used to counteract slow drifts due to temperature and seismic disturbances. With the optical table floated, the active stabilisation keeps the homodyne interferometer locked indefinitely.

Figure 3.4e shows an in-loop measurement of the apparent fluctuations in $\theta_{\rm hom}$ when the interferometer is locked (green trace), compared against the case where it is unlocked (red trace). The data is calibrated by using the fact that the peak-peak DC photocurrent, when the piezo stack is dithered, corresponds to $\theta_{\rm hom}$ varying by π. At low frequencies ($1 - 100$ Hz) the residual apparent fluctuations in $\theta_{\rm hom}$ is limited by electronic noise (grey trace) in the detection and feedback loops, whereas at frequencies above 1 kHz, the presence of piezo-mechanical resonances limit the applicable gain. Despite these technical limitations, Fig. 3.4e allows an upper-bound of, ${\rm Var}\left[\theta_{\rm hom}^2\right]^{1/2} < 100$ mrad, for the low frequency stability of the homodyne angle.

3.2.3.4 Detection of Conjugate Quadratures: Heterodyne

Contrary to a homodyne detector where the LO and signal share a common carrier frequency, a heterodyne detector (see Fig. 3.3b) is implemented by a frequency detuned LO.[24] For the case where the LO frequency is larger than the signal frequency by Ω_{IF}, the LO and signal fields may be represented by the ansatz (analogous to Eq. (3.2.28) for homodyne detection),

$$\hat{a}_{\mathrm{LO(sig)}}(t) = \left(\sqrt{\langle \hat{n}_{\mathrm{LO(sig)}}\rangle} + \delta\hat{a}_{\mathrm{LO(sig)}}(t)\right) e^{-i\omega_\ell t} \times \begin{cases} e^{-i(\Omega_{\mathrm{IF}} t + \theta_{\mathrm{LO}})} \\ e^{-i\theta_{\mathrm{sig}}} \end{cases} .$$

For the reasons detailed above for the case of homodyne detection (formally, $\Omega_{\mathrm{IF}} = 0$), it is technically useful to perform balanced detection, i.e. combine the LO and signal on a balanced beam-splitter, using a length-matched interferometer, i.e. the LO and signal arrive at the beam-splitter after acquiring equal phases.

In the case of a strong LO and balanced length, the mean and fluctuating parts of the heterodyne photocurrent (in the strong LO, i.e. $\langle \hat{n}_{\mathrm{LO}}\rangle \gg \langle \hat{n}_{\mathrm{sig}}\rangle$, and balanced, i.e. $\eta_t = \frac{1}{2}$, case),

$$\begin{aligned} \left\langle \hat{I}_{\mathrm{het}}\right\rangle &\approx -2q_e \sqrt{\langle \hat{n}_{\mathrm{LO}}\rangle \langle \hat{n}_{\mathrm{sig}}\rangle} \cos(\theta_{\mathrm{het}} + \Omega_{\mathrm{IF}} t) \\ \delta\hat{I}_{\mathrm{het}}(t) &\approx q_e \sqrt{2\langle \hat{n}_{\mathrm{LO}}\rangle}\, \delta\hat{q}_{\mathrm{sig}}^{\theta_{\mathrm{het}}+\Omega_{\mathrm{IF}} t}(t) \end{aligned} \tag{3.2.35}$$

where $\theta_{\mathrm{het}} := \theta_{\mathrm{sig}} - \theta_{\mathrm{LO}} + \pi/2$. Importantly, the photocurrent is not proportional to a unique signal quadrature, but in fact periodically samples all quadratures. In this sense, heterodyne detection may be identified as a realisation of a phase-insensitive linear amplifier of the signal.

Despite the fact that the heterodyne photocurrent operator samples all quadratures, the quadrature commutation relations (Eq. (3.2.25)) conspire to ensure that the heterodyne photocurrent commutes with itself, viz.[25]

$$\left[\delta\hat{I}_{\mathrm{het}}(t), \delta\hat{I}_{\mathrm{het}}(t')\right] = 4i\, q_e^2 \langle \hat{n}_{\mathrm{LO}}\rangle \cdot \delta(t-t') \sin(\Omega_{\mathrm{IF}}(t-t')) = 0,$$

rendering $\delta\hat{I}_{\mathrm{het}}$ a continuous observable.

The statistics of the heterodyne photocurrent is however quite unlike that of the homodyne. In fact, the photocurrent fluctuations (Eq. (3.2.35)) expressed in terms of the amplitude operators (using Eq. (3.2.24)),

$$\delta\hat{I}_{\mathrm{het}}(t) = q_e \sqrt{\langle \hat{n}_{\mathrm{LO}}\rangle} \left(\delta\hat{a}_{\mathrm{sig}}(t) e^{-i\theta_{\mathrm{het}} - i\Omega_{\mathrm{IF}} t} + \delta\hat{a}_{\mathrm{sig}}^\dagger(t) e^{i\theta_{\mathrm{het}} + i\Omega_{\mathrm{IF}} t}\right),$$

has the two-time correlator (omitting the proportionality factor $q_e^2 \langle \hat{n}_{\mathrm{LO}}\rangle$),

[24] A conceptually cleaner way to realise simultaneously detection of two conjugate quadratures is to split the incoming signal at a balanced beam-splitter, and subjecting either output to independent homodyne detectors sharing a common LO, but phase-shifted by $\pi/2$. This arrangement, called a multi-port homodyne detector [72], is clumsier to implement in practice.

[25] The second equality, by common abuse of notation, holds in the *sense of distribution*; i.e. it holds for any arbitrarily close approximation to $\delta(t)$.

$$\left\langle \delta \hat{I}_{\rm het}(t)\delta \hat{I}_{\rm het}(t')\right\rangle \propto \left\langle \delta\hat{a}_{\rm sig}(t)\delta\hat{a}^{\dagger}_{\rm sig}(t')\right\rangle e^{-i\Omega_{\rm IF}(t-t')} + \left\langle \delta\hat{a}^{\dagger}_{\rm sig}(t)\delta\hat{a}_{\rm sig}(t')\right\rangle e^{+i\Omega_{\rm IF}(t-t')}$$
$$- \left\langle \delta\hat{a}_{\rm sig}(t)\delta\hat{a}_{\rm sig}(t')\right\rangle e^{-2i\theta_{\rm het}} e^{-i\Omega_{\rm IF}(t+t')} - \left\langle \delta\hat{a}^{\dagger}_{\rm sig}(t)\delta\hat{a}^{\dagger}_{\rm sig}(t')\right\rangle e^{+2i\theta_{\rm het}} e^{+i\Omega_{\rm IF}(t+t')},$$

which is not stationary [73, 74]. The last two terms, being a periodic modulation of a stationary term, give rise to what is called *cyclostationary noise* [75, 76]. When the signal field carries excitations in a narrow band of frequencies much below the intermediate frequency $\Omega_{\rm IF}$, these non-stationary terms may be omitted.[26] The resulting photocurrent correlator,

$$\begin{aligned}\left\langle \delta \hat{I}_{\rm het}(t)\delta \hat{I}_{\rm het}(0)\right\rangle &\propto \left\langle \delta\hat{a}_{\rm sig}(t)\delta\hat{a}^{\dagger}_{\rm sig}(t')\right\rangle e^{-i\Omega_{\rm IF}t} + \left\langle \delta\hat{a}^{\dagger}_{\rm sig}(t)\delta\hat{a}_{\rm sig}(t')\right\rangle e^{+i\Omega_{\rm IF}t} \\ &= \left\langle \delta\hat{q}(t)\delta\hat{q}(0) + \delta\hat{p}(t)\delta\hat{p}(0)\right\rangle \cos\Omega_{\rm IF}t \\ &\quad + \left\langle \delta\hat{p}(t)\delta\hat{q}(0) + \delta\hat{q}(t)\delta\hat{p}(0)\right\rangle \sin\Omega_{\rm IF}t,\end{aligned} \tag{3.2.36}$$

is independent of the relative signal-LO phase $\theta_{\rm het}$. Note that due to simultaneous detection of conjugate quadratures, any mutual correlations between the two are reflected in the heterodyne photocurrent. Equation (3.2.36) together with the Wiener-Khinchine theorem (Eq. 2.1.6) gives the (single-sided) spectrum of the heterodyne photocurrent:

$$\bar{S}_I^{\rm het}[\Omega] = q_e^2 \left\langle \hat{n}_{\rm LO}\right\rangle \left(S_{aa}^{\rm sig}[\Omega + \Omega_{\rm IF}] + S_{a^\dagger a^\dagger}^{\rm sig}[\Omega - \Omega_{\rm IF}]\right),$$

expressed in terms of the unsymmetrised power spectral density of the (non-hermitian) flux amplitude operators (as defined in Eq. 2.1.10). The left-hand side, being a single-sided spectrum is defined only for $\Omega > 0$; in particular, fluctuations in the optical field originally about the optical carrier are translated to radio-frequencies about the intermediate frequency in the photocurrent. For a detector with bandwidth much less than $2\Omega_{\rm IF}$, the second term above can be neglected, resulting in the spectrum (centred about $\Omega_{\rm IF}$),

$$\bar{S}_I^{\rm het}[\Omega - \Omega_{\rm IF}] \approx q_e^2 \left\langle \hat{n}_{\rm LO}\right\rangle S_{aa}[\Omega]. \tag{3.2.37}$$

In this sense, a heterodyne detector measures the double-sided spectrum of the flux of the optical field, which reflects any correlations between amplitude and phase quadratures.

[26] In the contrary case, these terms give rise to cyclostationary shot noise [73, 77]—shot noise modulated at $\Omega_{\rm IF}$—in excess of the expectation from a stationary shot noise model. It is generally true that cyclostationary noise may be represented as a sum of correlated stationary noise processes [78]—therefore, it is possible to coherently cancel excess cyclostationary shot noise [73, 79, 80], or use the correlations for benefit [81].

Explicitly separating out the vacuum fluctuations, from the signal field, i.e. $\delta\hat{a}_{\mathrm{sig}} \rightarrow \delta\hat{a}_{\mathrm{vac}} + \delta\hat{a}_{\mathrm{sig}}$, and introducing the efficiencies for the detection, the spectrum of the heterodyne photocurrent is,

$$\bar{S}_I^{\mathrm{het}}[\Omega - \Omega_{\mathrm{IF}}] = \underbrace{\mathscr{R}^2 \bar{S}_P^{\mathrm{NE}}[\Omega]}_{\bar{S}_I^{\mathrm{het,det}}} + \underbrace{4\eta \cdot q_e \mathscr{R} \cdot P_{\mathrm{LO}}}_{\bar{S}_I^{\mathrm{het,shot}}} + \underbrace{\eta^2 q_e \mathscr{R} P_{\mathrm{LO}}\, S_{aa}^{\mathrm{sig}}[\Omega]}_{\bar{S}_I^{\mathrm{het,sig}}}. \quad (3.2.38)$$

Note that compared to homodyne detection Eq. (3.2.32), the shot noise contribution is twice larger, and the signal twice smaller—the former is due to the shot noise from both quadratures being detected, while the latter is due to the signal being spread symmetrically about the intermediate frequency (i.e. double-sidedness). In effect, heterodyne detection is four times less sensitive compared to a homodyne detector. The advantage however is that by detecting both quadratures of the signal field simultaneously, it furnishes sufficient information to determine the quantum state of the optical field in a single shot [82].

3.2.3.5 Design and Operation of a Realistic Heterodyne Detector

As illustrated by theoretical considerations, an experimentally practical heterodyne detector inherits all the characteristics of the homodyne detector in Fig. 3.4, except for a frequency shifted LO. Figure 3.5a depicts the essential layout of the balanced heterodyne interferometer constructed and employed in this thesis. The basic difference in the optics is the presence of an acousto-optic modulator (AOM, AA Optoelectronics MT110-B50A1) in the LO arm of the interferometer. The AOM was operated so as to maximise the diffracted optical power into the first order; at the chosen operation frequency $\Omega_{\mathrm{IF}} = 2\pi \cdot 78\,\mathrm{MHz}$, it was possible to attain a diffraction efficiency > 0.8.

Similar to the procedure followed to balance the homodyne detector, the input laser wavelength is modulated to induce interference fringes in the photocurrent. However, in the case of the heterodyne, the mean photocurrent Eq. (3.2.35), $\left\langle \hat{I}_{\mathrm{het}}(t) \right\rangle \propto \cos(\theta_{\mathrm{het}} + \Omega_{\mathrm{IF}} t)$ oscillates at the offset frequency Ω_{IF}. Therefore, to access the fringes resulting from a modulation of the phase θ_{het}, the photocurrent is mixed down using a RF local oscillator at the offset frequency Ω_{IF} (see schematic in Fig. 3.5a). The lengths are balanced by nullifying the fringe frequency. Unlike the homodyne, the phase θ_{het} need not be stabilised, since the photocurrent spectrum $\bar{S}_I^{\mathrm{het}}[\Omega]$ (Eq. (3.2.35)) is not sensitive to the mean phase.

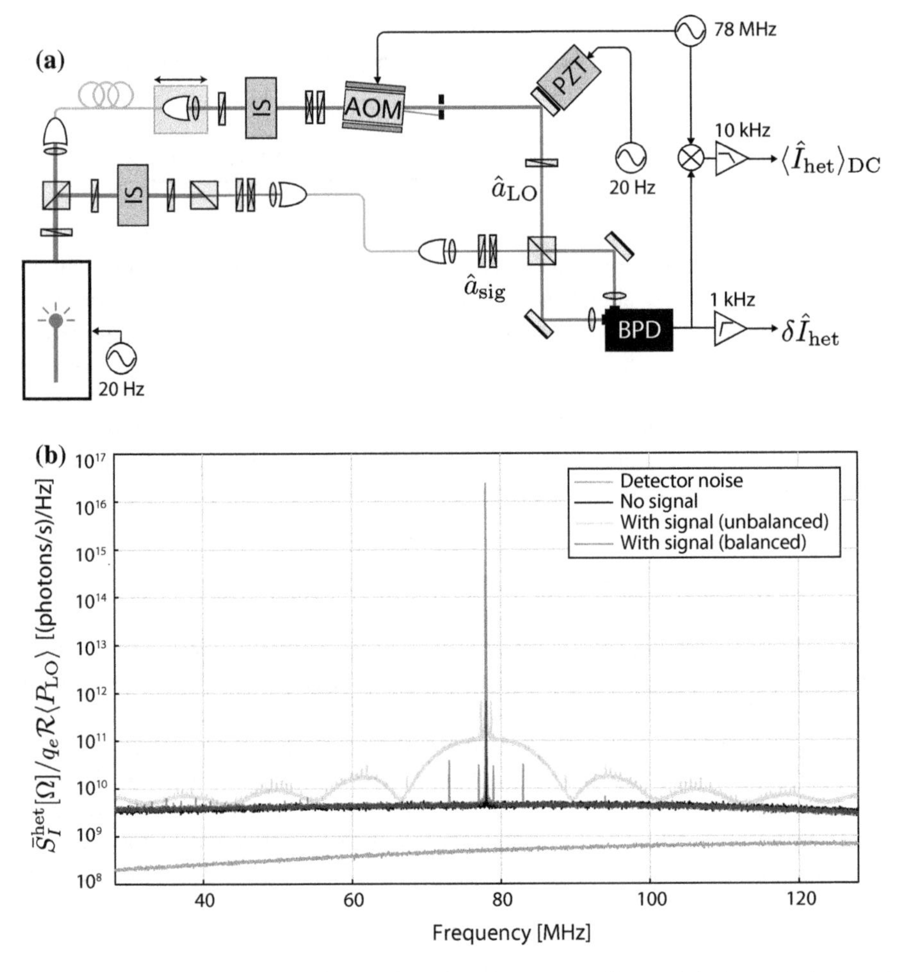

Fig. 3.5 Design and operation of heterodyne interferometer. a Essential design of the balanced heterodyne interferometer used in this thesis. An AOM in the LO path produces the desired frequency shift $\Omega_{\rm IF} = 2\pi \cdot 78$ MHz. **b** Heterodyne photocurrent spectrum for the interferometer unbalanced (light red) and balanced (red). Gray shows the electronic noise of the photodetector, and black the shot noise due to the LO. The spectrum is calibrated using the known DC optical power which is reflected as the variance of the carrier beat signal around the intermediate frequency $\Omega_{\rm IF} = 2\pi \cdot 78$ MHz

Figure 3.5b shows the cancellation of input laser noise achieved due to length balance. Shining a LO ($P_{\rm LO} \approx 1$ mW) alone gives rise to a shot noise contribution (black trace) $\bar{S}_I^{\rm het,shot} \gtrsim 10 \cdot \bar{S}_I^{\rm het,det}$. For an unbalanced interferometer driven by a (noisy) diode laser, the output photocurrent spectrum gives a direct measure of the laser phase noise transduced by the imbalance of the interferometer (see Appendix C). Indeed, the red trace in Fig. 3.5b, is consistent with diode laser frequency noise $\bar{S}_\omega[\Omega] = \Omega^2 \bar{S}_\phi[\Omega] \approx 2\pi(35\,{\rm Hz}^2/{\rm Hz})$, at Fourier frequencies $\Omega \approx 2\pi \cdot 4$ MHz from the carrier.

3.2.4 *From Propagating Modes to Standing Waves: Optical Cavity Coupled to a Waveguide*

The primary concern of the discussion so far has been optical fields that freely propagate, ultimately terminating at some detector. A qualitatively different sort of optical field exists—one that is "trapped" in a confined geometry. Such geometries, called *optical cavities*, are commonly employed in quantum optics to prolong (and thus enhance) the interaction between electromagnetic fields and other physical systems.

The typical optical fields we are interested in this thesis, also happen to be those that have been released after being stored in an optical cavity. Here we present a schematic of the optical cavity we are interested in—dielectric whispering-gallery mode optical microcavities [83, 84]—and the coupling of propagating optical fields in and out of such cavities via a waveguide (see [85] for a review of the theory of optical cavities).

Whispering-gallery mode (WGM) optical cavities, like the one shown in Fig. 3.6, support optical modes at specific frequencies ω_n roughly commensurate with standing waves resonating around the circumference. For example, for a spherical cavity of radius R and made of a dielectric material of refractive index ν, solutions of Maxwell's equations show that the mode wavelengths λ_n confirm to this intuition, i.e. $\lambda_n \approx 2\pi\nu R/n$ [86]. In general, when the free spectral range of the cavity, $\Delta\omega_{\mathrm{FSR}} := \omega_n - \omega_{n-1}$ is much larger than the energy decay rate κ_n, i.e. $\Delta\omega_{\mathrm{FSR}} \gg \kappa_n$, each mode may be treated independent of the others. Focusing on such a particular mode, at resonance frequency ω_c, the dynamics of its quantised standing wave amplitude $\hat{a}$ (normalised to photon number) is described by the hamiltonian [85],

$$\hat{H}_{\mathrm{c}} = \hbar\omega_c \hat{a}^\dagger \hat{a} + \hat{H}_{\mathrm{c},0} + \hat{H}_{\mathrm{c,ex}}, \tag{3.2.39}$$

where $\hat{H}_{\mathrm{c},0}$ models coupling to external sources responsible for the intrinsic cavity decay rate κ_0, and $\hat{H}_{\mathrm{c,ex}}$ models coupling to an external waveguide used to excite the cavity.

Light is coupled into the cavity using an optical waveguide—a tapered optical fiber—placed in the vicinity of the cavity evanescent field [89, 90]. Care is taken to ensure that the tapered fiber predominantly supports a single travelling mode[27] described by an amplitude flux $\hat{a}_{\mathrm{ex}}(z, t)$, along the (longitudinal) $z-$direction. Note that the travelling mode is normalised to a photon flux, and satisfies the commutation relation (see Eq. (3.2.2)),

$$[\hat{a}_{\mathrm{ex}}(z, t), \hat{a}^\dagger(z', t')] = \delta(t - t' - (z - z')/c). \tag{3.2.40}$$

[27]The tapered section, formed by adiabatically stretching a cylindrical optical fiber (780HP, 5μm mode waist), to a waist of $< 1\,\mu$m, supports degenerate TE, TM$_{00}$ modes (cutoff at $\approx$ 730 nm) [38, 91], with an evanescent part guided in free space. The optical fiber itself is excited using free space radiation in TE, TM modes, with a coupling efficiency $\gtrsim 80\%$.

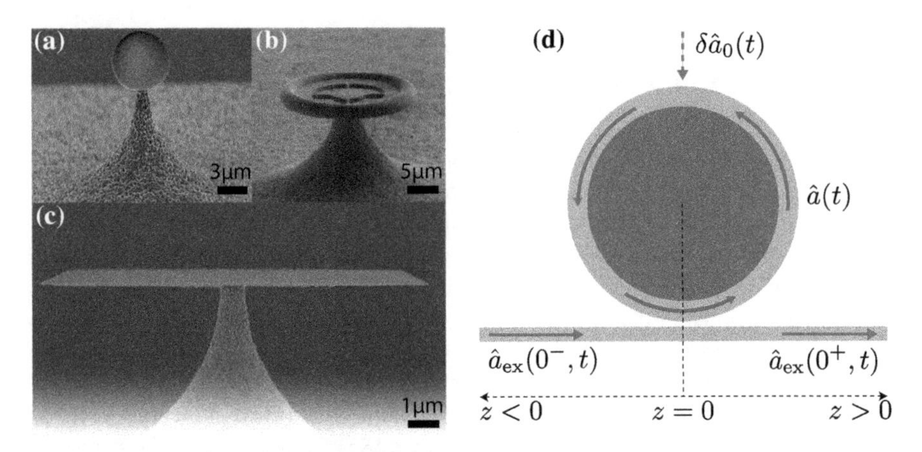

Fig. 3.6 Whispering-gallery mode cavities and coupling. a–c A smattering of whispering-gallery mode optical microcavities, where the resonant optical mode circulates along the circumference of the dielectric medium. **a, b** Spherical and toroidal cavities [87] made by CO_2 laser reflow of SiO_2. **c** Disk cavity fabricated by chemical-mechanical polishing [88]. **d** Schematic of waveguide (here, tapered optical fiber) coupling to a whispering gallery mode cavity. The cavity field $\hat{a}(t)$ is excited by the travelling wave field $\hat{a}_{ex}(z,t)$ of the waveguide through a beam-splitter type interaction at the point $z=0$. The cavity is also driven by vacuum fluctuations $\delta\hat{a}_0(t)$

The dynamics of this field is described by the hamiltonian,

$$\hat{H}_{ex} = \hbar\omega_{ex}\hat{a}^\dagger_{ex}(z,t)\hat{a}_{ex}(z,t) + \hat{H}_{c,ex} \tag{3.2.41}$$

where ω_{ex} is the frequency of the propagating mode. The cavity-waveguide coupling is modeled as an energy-conserving interaction localised at the point[28] $z=0$ (see Fig. 3.6b) [93], i.e.,

$$\hat{H}_{c,ex} = i\hbar\sqrt{\kappa_{ex}}\,\left(\hat{a}^\dagger(t)\hat{a}_{ex}(0,t) - \hat{a}(t)\hat{a}^\dagger_{ex}(0,t)\right). \tag{3.2.42}$$

Inserting this in the expression for the cavity hamiltonian in Eq. (3.2.39), and employing the commutation relation $[\hat{a}(t),\hat{a}^\dagger(t)] = 1$, gives the equation of motion for the cavity field,

$$\frac{\mathrm{d}\hat{a}}{\mathrm{d}t} = -i\omega_c\hat{a} + \sqrt{\kappa_{ex}}\,\hat{a}_{ex}(0,t) + \frac{i}{\hbar}[\hat{a},\hat{H}_{c,0}]. \tag{3.2.43}$$

Similarly, inserting Eq. (3.2.42) into the expression for the hamiltonian for the propagating field in Eq. (3.2.41), and employing the relevant commutation relation (Eq. (3.2.40)) gives,

[28] In a more realistic model where the coupling region has a finite extent, κ_{ex} effectively describes the detailed geometry of the coupling [92].

$$\frac{\mathrm{d}\hat{a}_{\mathrm{ex}}(z,t)}{\mathrm{d}t} = -i\omega_{\mathrm{ex}}\hat{a}_{\mathrm{ex}} + \sqrt{\kappa_{\mathrm{ex}}}\hat{a}(t)\delta(-z/c),$$
$$\text{or,} \quad -c\frac{\mathrm{d}\hat{a}_{\mathrm{ex}}(z,t)}{\mathrm{d}z} = -i\omega_{\mathrm{ex}}\hat{a}_{\mathrm{ex}} + \sqrt{\kappa_{\mathrm{ex}}}\hat{a}(t)\delta(-z/c), \tag{3.2.44}$$

where the second form is obtained by noting that for a propagating mode, satisfying $\hat{a}_{\mathrm{ex}}(z,t) = \hat{a}_{\mathrm{ex}}(0, t - z/c)$, time and the space derivative along the direction of propagation are related as, $\partial_t \hat{a}_{\mathrm{ex}}(z,t) = -c\partial_z \hat{a}_{\mathrm{ex}}(z,t)$, with c the propagation velocity in the waveguide. Integrating the latter equation within the coupling region, $z \in (0^-, 0^+)$, and employing the properties of the delta function,

$$\hat{a}_{\mathrm{ex}}(0^+, t) = \hat{a}_{\mathrm{ex}}(0^-, t) - \sqrt{\kappa_{\mathrm{ex}}}\,\hat{a}(t). \tag{3.2.45}$$

Defining the input (output) fields as the propagating field before (after) the coupling region:

$$\hat{a}_{\mathrm{in,out}}(t) := \hat{a}_{\mathrm{ex}}(0^{\mp}, t), \tag{3.2.46}$$

Eq. (3.2.45) takes the form of an input-output relation,

$$\hat{a}_{\mathrm{out}}(t) = \hat{a}_{\mathrm{in}}(t) - \sqrt{\kappa_{\mathrm{ex}}}\hat{a}(t), \tag{3.2.47}$$

between the waveguide and cavity modes in a scattering description of their coupling.

The equation of motion for the cavity field can be derived by returning Eq. (3.2.43): the discontinuity at $z = 0$ may be manipulated as,

$$\hat{a}_{\mathrm{ex}}(0,t) = \frac{1}{2}\left(\hat{a}_{\mathrm{ex}}(0^-, t) + \hat{a}_{\mathrm{ex}}(0^+, t)\right) = \hat{a}_{\mathrm{in}} - \frac{\sqrt{\kappa_{\mathrm{ex}}}}{2}\hat{a}(t);$$

here the first equality uses continuity of the field at the coupling point, and the second follows from the input-output relation (Eq. (3.2.47)). Inserting this back in the equation of motion for the cavity field, Eq. (3.2.43),

$$\frac{\mathrm{d}\hat{a}}{\mathrm{d}t} = -\left(i\omega_c + \frac{\kappa_{\mathrm{ex}}}{2}\right)\hat{a} + \sqrt{\kappa_{\mathrm{ex}}}\hat{a}_{\mathrm{in}}(t) + \frac{i}{\hbar}[\hat{H}_{\mathrm{c,0}}, \hat{a}].$$

Thus, coupling to the external waveguide opens a decay channel for the cavity mode, described by the external decay rate κ_{ex}. The explicit form of intrinsic losses, modelled by $\hat{H}_{\mathrm{c,0}}$, follow similar lines, and result in the equation of motion [94],

$$\frac{\mathrm{d}\hat{a}}{\mathrm{d}t} = -\left(i\omega_c + \frac{\kappa}{2}\right)\hat{a} + \sqrt{\kappa_0}\delta\hat{a}_0 + \sqrt{\kappa_{\mathrm{ex}}}\hat{a}_{\mathrm{in}}(t), \tag{3.2.48}$$

where κ_0 is the intrinsic decay rate, $\kappa = \kappa_0 + \kappa_{\mathrm{ex}}$ is the total decay rate, and $\delta\hat{a}_0$ is the zero-mean stochastic process driving the cavity through its intrinsic loss channel. Together with the input-output relation Eq. (3.2.47),

$$\hat{a}_{\text{out}}(t) = \hat{a}_{\text{in}}(t) - \sqrt{\kappa_{\text{ex}}}\,\hat{a}(t), \tag{3.2.49}$$

and the specification of the fluctuations associated with the intrinsic and external (waveguide) decay channel (here $n_{c,j}$ is the average thermal occupation of the channel $j \in \{0, \text{in}\}$),

$$\begin{aligned}\left\langle \delta\hat{a}_j^\dagger(t)\delta\hat{a}_j(0)\right\rangle &= n_{c,j}\delta(t)\\ \left\langle \delta\hat{a}_j(t)\delta\hat{a}_j^\dagger(t)\right\rangle &= (n_{c,j}+1)\delta(t),\end{aligned} \tag{3.2.50}$$

this completes the description of the optical cavity.

Typically, the optical cavity is excited using a coherent source at a definite optical frequency ω_ℓ, so that,

$$\hat{a}_{\text{in}}(t) = (|\bar{a}_{\text{in}}| + \delta\hat{a}_{\text{in}}(t))e^{-i\omega_\ell t}. \tag{3.2.51}$$

It proves convenient to adopt a description where the explicit time dependent factor $e^{-i\omega_\ell t}$ is implicit. At the level of the equation of motion Eq. (3.2.48) and the input-output relation Eq. (3.2.49), this is implemented by the transformation,[29] $\hat{a} \to \hat{a}e^{-i\omega_\ell t}$. Equations (3.2.48) and (3.2.49) then take the form,

$$\begin{aligned}\frac{\mathrm{d}\hat{a}}{\mathrm{d}t} &= \left(i\Delta - \frac{\kappa}{2}\right)\hat{a} + \sqrt{\kappa_0}\,\delta\hat{a}_0 + \sqrt{\kappa_{\text{ex}}}\,\delta\hat{a}_{\text{in}}(t)\\ \hat{a}_{\text{out}}(t) &= \hat{a} - \sqrt{\kappa_{\text{ex}}}\,\hat{a}_{\text{in}}(t),\end{aligned} \tag{3.2.52}$$

were Δ is the detuning between the input field in Eq. (3.2.51) and the cavity, viz.

$$\Delta := \omega_\ell - \omega_c.$$

3.2.4.1 Steady-State Cavity Spectroscopy

In a typical spectroscopy experiment, as shown in Fig. 3.7a, aiming to identify and characterise the whispering-gallery modes of the cavity, the cavity is pumped using a laser at frequency ω_ℓ (see Eq. (3.2.51)), and the transmitted power, $P_{\text{out}} = \hbar\omega_\ell\left\langle \hat{a}_{\text{out}}^\dagger\hat{a}_{\text{out}}\right\rangle$, is monitored as the laser frequency ω_ℓ is swept over the cavity resonance ω_c. In the experiment, we ensure that ω_ℓ is swept much slower than the cavity decay rate κ so that P_{out} is the steady-state transmission of the cavity.

In order to predict the outcome of transmission measurements, we start from the equation determining the intracavity field (Eq. (3.2.52)), expressed for the mean value of the field:

$$\frac{\mathrm{d}\langle\hat{a}\rangle}{\mathrm{d}t} = \left(i\Delta - \frac{\kappa}{2}\right)\langle\hat{a}\rangle + \sqrt{\kappa_{\text{ex}}}\,|\bar{a}_{\text{in}}|\,.$$

[29]Corresponding to a unitary transformation by the rotation operator, $\hat{R}(\phi) = e^{-i\phi\hat{a}^\dagger\hat{a}}$, of the hamiltonian $\hat{H}_c$ in Eq. (3.2.39); i.e. $\hat{H}_c \to \hat{R}(\omega_\ell t)\hat{H}_c\hat{R}^\dagger(\omega_\ell t)$.

The resulting stead-state intracavity field,

$$\bar{a} := \langle \hat{a} \rangle_{\text{ss}} = \frac{-\sqrt{\kappa_{\text{ex}}}}{i\Delta - \kappa/2} |\bar{a}_{\text{in}}| \,,$$

gives rise to the mean steady-state intracavity photon number,

$$\langle \hat{a}^\dagger \hat{a} \rangle_{\text{ss}} = |\bar{a}|^2 = \frac{\kappa_{\text{ex}} |\bar{a}_{\text{in}}|^2}{\Delta^2 + (\kappa/2)^2} = \frac{4}{\kappa} \left(\frac{\kappa_{\text{ex}}/\kappa}{1 + (4\Delta^2/\kappa^2)} \right) \frac{P_{\text{in}}}{\hbar\omega_\ell},$$

that describes the response of the cavity as a resonant build-up of the pump power $P_{\text{in}} := \hbar\omega_\ell \left\langle \hat{a}_{\text{in}}^\dagger \hat{a}_{\text{in}} \right\rangle$, depending on the relative detuning $\frac{\Delta}{\kappa/2}$ and the cavity-waveguide coupling efficiency,

$$\eta_c := \frac{\kappa_{\text{ex}}}{\kappa} = \frac{\kappa_{\text{ex}}}{\kappa_{\text{ex}} + \kappa_0}.$$

Finally, the steady-state transmission, $T_c(\Delta) := P_{\text{out}}/P_{\text{in}}$, for a given pump detuning,

$$T_c(\Delta) = 1 - \frac{4\eta_c(1 - \eta_c)}{1 + (4\Delta^2/\kappa^2)},$$

exhibits a Lorentzian suppression on approaching resonance ($|\Delta| \to 0$). However, the cavity only absorbs all the power on resonance, i.e. $T_c(0) = 0$, when the coupling is *critical*, i.e. $\eta_c = \frac{1}{2}$, corresponding to the case where the power coupled in by the waveguide exactly compensates for the power lost through the intrinsic decay channel.

These aspects are illustrated in Fig. 3.7c, which in fact depicts the transmission $T_c(\Delta)$. A widely tunable external cavity diode laser (NewFocus Velocity) is coupled into a fiber taper, which is brought within the evanescent field of the whispering-gallery mode cavity. The relative position of the fiber and cavity is controlled using a piezo-positioning stage (Attocube, ANPx101) which allows for sub-nm precision in taper-cavity gap. As the taper is brought closer into the evanescent field, the external coupling rate κ_{ex} increases [92], thereby allowing for control of the coupling efficiency η_c—a unique feature of this coupling technique. Control of input polarisation achieves perfect phase-matching into the resonant modes of the cavity. In order to obtain transmission signals as shown in Fig. 3.7c, the laser frequency is swept while the cavity transmission is monitored on a photodetector. In order to calibrate the laser frequency sweep, a part of the laser light is directed onto a fiber-loop cavity of known FSR[30] (≈ 250 MHz). This allows calibration of the relative detuning between the laser and the whispering gallery mode cavity. Figure 3.7b shows the variation in the resonant transmission as a function of the coupling efficiency η_c, obtained by

---

[30] The fiber-loop cavity is made by splicing together the ends of a 50:50 fiber beam-splitter using an approximately known length of fiber. In order to calibrate the FSR of this cavity, laser light is phase modulated using an EOM, as shown in Fig. 3.7a, imparting sidebands of known frequency separation. Thus the loop cavity FSR is calibrated to a known RF modulation frequency.

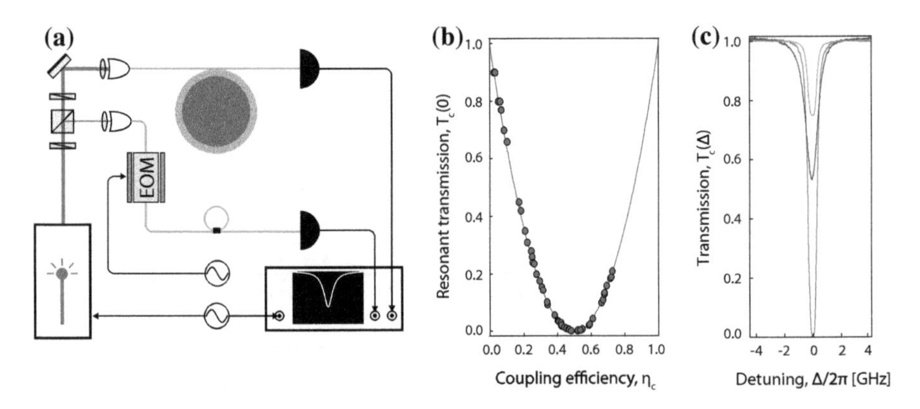

Fig. 3.7 Steady-state spectroscopy. a Experimental schematic of the spectroscopy scheme: laser light is coupled into a fiber taper which is then brought in close proximity to the whispering gallery mode cavity. **b** Transition from under-coupled to over-coupled regime, as the coupling efficiency η_c is varied. Relative position of the fiber taper and the cavity controls the waveguide coupling rate κ_{ex}. **c** Examples of cavity transmission when the cavity is under-coupled (light red), critically coupled (red), and over-coupled (dark red). See text for details regarding calibration of the detuning

varying the taper-cavity gap. The intrinsic decay rate of the cavity, κ_0, is obtained in the limit of heavy under-coupling ($\eta_c \to 0$). The data in Fig. 3.7c suggests that $\kappa_0 \approx 2\pi \cdot 450\,\text{MHz}$, a typical value observed in measurements of the optical cavity with a nanobeam coupled to it.[31]

[31]For this case, where $\kappa_0 \ll \omega_c$, the intrinsic decay rate may be understood as a combination of several effects [95], i.e.,

$$\kappa_0 = \kappa_{\text{rad}} + \kappa_{\text{vol}} + \kappa_{\text{surf}} + \kappa_{\text{mech}} = \omega_c \left(Q_{\text{rad}}^{-1} + Q_{\text{vol}}^{-1} + Q_{\text{surf}}^{-1} + Q_{\text{mech}}^{-1} \right),$$

here expressed as contributions to the optical Q-factor. The first term Q_{rad}^{-1} models the losses arising from imperfect confinement of light in the whispering-gallery mode; when the cavity is large (the cavity radius, $R_c > 10 \cdot \lambda$, for example) the losses due to radiation is expected to provide a limit [95] $Q_{\text{rad}} < 10^{11}$. In contrast, the data shown in Fig. 3.7 features $Q \approx 8 \cdot 10^5$. Losses in the cavity volume, leading to the contribution Q_{vol}^{-1}, arise from optical absorption in SiO_2; at the operating wavelength of 780 nm, estimates suggest $Q_{\text{vol}} < 10^{10}$ [95, 96]. Effects such as scattering off of surface inhomogeneities [95, 97] and/or contamination from water absorption [96], that depend on the surface area, go into Q_{surf}^{-1}. Water absorption at 1500 nm is known to provide the limit $Q_{\text{surf},\text{H}_2\text{0}} < 10^9$, which is known to be recoverable after sustained bake-out [96]; at 780 nm, the effect is expected to be smaller, mediated by harmonics of the absorption at 1500 nm. Unlike micro-cavities formed by surface reflow [87], for the disk geometry employed in this thesis (see Fig. 3.6c), surface scattering is known to play an important role in limiting the optical-Q [98]. A simple model suggests that [99] $Q_{\text{surf},\text{scat}} \propto R_c^{1/2} \langle \ell_{\text{surf}}^2 \rangle^{-1} \ell_{\text{surf},\text{corr}}^{-2}$, where R_c is the disk radius, $\langle \ell_{\text{surf}}^2 \rangle$ is the variance in the surface roughness and $\ell_{\text{surf},\text{corr}}$ the correlation length of the roughness pattern. Compared against the value of $Q_{\text{surf},\text{scat}} \approx 10^9$ reported in [99], our value of $Q \approx 10^6$ would imply a roughness pattern for which $\langle \ell_{\text{surf}}^2 \rangle^{1/2} \ell_{\text{surf},\text{corr}} \approx (150\,\text{nm})^2$. Given that this is an unusually large number, we believe that the contribution Q_{mech} due to optical losses in the presence of the beam is responsible for the observed linewidth.

The spectroscopic signature obtained by this technique provides access to the cavity's amplitude response. The response obtained, shown in Fig. 3.7c, can be used to stabilise the laser-cavity detuning at any point within the cavity bandwidth, except on resonance. The major technical disadvantage of locking a laser to the cavity using this technique is that the error signal for the control loop—essentially the transmission T_c—is susceptible to drifts in probe power and cavity coupling, causing the detuning to be affected by these factors.

3.2.4.2 Modulation Spectroscopy

Another spectroscopic technique relies not on using the cavity's response to a mean optical field, but rather to fluctuations in the input optical field. In order to describe it, it is therefore necessary to see how fluctuations in the input optical field manifest as intracavity fluctuations, and how they subsequently appear in the outgoing field.

Separating out the mean steady state intracavity amplitude from its fluctuating part, viz.

$$\hat{a}(t) = |\bar{a}| + \delta\hat{a}(t),$$
$$\text{where,} \quad |\bar{a}| := \sqrt{\langle \hat{a}^\dagger \hat{a} \rangle_{ss}} = \left(\frac{4\eta_c}{\kappa} \frac{|\bar{a}_{\text{in}}|^2}{1 + (4\Delta^2/\kappa^2)} \right)^{1/2},$$

and inserting it into Eq. (3.2.52), results in the equation of motion for the fluctuating part,

$$\delta\dot{\hat{a}}(t) = \left(i\Delta - \frac{\kappa}{2} \right) \delta\hat{a}(t) + \sqrt{(1-\eta_c)\kappa}\, \delta\hat{a}_0(t) + \sqrt{\eta_c \kappa}\, \delta\hat{a}_{\text{in}}(t).$$

Taking Fourier transforms of either side,

$$\delta\hat{a}[\Omega] = \chi_a[\Omega] \left(\sqrt{(1-\eta_c)\kappa}\, \delta\hat{a}_0[\Omega] + \sqrt{\eta_c\kappa}\, \delta\hat{a}_{\text{in}}[\Omega] \right),$$
$$\text{where,} \quad \chi_a[\Omega] = \left(-i(\Omega + \Delta) + \frac{\kappa}{2} \right)^{-1}, \tag{3.2.53}$$

is the susceptibility of the intracavity optical field to fluctuations in the input optical field. Note that the susceptibility encodes the cavity response as well as the laser-cavity detuning. The fluctuations in the outgoing field, given by the input-output relation Eq. (3.2.49),

$$\delta\hat{a}_{\text{out}}[\Omega] = (1 - \eta_c\kappa\, \chi_a[\Omega])\, \delta\hat{a}_{\text{in}}[\Omega] - \sqrt{\eta_c(1-\eta_c)}\, \kappa\, \chi_a[\Omega] \delta\hat{a}_0[\Omega], \tag{3.2.54}$$

carries this information, and may be retrieved by probing the cavity using an input of known spectral content.

One way to perform this, originally developed by Pound, Drever, Hall (PDH) and others [100, 101], involves frequency modulating the input field. In our experiment, depicted in Fig. 3.8a, the input field $\hat{a}_{\text{in}}(t)$ is passed through an electro-optic modulator (EOM), picking up a sinusoidal modulation of its phase, viz.

$$\begin{aligned} \hat{a}_{\text{in}}(t) &= |\bar{a}_{\text{in}}|\, e^{-i(\omega_\ell t - \xi_{\text{mod}} \sin \Omega_{\text{mod}} t)} \\ &\approx |\bar{a}_{\text{in}}|\, e^{-i\omega_\ell t} \left(1 + \frac{\xi_{\text{mod}}}{2} e^{i\Omega_{\text{mod}} t} - \frac{\xi_{\text{mod}}}{2} e^{-i\Omega_{\text{mod}} t} \right), \end{aligned} \tag{3.2.55}$$

approximated as sidebands at frequencies $\omega_\ell \pm \Omega_{\text{mod}}$ with depth $\xi_{\text{mod}} \ll 1$. The transmitted field is given by,

$$\hat{a}_{\text{out}}(t) = |\bar{a}_{\text{in}}|\, e^{-i\omega_\ell t} \left(\chi_a^{\text{out}}[0] + \frac{\xi_{\text{mod}}}{2} \chi_a^{\text{out}}[\Omega_{\text{mod}}] e^{i\Omega_{\text{mod}} t} - \frac{\xi_{\text{mod}}}{2} \chi_a^{\text{out}}[-\Omega_{\text{mod}}] e^{-i\Omega_{\text{mod}} t} \right),$$

where, $\chi_a^{\text{out}}[\Omega] := 1 - \eta_c \kappa\, \chi_a[\Omega]$, represents the susceptibility of the output field to the input field (see Eq. (3.2.54)). The transmitted sidebands now encode the cavity response, which appears in the detected photocurrent,

$$\begin{aligned} \left\langle \hat{I}_{\text{out}} \right\rangle \propto\, & |\bar{a}_{\text{in}}|^2 \left|\chi_a^{\text{out}}[0]\right|^2 + |\bar{a}_{\text{in}}|^2 \frac{\xi_{\text{mod}}^2}{4} \left(\left|\chi_a^{\text{out}}[\Omega_{\text{mod}}]\right|^2 + \left|\chi_a^{\text{out}}[-\Omega_{\text{mod}}]\right|^2 \right) \\ & + |\bar{a}_{\text{in}}|^2 \xi_{\text{mod}} \operatorname{Re} \left[\left(\chi_a^{\text{out}}[0] \chi_a^{\text{out}}[\Omega_{\text{mod}}] - \chi_a^{\text{out}}[0]^* \chi_a^{\text{out}}[-\Omega_{\text{mod}}] \right) e^{-i\Omega_{\text{mod}} t} \right] \\ & + |\bar{a}_{\text{in}}|^2 \frac{\xi_{\text{mod}}}{2} \operatorname{Re} \left[\chi_a^{\text{out}}[\Omega_{\text{mod}}] \chi_a^{\text{out}}[-\Omega_{\text{mod}}]^* e^{2i\Omega_{\text{mod}} t} \right], \end{aligned}$$

as a DC term, a term oscillating at the modulation frequency Ω_{mod}, and one oscillating at twice the modulation frequency. For a cavity with a symmetric response about resonance, the DC term does not carry unambiguous information regarding the laser-cavity detuning. The term oscillating at Ω_{mod} does furnish this information.

Figure 3.8a shows our implementation of this technique. A function generator provides a sinusoidal tone ($\Omega_{\text{mod}} = 2\pi \cdot 40$ MHz) that drives an EOM (NewFocus 4002), which imprints a phase-modulation tone on the laser before it is coupled into the cavity. Cavity transmission is detected on a sensitive avalanche photodetector (Thorlabs APD120); the resulting photocurrent is band-pass-filtered to isolate the component at Ω_{mod}, appropriately amplified (Minicircuits ZFL-500LN, or Femto HVA-200M) and mixed-down (Minicircuits ZP-3) with an electronic local oscillator (LO) at Ω_{mod}. Care is taken to ensure that the double-balanced mixer is operated in its linear regime and that its output passes through appropriate image-rejection filters. The LO is a phase-controlled copy of the signal used to drive the phase modulator, i.e. $V_{\text{LO}} = |V_{\text{LO}}| \sin(\Omega_{\text{mod}} t + \theta_{\text{LO}})$. The demodulated voltage at the output of the mixer is,

$$V_{\text{demod}} \propto |\bar{a}_{\text{in}}|^2 \xi_{\text{mod}} \cdot |V_{\text{LO}}| \cdot M_{\text{PDH}} \sin(\theta_{\text{LO}} + \theta_{\text{PDH}}). \tag{3.2.56}$$

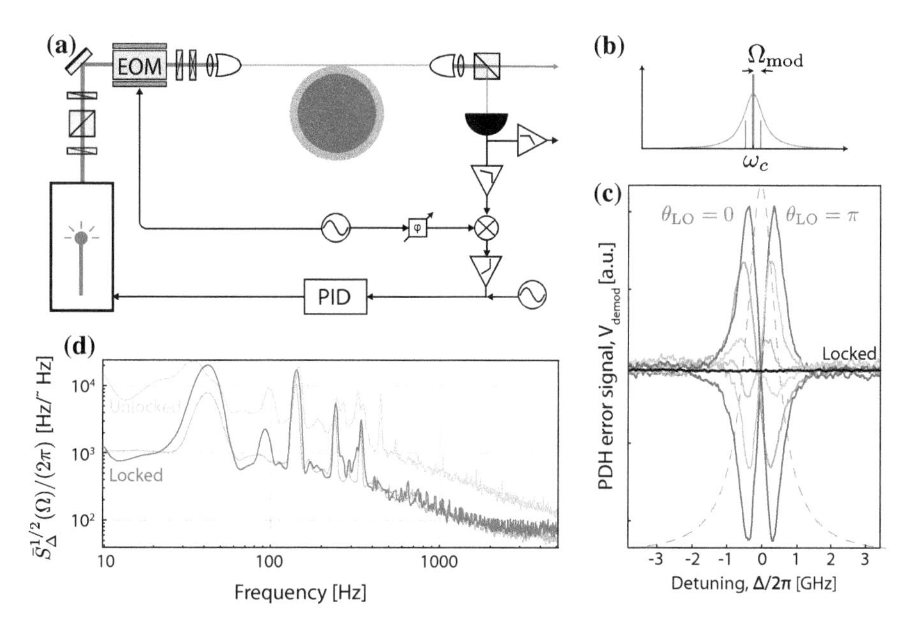

Fig. 3.8 Frequency modulation spectroscopy. **a** Experimental schematic of the spectroscopy scheme: an EOM driven by a RF generator produces frequency modulation sidebands on the detuned probe laser. The cavity transmission is detected using an avalanche photodetector, whose output is demodulated by a phase-tuned RF local oscillator. The mixer output may additionally be used to stabilise the probe on cavity resonance. **b** Schematic of the single-sideband technique: near resonance, the frequency modulation sidebands get transduced by the cavity phase response to amplitude modulation in transmission. Ideally, on resonance, the transmitted sidebands interfere destructively. **c** Demodulated output from the mixer as the laser is swept over cavity resonance. The various traces show the voltage trace as the electronic LO phase is tuned over half a cycle. The gray dashed line is an overlay of the cavity magnitude response plotted from the estimate of the linewidth obtained from the PDH signal. Black trace shows an in-loop signal once the laser is locked to cavity resonance. **d** Residual laser-cavity detuning noise estimated using the input into the laser frequency actuator. Gray is electronic noise in the control loop

Here, the magnitude (M_{PDH}) and phase (θ_{PDH}) of the PDH error signal is determined by the identity,

$$M_{\text{PDH}}\, e^{i\theta_{\text{PDH}}} = \chi_a^{\text{out}}[0]\chi_a^{\text{out}}[\Omega_{\text{mod}}] - \chi_a^{\text{out}}[0]^* \chi_a^{\text{out}}[-\Omega_{\text{mod}}] \approx \Omega_{\text{mod}} \left[\frac{\mathrm{d}\left|\chi_a^{\text{out}}[\Omega]\right|^2}{\mathrm{d}\Omega} \right]_{\Omega=0},$$

where the approximation is for the case where the modulation frequency is much smaller than the cavity bandwidth, i.e. $\Omega_{\text{mod}} \ll \frac{\kappa}{2}$, and therefore, $\chi_a^{\text{out}}[\Omega_{\text{mod}}] \approx \chi_a^{\text{out}}[0] + \Omega_{\text{mod}} \frac{\mathrm{d}\chi_a^{\text{out}}}{\mathrm{d}\Omega}$. In this *unresolved-sideband* case, $\theta_{\text{PDH}} \approx 0$, and the demodulated voltage, maximised for the choice $\theta_{\text{LO}} = 0$, provides the derivative of the magnitude response of the cavity transmission.

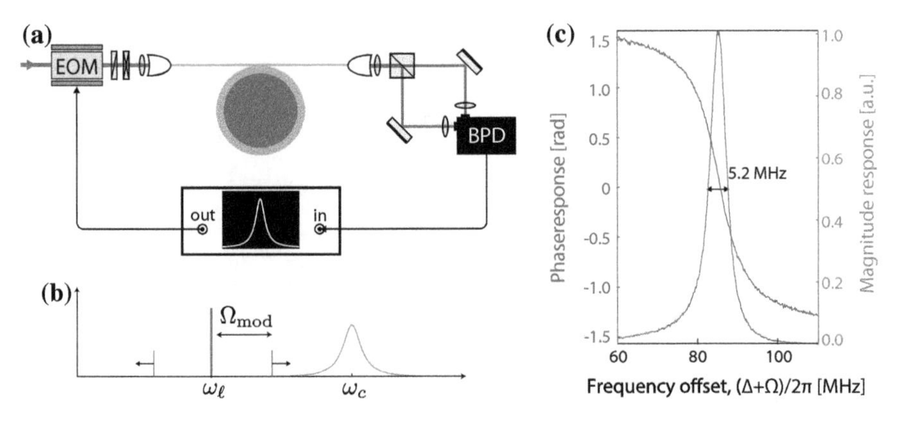

Fig. 3.9 Single-sideband modulation spectroscopy. a Experimental schematic of the spectroscopy scheme: an EOM driven by a network analyser imprints sidebands on the detuned probe laser. The cavity transmission is detected using a pair of balanced photodetectors, whose output is demodulated by the network analyser. **b** Schematic of the single-sideband technique: by being far detuned from the cavity resonance ω_c, specifically $|\Delta| \gg \kappa$, as the modulation frequency Ω_{mod} is swept, only one of the sidebands probes the cavity. **c** Example of a response taken on a microtoroid cavity, showing a linewidth of $\kappa = 2\pi \cdot 5.2\,\text{MHz}$

Figure 3.8c shows examples of the demodulated voltage as θ_{LO} is varied over half a cycle. At the optimal setting of the LO phase, $\theta_{\text{LO}} \approx 0$, the demodulated voltage provides the best error signal for stabilising the laser frequency to cavity resonance, i.e. $\Delta = 0$. This is realised by sending a copy of V_{demod} through a PI controller, appropriate filtering stages, and into the laser frequency controller actuating on the diode current. The black trace shows the actuator input when the laser is locked. Figure 3.8d shows the power spectral density of the actuator input, calibrated in units of laser-cavity detuning, providing an estimate of the residual detuning noise when the laser is nominally locked. Note that the PDH error signal voltage (Eq. (3.2.56)), $V_{\text{demod}} \propto \Omega_{\text{mod}}$, enabling calibration of the voltage noise in units of frequency. Low frequency broadband suppression of detuning noise $< 10^3\,\text{Hz}/\sqrt{\text{Hz}}$ at offset frequencies up to 1 kHz is easily achieved in our experiment; this is mainly limited by background electronic noise (grey trace) from the photodetector and control electronics. This level of detuning noise suppression proves crucial for experiment reported in Chap. 7, where this data will be revisited.

A variant of frequency-modulation spectroscopy which may be profitably employed when the cavity bandwidth is small compared to the accessible modulation frequency will be briefly described now. The central advantage of this technique, over standard PDH spectroscopy (as above), is that very little optical power actually enters the cavity; for exceptionally high-Q cavities suffering from low threshold for optical nonlinearities [102], this can be a technical boon.

As shown in Fig. 3.9b, the probe laser is far detuned from the cavity, i.e. $\Delta \ll -\frac{\kappa}{2}$, so that only one of the phase-modulation sidebands (here, upper sideband) probes the cavity response. In this effectively single-sideband modulation scenario, no extraneous interference occurs with the lower sideband, giving a demodulated voltage,

$$V_{\text{demod}} \propto |\bar{a}_{\text{in}}|^2 \, \xi_{\text{mod}} \cdot \left|\chi_a^{\text{out}}[\Omega_{\text{mod}}]\right| \sin\left(\theta_{\text{LO}} + \arg \chi_a^{\text{out}}[\Omega_{\text{mod}}]\right),$$

which, unlike the PDH voltage (Eq. (3.2.56)), directly provides access to the real ($\theta_{\text{LO}} = \frac{\pi}{2}$) and imaginary ($\theta_{\text{LO}} = 0$) parts of the cavity response.

In the experiment, as shown in Fig. 3.9a, both the modulation and demodulation are performed using a RF network analyser (Agilent, E5061B), so that the photodetected signal can be simultaneously demodulated over the two phases. Combining this quadrature-demodulated signal gives the magnitude and phase response of the cavity shown in Fig. 3.9c. Being a coherent detection technique, exceptionally low intracavity photon numbers ($|\bar{a}|^2| < 100$) may be reliably used, at the expense of longer averaging time.

In closing, we note that an additional frequency modulation on the sidebands allows for stabilisation of the probe laser at a variable offset detuning from cavity resonance.

References

1. W. Pauli, *Theory of Relativity* (Dover, 1958)
2. R.C. Tolman, *The Theory of the Relativity of Motion* (UC Berkeley Press, 1918)
3. B. Carter, H. Quintana, Proc. R. Soc. A **331**, 57 (1972)
4. J. Kijowski, G. Magli, J. Geom. Phys. **9**, 207 (1992)
5. D. Christodolou, Ann. Inst. Henri Poincare **69**, 335 (1998)
6. G. Maugin, *Continuum Mechanics Through the Twentieth Century* (Springer, 2013)
7. L.D. Landau, E.M. Lifshitz, *Theory of Elasticity* (4th ed.) (Butterworth-Heinemann, 1975)
8. R. Blandford, K.S. Thorne, *Modern Classical Physics* (Princeton University Press, 2016)
9. W. Kuhnel, *Differential Geometry: Curves—Surfaces—Manifolds* (AMS, 2006)
10. C. Lanczos, *The Variational Principles of Mechanics* (Dover, 1970)
11. E.P. Wigner, Proc. Natl. Acad. Sci **51**, 956 (1964)
12. R. Houtappel, H. van Dam, E.P. Wigner, Rev. Mod. Phys. **37**, 595 (1965)
13. J. Polchinski, arXiv:9210046 (1992)
14. H. Leutwyler, Phys. Rev. D **49**, 3033 (1994)
15. L.D. Landau, E.M. Lifshitz, *Mechanics* (3rd ed.) (Butterworth-Heinemann, 1976)
16. H. Bacry, J.-M. Levy-Leblond, J. Math. Phys. **9**, 1605 (1968)
17. M.J. Breuger, *Elementary Crystallography* (Wiley, 1956)
18. I.M. Gelfand, R. Minlos, Z. Shapiro, *Representations of the Rotation and Lorentz Groups and Their Applications* (Pergamon Press, 1963)
19. J. Jerphagnon, D. Chemla, R. Bonneville, Adv. Phys. **27**, 609 (1978)
20. P.G. Appleby, B.R. Duffy, R.W. Ogden, Glasgow Math. J. **29**, 185 (1987)
21. I.M. Gelfand, S.V. Fomin, *Calculus of Variations* (Prentice-Hall, 1963)
22. W. Heitler, *The Quantum Theory of Radiation*, 3rd edn. (Clarendon Press, 1954)
23. P.A.M. Dirac, *The Principles of Quantum Mechanics*, 4th edn. (Clarendon Press, 1982)

24. H. Ezawa, Ann. Phys. **67**, 438 (1971)
25. N. Akhiezer, I. Glazman, *Theory of Linear Operators in Hilbert Space* (Dover, 1993)
26. M. Pinard, Y. Hadjar, A. Heidmann, Eur. Phys. J. D **7**, 107 (1999)
27. N.W. Ashcroft, N.D. Mermin, *Solid State Physics* (Harcourt, 1976)
28. G. van Lear, G.E. Uhlenbeck, Phys. Rev **38**, 1583 (1931)
29. H.H. Szu, Phys. Rev. A **11**, 350 (1975)
30. G.W. Ford, M. Kac, P. Mazur, Math. Phys. **6**, 504 (1965)
31. E.B. Davies, Commun. Math. Phys. **33**, 171 (1973)
32. G.W. Ford, J.T. Lewis, R.F. O'Connell, Phys. Rev. A **37**, 4419 (1988)
33. V. Giovannetti, D. Vitali, Phys. Rev. A **63**, 023812 (2001)
34. G. Lindblad, Rep. Math. Phys. **10**, 393 (1976)
35. F. Haake, R. Reibold, Phys. Rev. A **32**, 2462 (1985)
36. S. Gnutzmann, F. Haake, Z. Phys. B **101**, 263 (1996)
37. R.F. Streater, J. Phys. A **15**, 1477 (1982)
38. J.A. Stratton, *Electromagnetic Theory* (McGraw Hill, 1941)
39. C. Cohen-Tannoudji, J. Dupont-Roc, G. Grynberg, *Photons and Atoms: Introduction to Quantum Electrodynamics* (Wiley, 1997)
40. C.M. Caves, B.L. Schumaker, Phys. Rev. A **31**, 3068 (1985)
41. L. Susskind, J. Glogower, Physics **1**, 49 (1964), reprinted in S. Barnett, J. Vaccaro, *The Quantum Phase Operator* (Taylor & Francis, 2007)
42. S.M. Barnett, D.T. Pegg, J. Phys. A **19**, 3849 (1986)
43. P. Busch, M. Grabowski, P.J. Lahti, Ann. Phys **237**, 1 (1995)
44. H.A. Haus, C.H. Townes, Proc. IRE **50**, 1544 (1962)
45. P.A.M. Dirac, Proc. R. Soc. A **114**, 243 (1927)
46. J. von Neumann, Compositio Mathematica **6**, 1 (1939)
47. R.J. Glauber, Phys. Rev. **131**, 2766 (1963a)
48. K.J. Blow, R. Loudon, S.J.D. Phoenix, T.J. Shepherd, Phys. Rev. A **42**, 4102 (1990)
49. G. Lachs, Phys. Rev. **138**, B1012 (1965)
50. A. Vourdas, Phys. Rev. A **34**, 3466 (1986)
51. M.O. Scully, W.E. Lamb, Phys. Rev **159**, 208 (1967)
52. S.J. van Enk, C.A. Fuchs, arXiv:quant-ph/0111157 (2001)
53. R. Simon, N. Mukunda, B. Dutta, Phys. Rev. A **49**, 1567 (1994)
54. U. Leonhardt, Rep. Prog. Phys. **66**, 1207 (2003)
55. R.J. Glauber, Phys. Rev. **130**, 2529 (1963b)
56. M.D. Srinivas, E.B. Davies, J. Mod. Opt **28**, 981 (1981)
57. H.J. Kimble, L. Mandel, Phys. Rev. A **30**, 844 (1984)
58. B. Yurke, Phys. Rev. A **32**, 311 (1985)
59. A.W. Hull, N.H. Williams, Phys. Rev. **25**, 147 (1925)
60. J.G. Graeme, *Photodiode Amplifiers* (McGraw Hill, 1996)
61. H.P. Yuen, V.W. Chan, Opt. Lett. **8**, 177 (1983)
62. B. Yurke, L.R. Corruccini, P.G. Kaminsky, L.W. Rupp, A.D. Smith, A.H. Silver, R.W. Simon, E.A. Whittaker, Phys. Rev. A **39**, 2519 (1989)
63. A. Roy, M. Devoret, Comp. Rend. Phys. **17**, 740 (2016)
64. C.M. Caves, Phys. Rev. D **26**, 1817 (1982)
65. B. Yurke, S.L. McCall, J.R. Klauder, Phys. Rev. A **33**, 4033 (1986)
66. V. Giovannetti, S. Lloyd, L. Maccone, Science **306**, 1330 (2004)
67. S. Steinlechner, B.W. Barr, A.S. Bell, S.L. Danilishin, A. Gläfke, C. Gräf, J.-S. Hennig, E.A. Houston, S.H. Huttner, S.S. Leavey, D. Pascucci, B. Sorazu, A. Spencer, K.A. Strain, J. Wright, S. Hild, Phys. Rev. D **92**, 072009 (2015)
68. J.H. Shapiro, IEEE, J. Quant. Electr. **21**, 237 (1985)
69. S. Machida, Y. Yamamoto, IEEE. J. Quant. Electr. **22**, 617 (1986)
70. K.E. Gillilland, H.D. Cook, K.D. Mielenz, R.B. Stephens, Metrologia **2**, 95 (1966)
71. A.G. McNish, Science **146**, 177 (1964)
72. N. Walker, J. Mod. Opt. **34**, 15 (1987)

73. T. Niebauer, R. Schilling, K. Danzmann, A. Rüdiger, W. Winkler, Phys. Rev. A **43**, 5022 (1991)
74. A. Buonanno, Y. Chen, N. Mavalvala, Phys. Rev. D **67**, 122005 (2003)
75. A.M. Yaglom, *Correlation Theory of Stationary and Related Random Functions*, vol. 2 (Springer, 1987)
76. H.L. Hurd, A. Miamee, *Periodically Correlated Random Sequences* (Wiley, 2007)
77. F. Quinlan, T.M. Fortier, H. Jiang, S.A. Diddams, J. Opt. Soc. Am. B **30**, 1775 (2013a)
78. H. Ogura, IEEE. Trans. Info. Theory **17**, 143 (1971)
79. M.B. Gray, A.J. Stevenson, H.-A. Bachor, D.E. McClelland, Opt. Lett. **18**, 759 (1993)
80. M. Rakhmanov, Appl. Opt. **40**, 6596 (2001)
81. F. Quinlan, T.M. Fortier, H. Jiang, A. Hati, C. Nelson, Y. Fu, J.C. Campbell, S.A. Diddams, Nat. Phot **7**, 290 (2013b)
82. Z.Y. Ou, H.J. Kimble, Phys. Rev. A **52**, 3126 (1995)
83. R.D. Richtmeyer, J. Appl. Phys. **10**, 391 (1939)
84. K.J. Vahala, Nature **424**, 839 (2003)
85. K. Ujihara, *Output Coupling in Optical Cavities and Lasers* (Wiley-VCH, 2010)
86. S. Schiller, Appl. Opt. **32**, 2181 (1993)
87. D.K. Armani, T.J. Kippenberg, S.M. Spillane, K.J. Vahala, Nature **421**, 925 (2003)
88. R. Schilling, H. Schütz, A. Ghadimi, V. Sudhir, D. Wilson, T. Kippenberg, Phys. Rev. Appl. **5**, 054019 (2016)
89. J.C. Knight, G. Cheung, F. Jacques, T.A. Birks, Opt. Lett. **22**, 1129 (1997)
90. M. Cai, O. Painter, K.J. Vahala, Phys. Rev. Lett. **85**, 74 (2000)
91. E. Snitzer, J. Opt. Soc. Am. **51**, 491 (1961)
92. M.L. Gorodetsky, V.S. Ilchenko, J. Opt. Soc. Am. B **16**, 147 (1999)
93. H.J. Carmichael, J. Opt. Soc. Am. B **4**, 1588 (1987)
94. C.W. Gardiner, A.S. Parkins, M.J. Collett, J. Opt. Soc. Am. B **4**, 1617 (1987)
95. V.B. Braginsky, M.L. Gorodetsky, V.S. Ilchenko, Phys. Lett. A **137**, 393 (1989)
96. M.L. Gorodetsky, A.A. Savchenkov, V.S. Ilchenko, Opt. Lett. **21**, 453 (1996)
97. M.L. Gorodetsky, A.D. Pryamikov, V.S. Ilchenko, J. Opt. Soc. Am. B **17**, 1051 (2000)
98. H. Lee, T. Chen, J. Li, K.Y. Yang, S. Jeon, O. Painter, K.J. Vahala, Nat. Photon **6**, 369 (2012)
99. D.W. Vernooy, V.S. Ilchenko, H. Mabuchi, E.W. Streed, H.J. Kimble, Opt. Lett. **23**, 247 (1998)
100. R.W. Drever, J.L. Hall, F. von Kowalski, J. Hough, G.M. Ford, A.J. Munley, H. Ward, Appl. Phys. B **31**, 97 (1983)
101. G.C. Bjorklund, M.D. Levenson, W. Lenth, C. Ortiz, Appl. Phys. B **32**, 145 (1983)
102. A.E. Fomin, M.L. Gorodetsky, I.S. Grudinin, V.S. Ilchenko, J. Opt. Soc. Am. B **22**, 459 (2005)

Chapter 4
Photon-Phonon Coupling: Cavity Optomechanics

Sections 3 and 4 brutally seize this formalism and mercilessly beat it to death...

Carlton Caves

The purpose of this chapter is to synthesise the developments of the previous two in a concrete setting. For example, in Sects. 3.1 and 3.2, the dynamics of an elastic medium, and that of light stored in an optical cavity has been considered; in the section below, we will see that elastic perturbations of the volume of an optical cavity lead to perturbations in the cavity frequency. Such a coupling between light and motion forms what is called a *cavity optomechanical* system. Perturbations in the cavity frequency trigger a feedback mechanism—shift in the cavity resonance frequency leads to changes in the stored optical energy, which exerts a radiation-pressure force that further displaces the elastic medium—that captures the basic dynamic consequence of light-motion interaction in cavity optomechanics; Sect. 4.2 describes this phenomenon. Section 4.3 discusses how the same interaction can be used to perform an ideal measurement of motion using light—one that can operate at the quantum limit that was derived in an abstract setting in Chap. 2.

4.1 Perturbing an Optical Cavity

An optical cavity is here taken to be some finite volume[1] V, with an appropriate boundary, that supports electromagnetic fields of the form,

[1]We assume that this domain is simply-connected so that all the usual manipulations of vector calculus hold.

V. Sudhir, *Quantum Limits on Measurement and Control of a Mechanical Oscillator*, Springer Theses, https://doi.org/10.1007/978-3-319-69431-3_4

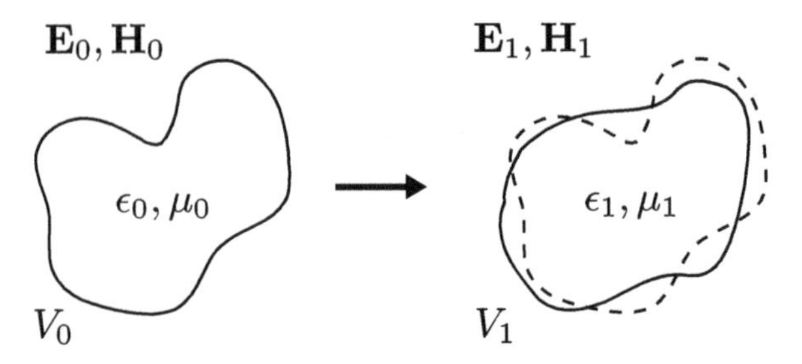

Fig. 4.1 Schematic of cavity perturbation. Left shows region V_0 forming a cavity of material characterised by dielectric constants ϵ_0, μ_0; right shows the region deformed to a new domain V_1, and with possibly different dielectric constants ϵ_1, μ_1

$$\mathbf{E}(\mathbf{r}, t) = \mathbf{E}(\mathbf{r})e^{-i\omega t}, \quad \mathbf{H}(\mathbf{r}, t) = \mathbf{H}(\mathbf{r})e^{-i\omega t}, \tag{4.1.1}$$

at the resonance frequency ω. The spatial variation of these fields, as well as the resonance frequencies, are determined by the Maxwell equations [1],

$$\nabla \times \mathbf{E}(\mathbf{r}) = i\omega\, \mu(\mathbf{r})\, \mathbf{H}(\mathbf{r}), \quad \nabla \times \mathbf{H}(\mathbf{r}) = -i\omega\, \epsilon(\mathbf{r})\, \mathbf{E}(\mathbf{r}); \tag{4.1.2}$$

here μ is the magnetic permeability (a strictly real number), and ϵ is the electric permittivity (a possibly complex number[2]). Note that the resonance frequency ω is allowed to be complex, so as to describe intrinsic losses of the cavity.

In this setting, the cavity may be perturbed only via a limited set of influences: (1) perturbation of the dielectric constants, μ, ϵ, through the introduction/removal of material, and/or (2) changes in the cavity domain V, for example by deformations of the boundary ∂V. Both perturbations lead to a change in the cavity frequency ω. When these effects are caused by an underlying elastic deformation field $\mathbf{u}$ (as defined in Sect. 3.1), a coupling between the cavity electromagnetic field and the elastodynamic field results. It is this coupling that we compute in the following via a perturbative treatment.

Following the work of Bethe and Schwinger [2] (see also [3]), consider the two configurations shown in Fig. 4.1, depicting a cavity perturbed in shape, i.e. $V_0 \to V_1$, and dielectric constants $(\epsilon_0(\mathbf{r}), \mu_0(\mathbf{r})) \to (\epsilon_1(\mathbf{r}), \mu_1(\mathbf{r}))$. Each situation is defined by equations analogous to Eq. (4.1.2), viz.

$$\begin{aligned} \nabla \times \mathbf{E}_0 &= i\omega_0\mu_0\, \mathbf{H}_0, \quad \nabla \times \mathbf{H}_0 = -i\omega_0\epsilon_0\, \mathbf{E}_0, \\ \nabla \times \mathbf{E}_1 &= i\omega_1\mu_1\, \mathbf{H}_1, \quad \nabla \times \mathbf{H}_1 = -i\omega_1\epsilon_1\, \mathbf{E}_1. \end{aligned} \tag{4.1.3}$$

[2]The second Maxwell equation follows from $\nabla \times \mathbf{H} = \mathbf{J} + \epsilon \frac{\partial \mathbf{E}}{\partial t}$, where ϵ is the real-valued electric permittivity. Assuming that losses in the cavity are modelled by the presence of a finite conductivity σ, i.e. the current density $\mathbf{J} = \sigma \mathbf{E}$, gives $\nabla \times \mathbf{H} = \left(\sigma + \epsilon \frac{\partial}{\partial t}\right) \mathbf{E}$. Inserting the ansatz Eq. (4.1.1) gives, $\nabla \times \mathbf{H} = -i\epsilon\omega \left(1 + i\frac{\sigma}{\omega\epsilon}\right) \mathbf{E}$. The identification, $\epsilon \left(1 + i\frac{\sigma}{\omega\epsilon}\right) \to \epsilon$, results in a phenomenological permittivity that is complex.

The general strategy is to relate the frequency shifts to mechanical quantities of the electromagnetic field, i.e. its energy and momentum. The change in these mechanical quantities can then be related to the motion of the elastic field.

Guided by the form of the energy and momentum of the field, Eq. (4.1.3) can be expressed as,

$$\begin{aligned}\mathbf{H}_1^* \cdot (\nabla \times \mathbf{E}_0) &= i\omega_0\mu_0 \mathbf{H}_1^* \cdot \mathbf{H}_0 \\ \mathbf{E}_0 \cdot (\nabla \times \mathbf{H}_1^*) &= i\omega_1\epsilon_1^* \mathbf{E}_1^* \cdot \mathbf{E}_0;\end{aligned}$$

which upon subtraction gives,

$$\nabla \cdot (\mathbf{H}_1^* \times \mathbf{E}_0) = i \left(\omega_1\epsilon_1^* \mathbf{E}_1^* \cdot \mathbf{E}_0 - \omega_0\mu_0 \mathbf{H}_1^* \cdot \mathbf{H}_0\right), \tag{4.1.4}$$

where the left-hand side is expressed using the identity $\nabla\cdot(\mathbf{f}\times\mathbf{g}) = \mathbf{g}\cdot\nabla\times\mathbf{f} - \mathbf{f}\cdot\nabla\times\mathbf{g}$. Similar manipulation of the remaining two equations in Eq. (4.1.3) gives,

$$\nabla \cdot (\mathbf{H}_0 \times \mathbf{E}_1^*) = -i \left(\omega_0\epsilon_0^* \mathbf{E}_1^* \cdot \mathbf{E}_0 - \omega_1\mu_1 \mathbf{H}_1^* \cdot \mathbf{H}_0\right). \tag{4.1.5}$$

Adding Eqs. (4.1.5) and (4.1.4), integrating the resulting equation over the perturbed domain V_1, and applying Gauss's theorem, gives the so-called Bethe-Schwinger equation [2],

$$\begin{aligned}\oint_{\partial V_1} \left(\mathbf{H}_1^* \times \mathbf{E}_0 + \mathbf{H}_0 \times \mathbf{E}_1^*\right) \cdot \mathrm{d}\mathbf{s} = i \int_{V_1} \big[&(\omega_1\epsilon_1^* - \omega_0\epsilon_0^*)\mathbf{E}_1^* \cdot \mathbf{E}_0 \\ &+ (\omega_1\mu_1 - \omega_0\mu_0)\mathbf{H}_1^* \cdot \mathbf{H}_0\big]\, \mathrm{d}^3 r,\end{aligned}$$

where $\mathbf{s}$ denotes the normal to the boundary surface ∂V_1. Expressing the above equation in a shorthand notation,

$$\begin{aligned}\mathscr{I}^{(V)} &= \omega_1\, \mathscr{I}_1^{(\epsilon,\mu)} - \omega_0\, \mathscr{I}_0^{(\epsilon,\mu)} \\ \text{where, } \mathscr{I}^{(V)} &:= -i \oint_{\partial V_1} \left(\mathbf{H}_1^* \times \mathbf{E}_0 + \mathbf{H}_0 \times \mathbf{E}_1^*\right) \cdot \mathrm{d}\mathbf{s} \\ \mathscr{I}_{0,1}^{(\epsilon,\mu)} &:= \int_{V_1} \left(\epsilon_{0,1}^* \mathbf{E}_1^* \cdot \mathbf{E}_0 + \mu_{0,1} \mathbf{H}_1^* \cdot \mathbf{H}_0\right) \mathrm{d}^3 r,\end{aligned} \tag{4.1.6}$$

the shifted frequency of the cavity is given by the formally exact recursion,

$$\omega_1 = \omega_0 + \frac{\mathscr{I}^{(V)}}{\mathscr{I}_0^{(\epsilon,\mu)}} - \omega_1 \left(\frac{\mathscr{I}_1^{(\epsilon,\mu)} - \mathscr{I}_0^{(\epsilon,\mu)}}{\mathscr{I}_0^{(\epsilon,\mu)}}\right). \tag{4.1.7}$$

A series of systematic approximations may now be performed. Firstly the recursion (in Eq. (4.1.7)) is iterated once to obtain the approximate fractional frequency shift,

$$\frac{\delta\omega}{\omega_0} := \frac{\omega_1 - \omega_0}{\omega_0} \approx \frac{\omega_0^{-1}\mathscr{I}^{(V)}}{\mathscr{I}_0^{(\epsilon,\mu)}} - \frac{\mathscr{I}_1^{(\epsilon,\mu)} - \mathscr{I}_0^{(\epsilon,\mu)}}{\mathscr{I}_0^{(\epsilon,\mu)}}. \tag{4.1.8}$$

Secondly, the integrals $\mathscr{I}$ may be estimated approximately, as follows. The surface integral $\mathscr{I}^{(V)}$ (see Eq. (4.1.6)) consists of two terms. Assuming that the boundaries $\partial V_{0,1}$ are essentially lossless (i.e. both the initial domain, and the perturbed domain have no intrinsic radiation loss), the electric field satisfies the boundary condition $\mathbf{E}_i|_{\partial V_i} = 0$ for each scenario $i = 0, 1$. Under this assumption,

$$\begin{aligned}\mathscr{I}^{(V)} &\approx -i \oint_{\partial V_1} \mathbf{H}_1^* \times \mathbf{E}_0 \cdot \mathbf{ds} \\ &= -i \left(\oint_{\partial V_1 - \partial V_0} + \oint_{\partial V_0} \right) \mathbf{H}_1^* \times \mathbf{E}_0 \cdot \mathbf{ds} \\ &= -i \oint_{\partial V_1 - \partial V_0} \mathbf{H}_1^* \times \mathbf{E}_0 \cdot \mathbf{ds},\end{aligned} \tag{4.1.9}$$

where by $\partial V_1 - \partial V_0$, we mean the surface enclosing the volume $\Delta V := V_1 - V_0$, defined as the set difference of the two domains. The final approximation consists of assuming that the perturbation does not appreciably alter the mode structure of the cavity, so that,

$$\mathbf{E}_1 \approx \mathbf{E}_0, \quad \mathbf{H}_1 \approx \mathbf{H}_0. \tag{4.1.10}$$

Inserting this in Eq. (4.1.9),

$$\mathscr{I}^{(V)} \approx -i \oint_{\partial V_1 - \partial V_0} \mathbf{H}_0^* \times \mathbf{E}_0 \cdot \mathbf{ds} = -\omega_0 \int_{\Delta V} \left(\epsilon_0 \, |\mathbf{E}_0|^2 - \mu_0 \, |\mathbf{H}_0|^2 \right) \mathrm{d}^3 r, \tag{4.1.11}$$

where the second equality is a consequence of Poynting's theorem [1]. Under the approximation in Eq. (4.1.10), the integrals $\mathscr{I}_{0,1}^{(\epsilon,\mu)}$ take the form,

$$\begin{aligned}\mathscr{I}_0^{(\epsilon,\mu)} &\approx \int_{V_0} \left(\epsilon_0^* \, |\mathbf{E}_0|^2 + \mu_0 \, |\mathbf{H}_0|^2 \right) \mathrm{d}^3 r \\ \mathscr{I}_1^{(\epsilon,\mu)} - \mathscr{I}_0^{(\epsilon,\mu)} &\approx \int_{V_0} \left(\Delta\epsilon^* \, |\mathbf{E}_0|^2 + \Delta\mu \, |\mathbf{H}_0|^2 \right) \mathrm{d}^3 r,\end{aligned} \tag{4.1.12}$$

where $\Delta\epsilon^* := \epsilon_1 - \epsilon_0$, and, $\Delta\mu := \mu_1 - \mu_0$, are the perturbations in the dielectric constants.

The frequency shift can now be estimated by inserting the approximate integrals in Eqs. (4.1.11) and (4.1.12) into the fractional frequency shift equation (Eq. (4.1.8)). At the level of perturbation theory carried out here, the frequency shift arises from two independent contributions, viz.

$$\frac{\delta\omega}{\omega_0} \approx \left(\frac{\delta\omega}{\omega_0}\right)_V + \left(\frac{\delta\omega}{\omega_0}\right)_{\epsilon,\mu}$$

$$\text{where,} \quad \left(\frac{\delta\omega}{\omega_0}\right)_V = -\frac{\int_{\Delta V}\left(\epsilon_0\,|\mathbf{E}_0|^2 - \mu_0\,|\mathbf{H}_0|^2\right)\,\mathrm{d}^3r}{\int_{V_0}\left(\epsilon_0^*\,|\mathbf{E}_0|^2 + \mu_0\,|\mathbf{H}_0|^2\right)\,\mathrm{d}^3r} \tag{4.1.13}$$

$$\text{and,} \quad \left(\frac{\delta\omega}{\omega_0}\right)_{\epsilon,\mu} = -\frac{\int_{V_0}\left(\Delta\epsilon^*\,|\mathbf{E}_0|^2 + \Delta\mu\,|\mathbf{H}_0|^2\right)\,\mathrm{d}^3r}{\int_{V_0}\left(\epsilon_0^*\,|\mathbf{E}_0|^2 + \mu_0\,|\mathbf{H}_0|^2\right)\,\mathrm{d}^3r}.$$

The first terms arises purely from volume deformation of the cavity domain, while the second arises purely from changes in dielectric constant(s) within the cavity. Note that for a predominantly dielectric cavity (i.e., where the electric field energy is much larger than the magnetic energy), an increase in the cavity dielectric constants and/or an increase in the cavity volume, both lead to a decrease in the resonance frequency. This analogy has been previously used to model moving boundary effects as an effective dielectric perturbation [4]. Finally we note that the conceptual ambiguities associated with identifying the force applied by an electromagnetic field on a moving body [5, 6] appear to be less severe in the approach outlined here, wherein the emphasis is on a well-defined observable —cavity frequency—and not on the details of the microscopic light-matter interaction.

4.2 Effective Description: Single-Mode Cavity Optomechanics

The effect of volume and/or dielectric perturbation, given in Eq. (4.1.13), may be related to the motion of a causative displacement field $\mathbf{u}$. The resulting variation of cavity volume and dielectric constant, can be related to the elastic field as,[3]

$$\Delta V \approx (\nabla\cdot\mathbf{u})V_0, \quad \Delta\epsilon^* \approx (\nabla\epsilon^*)\cdot\mathbf{u}, \quad \Delta\mu = (\nabla\mu)\cdot\mathbf{u}. \tag{4.2.1}$$

Inserting these in Eq. (4.1.13) provides a description of the cavity electromagnetic field interacting with the elastic field. Note however that other forms of coupling, for

[3]The volume change, ΔV, due to an elastic displacement $\mathbf{u}$ is that swept out by the surface element $\mathbf{ds}$ transverse to the displacement, i.e.,

$$\Delta V = \int \mathbf{u}\cdot\mathrm{d}\mathbf{s} = \int (\nabla\cdot\mathbf{u})\,\mathrm{d}^3\mathbf{r} \approx \nabla\cdot\mathbf{u}\int \mathrm{d}^3\mathbf{r} = (\nabla\cdot\mathbf{u})V_0.$$

For the dielectric constants, say ϵ^*, a Taylor expansion gives,

$$\Delta\epsilon^* \approx \frac{\partial\epsilon^*}{\partial\mathbf{r}}\cdot\mathrm{d}\mathbf{r} = (\nabla\epsilon^*)\cdot\mathbf{u}.$$

example through the elasto-optic effect [7, 8], which renders volume deformation and dielectric perturbation not independent, do not follow the above prescription.

Confining attention to the case of a dielectric cavity undergoing perturbation of its electric permittivity due to an elastic deformation (other cases are treated identically), the cavity frequency shift, now denoted $\delta\omega_c$, (from Eq. (4.1.13)) takes the form,

$$\delta\omega_c(t) \approx -\omega_c \frac{\int_{V_0} |\mathbf{E}_0|^2 \left(\mathbf{u}(\mathbf{r}, t) \cdot \nabla\epsilon^*\right) \mathrm{d}^3 r}{\int_{V_0} \left(\epsilon_0 |\mathbf{E}_0|^2 + \mu_0 |\mathbf{H}_0|^2\right) \mathrm{d}^3 r} =: -\sum_n G_n \hat{x}_n(t). \tag{4.2.2}$$

The second equality follows from inserting the elastic mode expansion (Eq. (3.1.21)),

$$\mathbf{u}(\mathbf{r}, t) = \sum_n \tilde{\mathbf{u}}_n(\mathbf{r}) \hat{x}_n(t),$$

and defining the *cavity frequency pull parameter*,

$$G_n := \omega_c \frac{\int_{V_0} |\mathbf{E}_0|^2 \left(\tilde{\mathbf{u}}_n(\mathbf{r}) \cdot \nabla\epsilon^*\right) \mathrm{d}^3 r}{\int_{V_0} \left(\epsilon_0 |\mathbf{E}_0|^2 + \mu_0 |\mathbf{H}_0|^2\right) \mathrm{d}^3 r} = \frac{\omega_c}{2} \frac{\int_{V_0} |\mathbf{E}_0|^2 \left(\tilde{\mathbf{u}}_n(\mathbf{r}) \cdot \nabla\epsilon^*\right) \mathrm{d}^3 r}{\int_{V_0} \epsilon_0 |\mathbf{E}_0|^2 \, \mathrm{d}^3 r}. \tag{4.2.3}$$

Here, the second equality is obtained by noticing that the denominator is the energy of the electromagnetic field in the volume V_0, which, for a stationary field configuration in a closed volume is divided equally between the electric and magnetic fields [2].

The cavity frequency shift describes a dispersive coupling between the electromagnetic and elastic degrees of freedom. In a quantised hamiltonian description (see Eqs. (3.1.25 and 3.2.39)), such an interaction manifests as a displacement-dependent modification of the cavity frequency, viz.,

$$\begin{aligned} \hat{H} &= \hbar\omega_c(\mathbf{u})\hat{a}^\dagger\hat{a} + \sum_n \hbar\Omega_n \left(\hat{b}_n^\dagger\hat{b}_n + \frac{1}{2}\right) \\ &= \hbar\left(\omega_c - \sum_n G_n \hat{x}_n\right)\hat{a}^\dagger\hat{a} + \sum_n \hbar\Omega_n \left(\hat{b}_n^\dagger\hat{b}_n + \frac{1}{2}\right) \\ &= \underbrace{\hbar\omega_c\hat{a}^\dagger\hat{a}}_{\hat{H}_c} + \sum_n \underbrace{\hbar\Omega_n \left(\hat{b}_n^\dagger\hat{b}_n + \frac{1}{2}\right)}_{\hat{H}_{\mathrm{m},n}} - \sum_n \underbrace{\hbar(G_n x_{\mathrm{zp},n})\hat{a}^\dagger\hat{a}(\hat{b}_n + \hat{b}_n^\dagger)}_{\hat{H}_{\mathrm{int},n}}. \end{aligned} \tag{4.2.4}$$

In terms of quantised excitations of the elastic field—phonons—and those of the quantised electromagnetic cavity field—photons —this hamiltonian may be read as the sum of their energies ($\hat{H}_c + \sum_n \hat{H}_{\mathrm{m},n}$), together with a nonlinear interaction term $\sum_n \hat{H}_{\mathrm{int},n}$. From the perspective of the photons, the interaction leads to a frequency shift, whereas from the perspective of the phonons, it leads to an additional force (here, $\hat{p}_n := i p_{\mathrm{zp},n}(\hat{b}_n - \hat{b}_n^\dagger)$ is the canonical momentum conjugate to $\hat{x}_n$, whose zero-point amplitude, $p_{\mathrm{zp},n}$ satisfies, $x_{\mathrm{zp},n} p_{\mathrm{zp},n} = \hbar/2$),

$$\hat{F}_{\mathrm{rad},n} := \frac{i}{\hbar}[\hat{H}_{\mathrm{int},n}, \hat{p}_n] = -\hbar G_n \, \hat{a}^\dagger \hat{a}.$$

Being a photon-number-dependent force, we identify it as being due to radiation pressure exerted on the phonon field by the intracavity photon field. Note that the unambiguously defined product,[4] the *vacuum optomechanical coupling rate*,

$$g_{0,n} := G_n x_{\mathrm{zp},n} = G_n \left(\frac{\hbar}{2 m_n \Omega_n} \right)^{1/2},$$

characterises the strength of the coupling between the photon and phonon degrees of freedom.

We now focus on a single mechanical mode, described by its position $\hat{x} = x_{\mathrm{zp}}(\hat{b} + \hat{b}^\dagger)$, interacting with a cavity field at the rate $g_0 = G x_{\mathrm{zp}}$. The equations of motion for the intracavity optical field $\hat{a}$ and the mechanical oscillator position $\hat{x}$, that follow from the hamiltonian in Eq. (4.2.4), together with the cavity-waveguide coupling and loss terms of either subsystem (see Eqs. (3.1.28) and (3.2.48)), are [9],

$$\begin{aligned} \dot{\hat{a}} &= \left(i(\Delta - G\hat{x}) - \frac{\kappa}{2} \right) \hat{a} + \sqrt{(1 - \eta_c)\kappa}\, \delta \hat{a}_0 + \sqrt{\eta_c \kappa}\, \hat{a}_{\mathrm{in}} \\ \ddot{\hat{x}} + \Gamma_{\mathrm{m}} \dot{\hat{x}} + \Omega_{\mathrm{m}}^2 \hat{x} &= m^{-1} \left(\delta \hat{F}_{\mathrm{th}} - \hbar G\, \hat{a}^\dagger \hat{a} \right). \end{aligned} \tag{4.2.5}$$

In the rest of this section we use these equations of motion to explore the basic consequences of the radiation pressure optomechanical interaction for the dynamics of the photon and phonon modes.

4.2.1 Steady-State Shifts

The above equations of motion, due to their nonlinearity, sustain several steady-states. Employing the ansatz,

$$\hat{a}(t) = \bar{a} + \delta \hat{a}(t), \quad \hat{x}(t) = \bar{x} + \delta \hat{x}(t),$$

the steady-state amplitudes are determined by the nonlinear algebraic relations,

---

[4]From the discussion in footnote 3.8 (in page 42), a scale transformation of the elastic displacement function by a factor λ, i.e. $\tilde{u}_n(\mathbf{r}) \to \lambda \tilde{u}_n(\mathbf{r})$, implies a corresponding transformation of the mode mass, $m_n \to \lambda^2 m_n$, and therefore a transformation of the zero-point motion, $x_{\mathrm{zp},n} \to \lambda^{-1} x_{\mathrm{zp},n}$. From the definition of the frequency-pull parameter in Eq. (4.2.3), the same scale transformation induces the change, $G_n \to \lambda G_n$, meaning that it is dependent on the definition of the displacement function (but not on the cavity electric field). However, the product, $G_n x_{\mathrm{zp},n}$ is invariant to any scale transformation of the displacement field, and is thus independent of its choice.

$$\bar{a} = \frac{-\sqrt{\eta_c \kappa}\, \bar{a}_{\mathrm{in}}}{i(\Delta - G\bar{x}) - \kappa/2}, \quad \bar{x} = -\frac{\hbar G}{m\Omega_{\mathrm{m}}^2} |\bar{a}|^2 . \tag{4.2.6}$$

Schematically, this leads to a static shift in the mean laser-cavity detuning due to the mean position of the mechanical element, $\bar{x}$, leading to a change in the mean intracavity photon number. The resulting delayed radiation pressure force causes the mean position of the oscillator to change; ultimately, the steady-state is stable when these changes are self-consistent [10].

In order to investigate the potential instability, Eq. (4.2.6) may be expressed as the cubic equation,

$$\frac{\bar{x}}{x_{\mathrm{zp}}} + \left(\frac{2\Delta}{\kappa} - \frac{2g_0}{\kappa}\frac{\bar{x}}{x_{\mathrm{zp}}}\right)^2 \frac{\bar{x}}{x_{\mathrm{zp}}} = -\frac{g_0}{\Omega_{\mathrm{m}}} n_{c,\mathrm{res}},$$

for the normalised steady-state displacement $\bar{x}/x_{\mathrm{zp}}$. Here $n_{c,\mathrm{res}} = \frac{4\eta_c}{\kappa} |\bar{a}_{\mathrm{in}}|^2$, the mean intracavity photon number on resonance, proxies the injected power in this equation. The necessary condition for bistability is that at least two roots of this equation are real. This happens when [11],

$$\frac{2\Delta}{\kappa} < -\sqrt{3}, \quad \text{and,} \quad n_{c,\mathrm{res}} > n_{c,\mathrm{thresh}} := \left(\frac{1}{3\sqrt{3}}\right) \frac{\kappa \Omega_{\mathrm{m}}}{g_0^2}.$$

However these are not sufficient conditions [12]. For the system employed in this thesis, the static optomechanical bistability threshold, $n_{c,\mathrm{thresh}} \approx 2 \cdot 10^6$, while the maximum mean intracavity photon numbers employed are ~ 50 times smaller.

Static instabilities due to an inconsistent steady-state were some of the first effects of radiation pressure to have been observed in high-power interferometers. In the 1980s optical bistability due to very low frequency ($\Omega_{\mathrm{m}} \approx 2\pi \cdot 1\,\mathrm{Hz}$) modes was observed [13, 14]. The steady-state, if stable, leads to a static radiation pressure force, $\hbar G\, |\bar{a}|^2$, that causes a shift in the mechanical oscillator resonance frequency,

$$\Omega_{\mathrm{m,stat}}^2 = m^{-1} \frac{\partial}{\partial \bar{x}} (\hbar G\, |\bar{a}|^2),$$

which has also been observed in early experiments with low frequency oscillators [15].

4.2.2 *Dynamical Back-Action*

Assuming a stable steady-state, fluctuations $\delta\hat{a}$, $\delta\hat{x}$ are governed by linearised equations of motion [9],

$$\begin{aligned}
\delta\dot{\hat{a}} &= \left(i\Delta - \frac{\kappa}{2}\right)\delta\hat{a} + ig_0\sqrt{n_c}\frac{\delta\hat{x}}{x_{\text{zp}}} + \sqrt{(1-\eta_c)\kappa}\,\delta\hat{a}_0 + \sqrt{\eta_c\kappa}\,\delta\hat{a}_{\text{in}} \\
\delta\ddot{\hat{x}} + \Gamma_{\text{m}}\delta\dot{\hat{x}} + \Omega_{\text{m}}^2\delta\hat{x} &= m^{-1}\left(\delta\hat{F}_{\text{th}} - \hbar\frac{g_0\sqrt{n_c}}{x_{\text{zp}}}\left(e^{-i\theta_c}\delta\hat{a} + e^{i\theta_c}\delta\hat{a}^\dagger\right)\right)
\end{aligned} \tag{4.2.7}$$

Here we define the magnitude, n_c, and phase, θ_c, of the steady-state intracavity amplitude $\bar{a}$, viz.,

$$n_c := \frac{4\eta_c}{\kappa}\frac{|\bar{a}_{\text{in}}|^2}{1+(4\Delta^2/\kappa^2)}, \quad \theta_c := \arctan\frac{2\Delta}{\kappa}.$$

The former quantity, n_c, is the mean intracavity photon number. In the case of a single cavity mode, the phase θ_c is redundant, since it may be accommodated for by using an appropriately retarded input field (i.e. $\hat{a}_{\text{in}} \to \hat{a}_{\text{in}}e^{-i\theta_c}$). Note however that when the cavity is embedded in an interferometer, as for example in our experiments, this corresponds to an equivalent advancement of the LO phase. Keeping this in mind we henceforth set $\theta_c = 0$.

Note finally that the vacuum interaction strength g_0, is enhanced by the presence of a large ($n_c \gg 1$) intracavity photon number, to the dressed interaction strength,

$$g := g_0\sqrt{n_c}.$$

It is in fact this renormalisation of the interaction strength that makes it possible to observe dynamic optomechanical phenomena despite the fact that the cavity frequency shift caused by the per-photon force, $\hbar G$, is small compared to the cavity linewidth κ.

Expressed in terms of the Fourier transform, $\delta\hat{a}[\Omega]$, $\delta\hat{x}[\Omega]$, Eq. (4.2.7) takes the form,

$$\begin{aligned}
\delta\hat{a}^{(g)}[\Omega] &= \delta\hat{a}^{(0)}[\Omega] + \chi_a^{(0)}[\Omega]\cdot ig\,\frac{\delta\hat{x}^{(g)}[\Omega]}{x_{\text{zp}}} \\
\delta\hat{x}^{(g)}[\Omega] &= \delta\hat{x}^{(0)}[\Omega] - \chi_x^{(0)}[\Omega]\cdot\frac{\hbar g}{x_{\text{zp}}}\left(\delta\hat{a}^{(g)}[\Omega] + \delta\hat{a}^{(g)}[-\Omega]^\dagger\right),
\end{aligned} \tag{4.2.8}$$

where the intrinsic susceptibilities,

$$\chi_a^{(0)}[\Omega] = \left(-i(\Omega+\Delta) + \frac{\kappa}{2}\right)^{-1}, \quad \chi_x^{(0)}[\Omega] = m^{-1}\left(\Omega_{\text{m}}^2 - \Omega^2 - i\Omega\Gamma_{\text{m}}\right)^{-1}, \tag{4.2.9}$$

dictate the response of the cavity and the phonon modes to their respective generalised forces in the absence of optomechanical coupling. In the above equations, $\delta\hat{a}^{(g)}$, $\delta\hat{x}^{(g)}$ denote the cavity field and mechanical position modified by the interaction; their intrinsic motion—in the absence of interaction ($g = 0$)—is given by,

$$\begin{aligned}\delta\hat{a}^{(0)}[\Omega] &:= \chi_a^{(0)}[\Omega]\left(\sqrt{(1-\eta_c)\kappa}\,\delta\hat{a}_0[\Omega] + \sqrt{\eta_c\kappa}\,\delta\hat{a}_{\text{in}}[\Omega]\right)\\ \delta\hat{x}^{(0)}[\Omega] &:= \chi_x^{(0)}[\Omega]\,\delta\hat{F}_{\text{th}}[\Omega].\end{aligned} \tag{4.2.10}$$

The generalised forces that drive the optomechanical system in the presence of a finite coupling come in two forms. From the right-hand side of Eq. (4.2.8), these are: ones due to the dynamics of either mode, and those due to stochastic fluctuations from the environment (interpreted broadly, including the other sub-system). The former effectively leads to a renormalisation of the susceptibilities—an effect called *dynamical back-action* [9, 16, 17]; the latter leads to *stochastic (measurement) back-action* [9, 18–20].[5] When the fluctuations due to the environment are limited to the level allowed by quantum mechanics, stochastic back-action is called *quantum back-action.*

In a classical picture, dynamical back-action may be described as a form of autonomous feedback between the mechanical oscillator and the cavity field [10, 17, 22], which leads to a renormalisation of the response of either sub-system to their respective external forces. Dynamical back-action of the cavity field on the mechanical oscillator becomes prominent when the photon lifetime in the cavity is comparable to the mechanical oscillator period. In this conventional regime ($\kappa \gtrsim \Omega_{\text{m}} \gg \Gamma_{\text{m}}$), the modified dynamics of the mechanical oscillator,

$$\delta\hat{x}^{(g)}[\Omega] = \chi_x^{(g)}[\Omega]\left(\delta\hat{F}_{\text{th}}[\Omega] + \delta\hat{F}_{\text{BA}}[\Omega]\right),$$

obtained by solving Eq. (4.2.8), features a renormalised susceptibility,

$$\chi_x^{(g)}[\Omega]^{-1} = \chi_x^{(0)}[\Omega]^{-1} + i\hbar\left(\frac{g}{x_{\text{zp}}}\right)^2\left(\chi_a^{(0)}[\Omega] - \chi_a^{(0)}[-\Omega]^*\right) \tag{4.2.11}$$

that modifies the response of the oscillator. The renormalised mechanical susceptibility [9],

$$\begin{aligned}\frac{\chi_x^{(g)}[\Omega]^{-1}}{m} &\approx \frac{\chi_x^{(0)}[\Omega]^{-1}}{m} + 2ig^2\Omega_{\text{m}}(\chi_a^{(0)}[\Omega_{\text{m}}] - \chi_a^{(0)}[-\Omega_{\text{m}}]^*)\\ &= \left(-\Omega^2 + \Omega_{\text{m}}^2 - i\Omega\Gamma_{\text{m}}\right) + 2ig^2\Omega_{\text{m}}(\chi_a^{(0)}[\Omega_{\text{m}}] - \chi_a^{(0)}[-\Omega_{\text{m}}]^*)\\ &= -\Omega^2 + \left[\Omega_{\text{m}}^2 + 2g^2\Omega_{\text{m}}\,\text{Re}\,i(\chi_a^{(0)}[\Omega_{\text{m}}] - \chi_a^{(0)}[-\Omega_{\text{m}}]^*)\right]\\ &\quad - i\Omega\left[\Gamma_{\text{m}} - 2g^2\frac{\Omega_{\text{m}}}{\Omega}\,\text{Im}\,i(\chi_a^{(0)}[\Omega_{\text{m}}] - \chi_a^{(0)}[-\Omega_{\text{m}}]^*)\right]\end{aligned}$$

[5] Although widely credited to the investigations of Braginsky's group in Moscow in the 1970s [16, 18], it appears that dynamical back-action was observed and described in the 1930s by a group led by Hartley at Bell Labs [21]. It is however unclear whether the effect was solely due to radiation pressure. It is interesting to note Hartley's recognition of the mechanism being analogous to Raman scattering, an effect that was only described in the previous decade.

describes a harmonic oscillator with a shifted frequency,

$$\begin{aligned}\Omega_{\text{eff}} &:= \left[\Omega_{\text{m}}^2 + 2g^2\Omega_{\text{m}} \operatorname{Re} i(\chi_a^{(0)}[\Omega_{\text{m}}] - \chi_a^{(0)}[-\Omega_{\text{m}}]^*)\right]^{1/2} \\ &\approx \Omega_{\text{m}} + \frac{2g^2}{\kappa}\left(\frac{2(\Delta+\Omega_{\text{m}})/\kappa}{1+(2(\Delta+\Omega_{\text{m}})/\kappa)^2} + \frac{2(\Delta-\Omega_{\text{m}})/\kappa}{1+(2(\Delta-\Omega_{\text{m}})/\kappa)^2}\right),\end{aligned} \quad (4.2.12)$$

and modified linewidth [9],

$$\begin{aligned}\Gamma_{\text{eff}} &:= \Gamma_{\text{m}} - 2g^2\frac{\Omega_{\text{m}}}{\Omega} \operatorname{Im} i(\chi_a^{(0)}[\Omega_{\text{m}}] - \chi_a^{(0)}[-\Omega_{\text{m}}]^*) \\ &\approx \Gamma_{\text{m}} + \frac{4g^2}{\kappa}\left(\frac{1}{1+(2(\Delta+\Omega_{\text{m}})/\kappa)^2} - \frac{1}{1+(2(\Delta-\Omega_{\text{m}})/\kappa)^2}\right).\end{aligned} \quad (4.2.13)$$

Figure 4.2 shows the modification of the mechanical frequency—*optical spring*—against the possible values of the linewidth modification—*optical damping*—as the laser-cavity detuning is varied from negative values, into resonance, and large positive values. The various traces show the effect of sideband resolution, $\frac{2\Omega_{\text{m}}}{\kappa}$, which dictates the efficiency of the autonomous feedback. Dynamical back-action induced modification of the mechanical response, leading to optical spring [23, 24], linewidth narrowing (amplification) [25–27], or damping (cooling) [28–32] have been observed.

Ultimately, dynamical back-action modification of the mechanical response is limited by the quantum back-action force,

$$\delta\hat{F}_{\text{BA}}[\Omega] := -\hbar\frac{g}{x_{\text{zp}}}\left(\delta\hat{a}^{(0)}[\Omega] + \delta\hat{a}^{(0)}[-\Omega]^\dagger\right) = -\hbar\frac{g}{x_{\text{zp}}}\sqrt{2}\delta\hat{q}^{(0)}[\Omega], \quad (4.2.14)$$

due to vacuum fluctuations in the optical amplitude, acting on the oscillator. For dynamical back-action amplification, this is predicted to lead to a fundamental phase diffusion of the oscillator [33], while for cooling, quantum fluctuations lead to a minimum accessible temperature [34, 35]. In the resolved-sideband regime, characterised by $\kappa \ll \Omega_{\text{m}}$, it becomes possible to cool the mechanical oscillator to a level where its energy is comparable to the energy in its ground state [36–40].

Dynamical back-action can also be used to modify the optical susceptibility, viz.,

$$\chi_a^{(g)}[\Omega]^{-1} = \chi_a^{(0)}[\Omega]^{-1} + i\hbar\frac{g^2}{x_{\text{zp}}^2}\chi_x^{(0)}[\Omega]$$

In the conventional regime this manifests as an optomechanically-induced absorption/transparency [41–43]. In an unconventional regime, characterised by $\Gamma_{\text{m}} \gtrsim \kappa$ [44], modification of the optical susceptibility leads to phenomena analogous to lasing—i.e. optical amplification with added noise limited by quantum fluctuations.

Thus, radiation pressure optomechanical coupling provides an opportunity to control the dynamics of the mechanical mode, or of the optical field, by modifying its

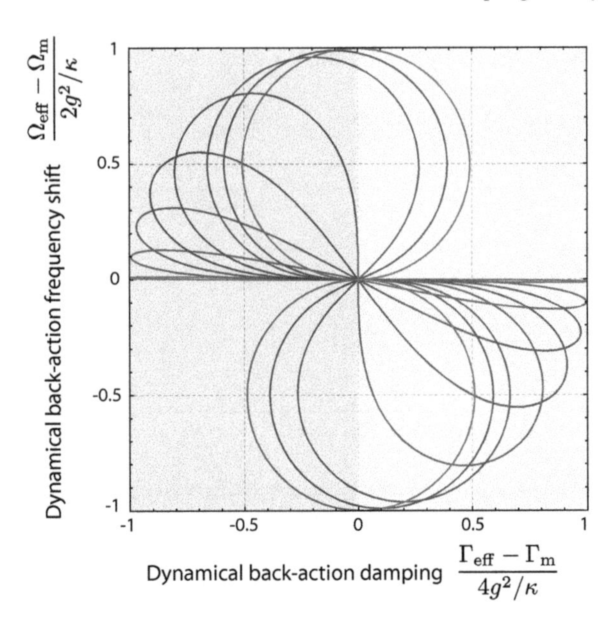

Fig. 4.2 Dynamical back-action. Parametric plot of the normalised optical spring shift versus the normalised damping, due to dynamic back-action. Each trace is plotted as the normalised detuning, $\frac{2\Delta}{\kappa}$, varies from $-\infty \ldots + \infty$. The different traces represent values of the sideband resolution factor, $\frac{2\Omega_{\text{m}}}{\kappa}$—red represents $\Omega_{\text{m}} = 10^{-2} \cdot \frac{\kappa}{2}$—deep in the sideband unresolved regime, while blue represents, $\Omega_{\text{m}} = 10^{2} \cdot \frac{\kappa}{2}$—deep in the sideband resolved regime. The shaded orange region (corresponding to $\Gamma_{\text{eff}} - \Gamma_{\text{m}} < 0$) represents regions of potential instability (corresponding to $\Gamma_{\text{eff}} < 0$), whereas the blue region is unconditionally stable

susceptibility. The fundamental limit to autonomous control of this kind is set by quantum back-action.

However, it is the alternate provision—that of being able to precisely measure mechanical motion—that a major part of this thesis is concerned with. In particular, in the unresolved sideband regime (i.e. $\kappa \gg \Omega_{\text{m}}$), dynamical back-action is a weak effect, and therefore it becomes possible to measure the intrinsic mechanical motion. The fundamental limit of how well the measurement can be performed is again set by quantum back-action. The following section explores the measurement aspect of optomechanics in detail.

4.3 Continuous Linear Measurement Using Cavity Optomechanics

The cavity field—henceforth the *meter*—by being coupled to the mechanical oscillator—the *system*—can realise the ideal linear measurement model considered in Chap. 2. The task of any such measurement is to allow for the inference of the intrinsic mechanical motion (defined in Eq. (4.2.10)),

$$\delta\hat{x}^{(0)} = \chi_x^{(0)}[\Omega]\,\delta\hat{F}_{\text{th}}[\Omega].$$

However, the linear coupling between the cavity field and the oscillator precludes the possibility of having access to just the intrinsic motion. From Eq. (4.2.8), the intracavity field consists of three terms,

$$\begin{aligned}\delta\hat{a}^{(g)}[\Omega] = \delta\hat{a}^{(0)}[\Omega] &+ \left(\frac{ig}{x_{\text{zp}}}\right)\chi_a^{(0)}[\Omega]\chi_x^{(g)}[\Omega]\chi_x^{(0)}[\Omega]^{-1}\,\delta\hat{x}^{(0)}[\Omega] \\ &+ \left(\frac{ig}{x_{\text{zp}}}\right)\chi_a^{(0)}[\Omega]\chi_x^{(g)}[\Omega]\delta\hat{F}_{\text{BA}}[\Omega];\end{aligned} \tag{4.3.1}$$

the first due to the unavoidable intrinsic fluctuations of the intracavity field, the second is the transduced motion of the oscillator (possibly modified by dynamical back-action), and the third, the motion of the oscillator driven by the fluctuations of the intracavity field (i.e. quantum back-action).

Since we are interested in retrieving the intrinsic motion, it is desirable that the measurement has no dynamical back-action, i.e. $\chi_x^{(g)}[\Omega] = \chi_x^{(0)}[\Omega]$. From the expressions for the optical spring and damping in Eqs. (4.2.12) and (4.2.13), it follows that for resonant probing, i.e. $\Delta = 0$, dynamic back-action is absent. In this case, the cavity transmission, given by the input-output relation (Eq. (3.2.49)), carries all three components:

$$\begin{aligned}\delta\hat{a}_{\text{out}}[\Omega] = &\left(1 - \frac{2\eta_c}{1 - 2i\Omega/\kappa}\right)\delta\hat{a}_{\text{in}}[\Omega] - \left(\frac{2\sqrt{\eta_c(1-\eta_c)}}{1 - 2i\Omega/\kappa}\right)\delta\hat{a}_0[\Omega] \\ &- i\left(\frac{\sqrt{\eta_c}}{1 - 2i\Omega/\kappa}\right)\left(\frac{2g}{\sqrt{\kappa}}\right)\frac{\delta\hat{x}^{(0)}[\Omega] + \delta\hat{x}_{\text{BA}}[\Omega]}{x_{\text{zp}}};\end{aligned} \tag{4.3.2}$$

the first two terms represent the intrinsic cavity field leaking out in transmission, while the second line shows the position fluctuations transduced by the measurement interaction, contaminated by the back-action driven position fluctuations,

$$\delta\hat{x}_{\text{BA}}[\Omega] := \chi_x^{(0)}[\Omega]\,\delta\hat{F}_{\text{BA}}[\Omega].$$

From Eq. (4.3.2), it is apparent that the role of the cavity is to enhance measurement sensitivity. For fixed intracavity photon number, the transmitted photon flux (per frequency band at an offset of Ω_{m} from the carrier) due to the zero-point motion is $\left(1 + 4\Omega_{\text{m}}^2/\kappa^2\right)^{-1} \cdot \eta_c \cdot \frac{4g^2}{\kappa}$. The frequency dependent pre-factor $(1 + 4\Omega_{\text{m}}^2/\kappa^2)^{-1}$ may be interpreted as an additional efficiency penalty arising from the finite bandwidth of the cavity acting as a low-pass filter with respect to the mechanical motion. In the deep unresolved-sideband regime, $\Omega_{\text{m}} \ll \kappa$, this efficiency factor is unity. Theoretically, therefore, probing on resonance with a sufficiently large bandwidth cavity provides the ideal meter to sense the oscillator motion.[6] Working in the broadband approximation, i.e. $\Omega_m/\kappa \ll 1$, fluctuations of the mechanical oscillator position are confined to the phase quadrature of the output field, viz.,

$$\begin{aligned} \delta\hat{p}_{out}[\Omega] &= (1-2\eta_c)\delta\hat{p}_{in}[\Omega] - 2\sqrt{\eta_c(1-\eta_c)}\,\delta\hat{p}_0[\Omega] \\ &\quad - \sqrt{2\eta_c}\left(\frac{2g}{\sqrt{\kappa}}\right)\left(\frac{\delta\hat{x}^{(0)}[\Omega]+\delta\hat{x}_{BA}[\Omega]}{x_{zp}}\right). \end{aligned}$$

A homodyne detector tuned to the phase quadrature directly detects $\delta\hat{p}_{out}$. We define the observable $\delta\hat{y}_{hom}$ that models the position-equivalent record of the homodyne detector,

$$\delta\hat{y}_{hom}[\Omega] := \frac{\delta\hat{p}_{out}[\Omega]}{-\sqrt{2\eta_c}(2g/\sqrt{\kappa})/x_{zp}} = \delta\hat{x}_{imp,hom}[\Omega] + \delta\hat{x}^{(0)}[\Omega] + \delta\hat{x}_{BA}[\Omega].$$

Here, we have defined,

$$\delta\hat{x}_{imp,hom}[\Omega] := x_{zp}\left(\frac{\sqrt{\kappa}}{2g}\right)\left(\frac{2\eta_c-1}{\sqrt{2\eta_c}}\,\delta\hat{p}_{in}[\Omega] + \sqrt{2(1-\eta_c)}\,\delta\hat{p}_0[\Omega]\right), \quad (4.3.3)$$

the position-equivalent imprecision in the measurement record. For the relevant operating conditions ($\Delta = 0, \Omega_m \ll \kappa$), the motion induced by the coupling to the meter—*measurement back-action*—is given by,

$$\delta\hat{x}_{BA}[\Omega] = \left(\frac{2g}{\sqrt{\kappa}}\right)\frac{\hbar\chi_x^{(0)}[\Omega]}{x_{zp}}\left(\sqrt{2\eta_c}\,\delta\hat{q}_{in}[\Omega] + \sqrt{2(1-\eta_c)}\,\delta\hat{q}_0[\Omega]\right). \quad (4.3.4)$$

Note that the definition of the back-action motion depends on the system-meter coupling alone, while that of the measurement imprecision depends also on the meter-detector coupling (i.e. the choice of detector used to measure the meter state—here, homodyne detection).

For quantum-noise-limited optical fields, the measurement imprecision and measurement-back action spectral densities take the form (following from Eqs. (4.3.3) and (4.3.4)),

$$\begin{aligned} \bar{S}_{xx}^{imp,hom}[\Omega] &= \frac{x_{zp}^2}{4\eta_c}\left(\frac{\kappa}{4g^2}\right) \\ \bar{S}_{xx}^{BA}[\Omega] &= \frac{\hbar^2}{x_{zp}^2}\left|\chi_x^{(0)}[\Omega]\right|^2\left(\frac{4g^2}{\kappa}\right) = \left|\chi_x^{(0)}[\Omega]\right|^2\bar{S}_{FF}^{BA}[\Omega]. \end{aligned} \quad (4.3.5)$$

Their product—an expression of the uncertainty principle—is given by,

[6] Effectively, the cavity needs to be broadband, and high-finesse.A micro-cavity meets these conflicting requirements owing to the small round-trip time.

$$\bar{S}_{FF}^{\mathrm{BA}}[\Omega]\bar{S}_{xx}^{\mathrm{imp,hom}}[\Omega] = \frac{\hbar^2}{4\eta_c}; \tag{4.3.6}$$

the ideal bound of $\hbar^2/4$ is achieved when the cavity is heavily over-coupled, corresponding to unit coupling efficiency $\eta_c = 1$. Therefore, in principle, the cavity optomechanical interaction, together with high-efficiency homodyne detection of the optical field, is capable of achieving the ideal performance of a linear quantum measurement chain.

The standard quantum limit (SQL) for such a measurement strategy refers to the absolute minimum apparent motion seen at the detector output. The total noise power in the homodyne measurement record, described in terms of the spectral density of the observable $\delta\hat{y}_{\mathrm{hom}}$,

$$\begin{aligned}\bar{S}_{yy}^{\mathrm{hom}}[\Omega] &= \bar{S}_{xx}^{(0)}[\Omega] + \bar{S}_{xx}^{\mathrm{imp,hom}}[\Omega] + \bar{S}_{xx}^{\mathrm{BA}}[\Omega] \\ &\geq \bar{S}_{xx}^{(0)}[\Omega] + 2\sqrt{\bar{S}_{xx}^{\mathrm{imp,hom}}[\Omega]\cdot\left|\chi_x^{(0)}[\Omega]\right|^2 \bar{S}_{FF}^{\mathrm{BA}}[\Omega]} \\ &\geq \bar{S}_{xx}^{(0)}[\Omega] + \frac{\hbar}{\sqrt{\eta_c}}\left|\chi_x^{(0)}[\Omega]\right|,\end{aligned} \tag{4.3.7}$$

contains, in addition to the intrinsic motion $\bar{S}_{xx}^{(0)}$, an excess noise at each frequency arising from the quantum fluctuations in the optical field, commensurate with the general considerations of Chap. 2. These excess fluctuations arise from the vacuum fluctuations in the two degrees of freedom of the meter: phase vacuum fluctuations leading to measurement imprecision in the homodyne detector, and amplitude vacuum fluctuations leading to apparent motion due to quantum back-action. The absolute minimum of the total apparent motion $\bar{S}_{yy}^{\mathrm{hom}}$—the SQL—is realized when the intrinsic fluctuations in the mechanical oscillator are also quantum-noise-limited, i.e. when its thermal occupation is zero ($n_{\mathrm{m,th}} = 0$), and when the detection efficiency is unity ($\eta_c = 1$). In this pristine case, the peak spectral density of apparent position fluctuations given by Eq. (4.3.7),

$$\bar{S}_{xx}^{\mathrm{SQL}}[\Omega_{\mathrm{m}}] := \min\, \bar{S}_{yy}^{\mathrm{hom}}[\Omega_{\mathrm{m}}] = 2\bar{S}_{xx}^{\mathrm{zp}}[\Omega_{\mathrm{m}}] = \frac{4x_{\mathrm{zp}}^2}{\Gamma_{\mathrm{m}}}, \tag{4.3.8}$$

is twice the intrinsic zero-point motion of the oscillator (defined in Eq. (3.1.30)).

The apparent motion at the SQL provides a fundamental and natural scale to compare the performance of any linear measurement of the position fluctuations of a mechanical oscillator. In fact, the apparent position fluctuations, $\bar{S}_{yy}^{\mathrm{hom}}$, referred to the standard quantum limit,

$$\frac{\bar{S}_{yy}^{\mathrm{hom}}[\Omega]}{\bar{S}_{xx}^{\mathrm{SQL}}[\Omega_{\mathrm{m}}]} = n_{\mathrm{imp,hom}} + \frac{(\Omega_{\mathrm{m}}\Gamma_{\mathrm{m}})^2}{(\Omega^2-\Omega_{\mathrm{m}}^2)^2 + (\Omega\Gamma_{\mathrm{m}})^2}\left(n_{\mathrm{m,th}} + n_{\mathrm{m,BA}} + \frac{1}{2}\right), \tag{4.3.9}$$

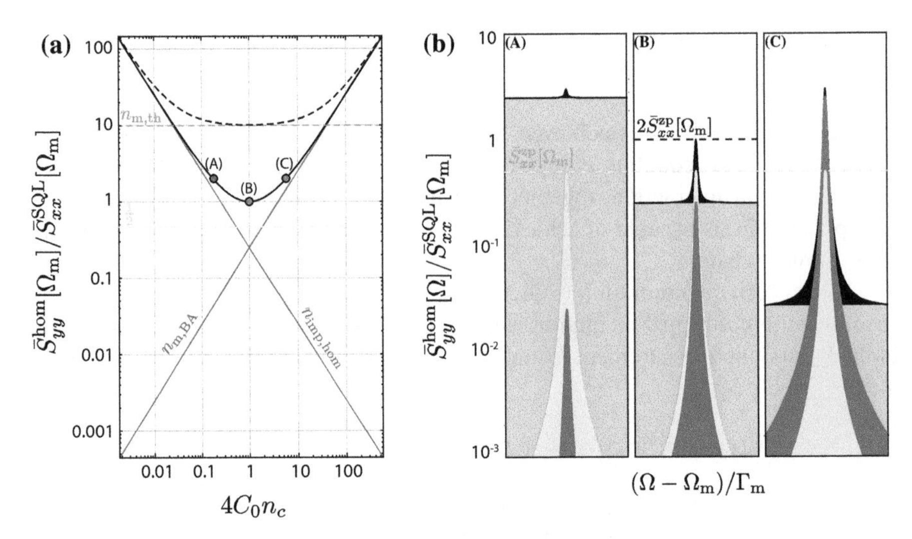

Fig. 4.3 Anatomy of the standard quantum limit of position measurement. a Plot shows the phonon-equivalent total signal and its noise budget. Imprecision noise (blue) decreases with measurement strength, while back-action noise increases proportionately so as to maintain the imprecision-back-action product (Eq. (4.3.12)). Yellow is the zero-point motion of the oscillator. Black shows the total phonon-equivalent noise at the output of the detector. Orange dashed shows intrinsic thermal motion when the oscillator is in a thermal state; black dashed shows the total noise in this case. **b** Spectra for three different points along the thick black curve in (**a**). Here, black shows the spectrum of the observable $\delta\hat{y}_{\rm hom}$ normalised to the SQL. Blue shows the contribution from imprecision, which goes down as the measurement strength is increased; red is the back-action due to the measurement, going up with measurement strength. Yellow is the zero-point motion

may be simply parametrised in terms of equivalent thermal quanta describing the three different contributions to the measurement record. Here, $n_{\rm m,th}+\frac{1}{2}$, is due to the intrinsic motion (thermal and zero-point) of the oscillator, while the measurement imprecision and back-action contributions are encoded as,

$$\begin{aligned} n_{\rm imp,hom} &:= \frac{\bar{S}_{xx}^{\rm imp,hom}[\Omega_{\rm m}]}{\bar{S}_{xx}^{\rm SQL}[\Omega_{\rm m}]} = \frac{1}{16\eta C_0 n_c}, \\ n_{\rm m,BA} &:= \frac{\bar{S}_{xx}^{\rm BA}[\Omega_{\rm m}]}{\bar{S}_{xx}^{\rm SQL}[\Omega_{\rm m}]} = C_0 n_c. \end{aligned} \tag{4.3.10}$$

The dimensionless parameter defined here, the *single-photon cooperativity*,

$$C_0 := \frac{4g_0^2}{\kappa\Gamma_{\rm m}}, \tag{4.3.11}$$

describes the merits of the linearised optomechanical system as a position sensor; η describes the total efficiency of the measurement chain, including input-output coupling due to the cavity and losses in the homodyne interferometer. The imprecision-back-action product in Eq. (4.3.6) takes the form,

$$n_{\mathrm{m,BA}} n_{\mathrm{imp,hom}} = \frac{1}{4\hbar^2} \bar{S}_{FF}^{\mathrm{BA}}[\Omega_{\mathrm{m}}] \bar{S}_{xx}^{\mathrm{imp,hom}}[\Omega_{\mathrm{m}}] \geq \frac{1}{16\eta_c}, \tag{4.3.12}$$

in terms of the phonon-equivalent quantities. Here, we employ an inequality to convey the notion that other sources of classical noise—either in the imprecision, or in the back-action—may preclude the lower limit prescribed by quantum mechanics.

Figure 4.3 depicts the anatomy of the standard quantum limit. Figure 4.3a shows the peak spectral density $\bar{S}_{yy}[\Omega_{\mathrm{m}}]$ normalised to its value at the SQL, as the measurement strength, $4C_0 n_c$, varies. As the measurement strength is increased, typically by increasing the photon number n_c, the imprecision contribution (blue) goes down due to the properties of phase shot-noise of the coherent-state meter field, while the back-action contribution (red) increases, as the oscillator experiences an increased photon recoil. Together with the zero-point motion (yellow), this gives the ideal characteristic of the total noise power in the detector output (black). The standard quantum limit is the point marked (B), where the detector output exhibits a minimum, achieved at the intracavity photon number,

$$n_c^{\mathrm{SQL}} := \frac{1}{4C_0}.$$

As shown in Fig. 4.3b sub-panel (B): at the SQL, the total output (black) is twice the zero-point motion (yellow). In fact, from Eqs. (4.3.9) and (4.3.10), it follows that at the SQL, the total phonon-equivalent noise at the detector output,

$$n_{\mathrm{m}}^{\mathrm{SQL}} := \left[n_{\mathrm{imp,hom}} + n_{\mathrm{BA}} + \frac{1}{2} \right]_{n_c = n_c^{\mathrm{SQL}}} = \frac{1}{4} + \frac{1}{4} + \frac{1}{2} = 1,$$

is such that imprecision and back-action contribute equally and each exactly half the zero-point motion [45]. In the non-ideal case, where the oscillator has a finite thermal occupation (i.e. $n_{\mathrm{m,th}} > 0$), the excess thermal motion (orange dashed in Fig. 4.3a) adds a constant excess to the detector output spectrum $\bar{S}_{yy}^{\mathrm{hom}}$ (shown as black dashed line) precluding the possibility of realising an ideal measurement with the above noise budget.

In the presence of thermal motion of the oscillator, i.e. $n_{\mathrm{m,th}} > 0$, it is convenient to amend the ideal imprecision-back-action equality in Eq. (4.3.6), to an inequality of the form

$$\begin{aligned} \bar{S}_{FF}^{\mathrm{tot}}[\Omega] \bar{S}_{xx}^{\mathrm{imp,hom}}[\Omega] &\geq \frac{\hbar^2}{4\eta_c} \geq \frac{\hbar^2}{4}, \\ \text{equivalently,} \quad \left(n_{\mathrm{m,th}} + n_{\mathrm{m,BA}}\right) n_{\mathrm{imp,hom}} &\geq \frac{1}{16\eta_c} \geq \frac{1}{16}, \end{aligned} \tag{4.3.13}$$

that takes into account the total force acting on the oscillator, $\bar{S}_{FF}^{\mathrm{tot}}[\Omega] := \bar{S}_{FF}^{\mathrm{th}}[\Omega] + \bar{S}_{FF}^{\mathrm{BA}}[\Omega]$, including the non-negligible thermal force from the environment. Despite the presence of thermal motion that precludes the possibility of achieving the absolute minimum, $\bar{S}_{xx}^{\mathrm{SQL}}$, the total-force-imprecision product characterises the ideality of the detector subject to the constraints of quantum mechanics.

References

1. J.A. Stratton, *Electromagnetic Theory* (McGraw Hill, 1941)
2. H.A. Bethe, J. Schwinger, Perturbation theory for cavities. Technical Report No. D1-117 (MIT Radiation Laboratory, 1943)
3. G.D. Wassermann, Math. Proc. Camb. Phil. Soc. **44**, 251 (1948)
4. S.G. Johnson, M. Ibanescu, M.A. Skorobogatiy, O. Weisberg, J.D. Joannopoulos, Y. Fink, Phys. Rev. E **65**, 066611 (2002)
5. P. Penfield, H.A. Haus, *Electrodynamics of moving media*, (MIT Press, 1967)
6. S.M. Barnett, R. Loudon, Phil. Trans. Roy. Soc. A **368**, 927 (2010)
7. A. Feldman, Phys. Rev. B **11**, 5112 (1975)
8. P.T. Rakich, P. Davids, Z. Wang, Opt. Exp. **18**, 14439 (2010)
9. M. Aspelmeyer, T.J. Kippenberg, F. Marquardt, Rev. Mod. Phys. **86**, 1391 (2014)
10. L. Bel, J.-L. Boulanger, N. Deruelle, Phys. Rev. A **37**, 1563 (1988)
11. S. Mancini, P. Tombesi, Phys. Rev. A **49**, 4055 (1994)
12. S. Aldana, C. Bruder, A. Nunnenkamp, Phys. Rev. A **88**, 043826 (2013)
13. A. Dorsel, J. McCullen, P. Meystre, E. Vignes, H. Walther, Phys. Rev. Lett. **51**, 1550 (1983)
14. A. Gozzini, I. Longo, S. Barbarino, F. Maccarrone, F. Mango, J. Opt. Soc. Am. B **2**, 1841 (1985)
15. B.S. Sheard, M.B. Gray, C.M. Mow-Lowry, D.E. McClelland, S.E. Whitcomb, Phys. Rev. A **69**, 051801 (2004)
16. V. Braginsky, A. Manukin, Sov. Phys. JETP **25**, 653 (1967)
17. T.J. Kippenberg, K.J. Vahala, Science **321**, 1172 (2008)
18. V.B. Braginsky, Y.I. Vorontsov, Sov. Phys. Usp. **17**, 644 (1975)
19. C.M. Caves, Phys. Rev. Lett. **45**, 75 (1980)
20. A. Pace, M. Collett, D. Walls, Phys. Rev. A **47**, 3173 (1993)
21. R.V.L. Hartley, Bell. Sys. Tech. J. **15**, 424 (1936)
22. T. Botter, D.W.C. Brooks, N. Brahms, S. Schreppler, D.M. Stamper-Kurn, Phys. Rev. A **85**, 013812 (2012)
23. T. Corbitt, Y. Chen, E. Innerhofer, H. Müller-Ebhardt, D. Ottaway, H. Rehbein, D. Sigg, S. Whitcomb, C. Wipf, N. Mavalvala, Phys. Rev. Lett. **98**, 11 (2007)
24. K.K. Ni, R. Norte, D.J. Wilson, J.D. Hood, D.E. Chang, O. Painter, H.J. Kimble, Phys. Rev. Lett. **108**, 214302 (2012)
25. H. Rokhsari, T.J. Kippenberg, T. Carmon, K.J. Vahala, Opt. Exp. **13**, 5293 (2005)
26. K. Vahala, M. Herrmann, S. Knunz, V. Batteiger, G. Saathoff, T.W. Haensch, T. Udem, Nature Phys. **5**, 682 (2009)
27. M. Evans et al., Phys. Rev. Lett. **114**, 161102 (2015)
28. S. Gigan, H.R. Böhm, M. Paternostro, F. Blaser, G. Langer, J.B. Hertzberg, K.C. Schwab, D. Bäuerle, M. Aspelmeyer, A. Zeilinger, Nature **444**, 67 (2006)
29. O. Arcizet, P.-F. Cohadon, T. Briant, M. Pinard, A. Heidmann, Nature **444**, 71 (2006)
30. A. Schliesser, P. Del'Haye, N. Nooshi, K. Vahala, T. Kippenberg, Phys. Rev. Lett. **97**, 243905 (2006)
31. T. Corbitt, D. Ottaway, E. Innerhofer, J. Pelc, N. Mavalvala, Phys. Rev. A **74**, 021802(R) (2006)
32. J.D. Teufel, J.W. Harlow, C. Regal, K.W. Lehnert, Phys. Rev. Lett. **101**, 197203 (2008)
33. K. Vahala, Phys. Rev. A **78**, 023832 (2008)
34. F. Marquardt, J. Chen, A.A. Clerk, S.M. Girvin, Phys. Rev. Lett. **99**, 093902 (2007)
35. I. Wilson-Rae, N. Nooshi, W. Zwerger, T.J. Kippenberg, Phys. Rev. Lett. **99**, 093901 (2007)

36. T. Rocheleau, T. Ndukum, C. Macklin, J.B. Hertzberg, A.A. Clerk, K.C. Schwab, Nature **463**, 72 (2010)
37. J.D. Teufel, T. Donner, D. Li, J.W. Harlow, M.S. Allman, K. Cicak, A.J. Sirois, J.D. Whittaker, K.W. Lehnert, R.W. Simmonds, Nature **475**, 359 (2011)
38. J. Chan, T.P.M. Alegre, A.H. Safavi-Naeini, J.T. Hill, A. Krause, S. Gröblacher, M. Aspelmeyer, O. Painter, Nature **478**, 89 (2011)
39. E. Verhagen, S. Deleglise, S. Weis, A. Schliesser, T.J. Kippenberg, Nature **482**, 63 (2012)
40. R.W. Peterson, T.P. Purdy, N.S. Kampel, R.W. Andrews, K.W. Yu, P.-L. Lehnert, C.A. Regal, Phys. Rev. Lett. **116**, 063601 (2016)
41. S. Weis, R. Rivière, S. Deléglise, E. Gavartin, O. Arcizet, A. Schliesser, T.J. Kippenberg, Science **330**, 1520 (2010)
42. A.H. Safavi-Naeini, T.P. Mayer Alegre, J. Chan, M. Eichenfield, M. Winger, Q. Lin, J.T. Hill, D.E. Chang, O. Painter, Nature **472**, 69 (2011)
43. X. Zhou, F. Hocke, A. Schliesser, A. Marx, H. Heubl, R. Gross, T.J. Kippenberg, Nature Phys. **9**, 179 (2013)
44. A. Nunnenkamp, V. Sudhir, A. Feofanov, A. Roulet, T. Kippenberg, Phys. Rev. Lett. **113**, 023604 (2014)
45. A.A. Clerk, M.H. Devoret, S.M. Girvin, F. Marquardt, R.J. Schoelkopf, Rev. Mod. Phys. **82**, 1155 (2010)

Chapter 5
Experimental Platform: Cryogenic Near-Field Cavity Optomechanics

Quantum phenomena do not occur in a Hilbert space, they occur in a laboratory.

Asher Peres

Radiation-pressure coupling between a mechanical oscillator and an electromagnetic cavity allows, in principle, for an exquisitely sensitive measurement of the oscillator's motion. If the oscillator is assumed to be in thermal equilibrium, such that its motion is driven by a thermal force with spectral density $\bar{S}_{FF}^{\text{th}}$, then quantum mechanics imposes a fundamental tradeoff between the measurement precision, expressed as a position-equivalent spectral density $\bar{S}_{xx}^{\text{imp}}$, and measurement back-action, expressed as a force noise $\bar{S}_{FF}^{\text{BA}}$. This constraint is codified by the statement of the uncertainty principle (Eq. (4.3.13)),

$$\left(\bar{S}_{FF}^{\text{th}}[\Omega] + \bar{S}_{FF}^{\text{BA}}[\Omega]\right) \bar{S}_{xx}^{\text{imp}}[\Omega] \geq \frac{\hbar^2}{4\eta_c} \geq \frac{\hbar^2}{4},$$

$$\text{equivalently,} \quad \left(n_{\text{m,th}} + n_{\text{m,BA}}\right) n_{\text{imp}} \geq \frac{1}{16\eta_c} \geq \frac{1}{16}.$$

Here, the second line expresses the first inequality in terms of phonon-equivalent thermal force, back-action, and imprecision (introduced in Eq. (4.3.10)). A quantum-ideal measurement is realised when,

(a) $n_{\text{m,BA}} n_{\text{imp}} \to \frac{1}{16}$, i.e. a quantum-noise-limited electromagnetic field is used as the meter
(b) $n_{\text{m,th}} n_{\text{imp}} \to 0$, i.e. sufficiently low measurement imprecision is achieved.

The last requirement, somewhat counter-intuitively,[1] is equivalent to the oscillator motion being back-action dominated, i.e.

[1]In this case, intuition is garnered from a different perspective on the measurement process: ideally, the meter and the system gets tightly entangled in the course of the measurement interaction;

V. Sudhir, *Quantum Limits on Measurement and Control of a Mechanical Oscillator*, Springer Theses, https://doi.org/10.1007/978-3-319-69431-3_5

$$\frac{n_{\mathrm{m,BA}}}{n_{\mathrm{m,th}}} = \frac{C_0 n_c}{n_{\mathrm{m,th}}} \gtrsim 1.$$

Thus, performing a quantum-ideal measurement relies on having an optomechanical system with a large single-photon cooperativity C_0, a low thermal phonon occupation $n_{\mathrm{m,th}}$ and the ability to sustain sufficiently large intracavity photon numbers to dominate thermal motion with back-action while remaining a linear transducer. The *quantum cooperativity* [1],

$$\frac{C_0}{n_{\mathrm{m,th}}} = \frac{4g_0^2}{\kappa(n_{\mathrm{m,th}}\Gamma_{\mathrm{m}})} \approx \frac{(\hbar G)^2}{k_B T} \cdot \frac{1}{\kappa} \cdot \frac{1}{m\Gamma_{\mathrm{m}}},$$

characterises the requirement for fixed intracavity photon number. Clearly, the experimental system must consist of a low-mass low-loss mechanical oscillator integrated with a high-Q optical cavity imparting a large per-photon radiation pressure force operating at sufficiently low temperature. The rest of this chapter details the optomechanical system developed in our group, which in tandem with cryogenic operation, allow us to meet these requirements.

5.1 Stressed Nanostring Coupled to an Optical Microcavity

The radiation pressure of light, exerting a very feeble force, requires a specially engineered mechanical object so as to induce appreciable motion. The field of contemporary cavity optomechanics [1] achieves this goal by employing low-mass mechanical objects coupled to intense optical fields in a small mode-volume cavity.

Practical requirements—the need for a highly stable and miniature frequency reference for integrated circuits—led researchers towards high-Q nanomechanical oscillators in the 1960s [2]. By the late 1990s, these initial ideas had given birth to a host of electronic devices and sensors based on nanomechanical oscillators [3, 4]. It was hoped that by simultaneously shrinking the size of the oscillator (low m), while maintaining the mechanical quality (low Γ_{m}), quantum back-action on nanomechanical objects could be witnessed and exploited [5]. Phenomenologically however, the mechanical quality factor,

$$Q_{\mathrm{m}} = 2\pi \frac{\text{energy stored in elastic motion}}{\text{energy lost per cycle}} = \frac{\Omega_{\mathrm{m}}}{\Gamma_{\mathrm{m}}}, \tag{5.1.1}$$

was found to obey the rough scaling with volume [6–8], $Q_{\mathrm{m}} \propto V_{\mathrm{m}}^{1/3}$, suggesting that conventional nanomechanical oscillators may not simultaneously achieve low mass and high Q_{m}. A welcome break from this trend was observed in nanostring oscillators

monogamy of entanglement then requires that the system has to be exclusively entangled with the meter.

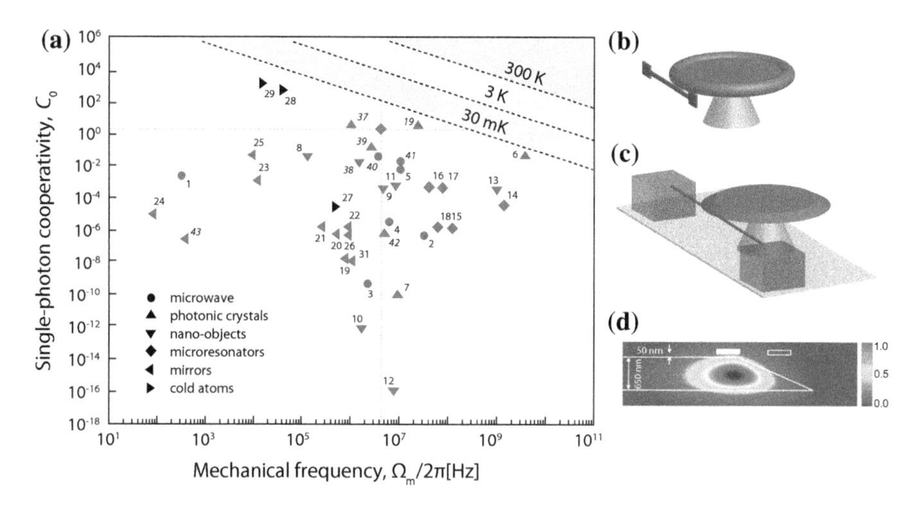

Fig. 5.1 Unity single-photon cooperativity with near-field gradient coupling. a Survey of the contemporary landscape of cavity optomechanics in terms of the typically achieved single-photon cooperativity C_0 (Figure adapted with permission from Ref. [13]. Copyrighted by the American Physical Society) **b** In the original near-field gradient force optomechanics architecture developed in our group [14, 15], a high-stress Si_3N_4 nanobeam (red) is manually placed in the evanescent field of an SiO_2 optical microcavity. **c** In the approach followed in this thesis, building upon an earlier generation of work [16], the nanobeam and micro-cavity are heterogeneously integrated on a chip for improved technical stability, making it convenient for cryogenic deployment. **d** An example finite-element simulation showing the difference between the current generation device and the previous generation. In the previous generation, the nanobeam (white outline) was placed far outside the maximum of the evanescent field gradient, whereas in the current generation (white solid), the beam is at the optimal position

with intrinsic tensile stress [9, 10], exhibiting $Q_m \approx 10^6$ at an oscillation frequency of $\Omega_m \approx 2\pi \cdot 1$ MHz at room temperature. The conventional understanding is that the pre-stress of the elastic medium increases the stored elastic energy without affecting intrinsic loss mechanisms (see Sect. 5.1.2) [11, 12]. Thus, stressed nanobeams were discovered to have unique mechanical properties that make them appropriate subjects of quantum-limited measurements.

However, due to their small transverse dimensions ($\sim$10–100 nm), the nanobeam geometry does not have a large cross-section for electromagnetic scattering, making them indifferent to the conventional scattering type radiation pressure force. Therefore, nanobeams have been coupled to optical fields via gradient forces due to the evanescent field of optical cavities [14, 15, 17–20].

The approach followed in our group [14–16] relies on coupling a high-Q nanostring to the evanescent field of a high-finesse whispering gallery optical cavity. The low mass ($m \lesssim 5$ pg) of the oscillator gives a large zero-point motion ($x_{zp} \gtrsim 100$ pm). In order to realise a large frequency pull parameter G, the first generation of devices [14, 15], shown in Fig. 5.1a, were realised by manually positioning the nanobeam in the evanescent field of a whispering-gallery micro-toroid cavity, realising $G \approx 2\pi \cdot 1\,\text{GHz/nm}$. Operating at room temperature, this system could realise a measurement imprecision at the standard quantum limit [14]. In a second generation [16], shown in Fig. 5.1, the nanobeam and microcavity were integrated on-chip, to improve operational stability. However, in the second generation device, the placement of the nanobeam relative to the cavity was not optimal—essentially having to do with the fabrication procedure employed. The devices used in this thesis, while nominally similar to Fig. 5.1b, are based on a vastly improved fabrication procedure [13] capable of placing the nanobeam at the optimal position in the evanescent optical field so as to maximise the gradient force. Future theses from the group will discuss the fabrication process in depth, see [13] for details.

Ultimately, the optomechanical system employed here features single photon cooperativites,

$$C_0 \approx 1 \cdot \left(\frac{g_0/2\pi}{25\,\text{kHz}}\right)^2 \left(\frac{0.5\,\text{GHz}}{\kappa/2\pi}\right) \left(\frac{5\,\text{Hz}}{\Gamma_\text{m}/2\pi}\right), \tag{5.1.2}$$

of order unity; Fig. 5.1a shows a survey of the contemporary landscape of cavity optomechanics in terms of C_0. Despite the fact that the quantum cooperativity, $C_0/n_{\text{m,th}} \ll 1$ at any technologically feasible cryogenic temperature, by being able to probe the cavity with a sufficiently large intracavity photon number ($n_c \lesssim 10^5$), and operating at 4 K, we are able to achieve $(C_0/n_{\text{m,th}})n_c \approx 1$.

The remainder of this section details the prevailing understanding of the various parameters that go into the single photon cooperativity, and technical details regarding the cryogenic experiment. Extraneous (classical) sources of imprecision and back-action will be considered in Chap. 6.

5.1.1 Near-Field Coupling

The expression for the cavity frequency pull parameter in terms of the perturbation caused by the presence of the dielectric nanobeam is given by Eq. (4.2.3),

$$G = \frac{\omega_c}{2} \frac{\int |\mathbf{E}(\mathbf{r})|^2 \left(\tilde{\mathbf{u}}(\mathbf{r}) \cdot \nabla\epsilon\right) \mathrm{d}^3 r}{\int \epsilon\, |\mathbf{E}(\mathbf{r})|^2 \,\mathrm{d}^3 r}; \tag{5.1.3}$$

here the integrals are taken over the entire volume of the cavity optical field including the evanescent field. The integral in the numerator gives the cavity frequency shift due to the presence of the nanobeam within the evanescent field of the cavity mode. Application of straightforward vector identities (and neglecting surface terms) reveal that this perturbation may be expressed as the sum of two contributions, i.e.,

$$\int |\mathbf{E}(\mathbf{r})|^2 \left(\tilde{\mathbf{u}}(\mathbf{r}) \cdot \nabla\epsilon\right) \mathrm{d}^3 r = \int \epsilon\tilde{\mathbf{u}} \cdot \left(\nabla |\mathbf{E}|^2\right) \mathrm{d}^3 r + \int \epsilon\, |\mathbf{E}|^2 \left(\nabla \cdot \tilde{\mathbf{u}}\right) \mathrm{d}^3 r;$$

the first couples the beam displacement to a generalised force due to the gradient in the electric field intensity (i.e. a gradient force), while the second arises from volume deformations of the beam due to the incident field intensity (i.e. a scattering-type force). For nanobeams with dimensions comparable to the optical wavelength, the gradient force dominates.

In order to estimate the coupling, we now make a few simplifying assumptions. Firstly, although the frequency pulling parameter depends on the geometric overlap between the electric field and the beam displacement profile, it is acceptable to neglect this dependence, and later accommodate it in the definition of the single-photon coupling g_0, via an effective zero-point motion. Thus we may redefine (equivalent to the identification $\tilde{\mathbf{u}} \cdot \nabla\epsilon = \epsilon_{\text{beam}} - 1$),

$$G = \frac{\omega_c}{2}\frac{\partial}{\partial z}\left(\frac{\int_{\text{beam}}(\epsilon_{\text{beam}}(\mathbf{r}) - 1)\,|\mathbf{E}(\mathbf{r})|^2\,\mathrm{d}^3 r}{\int_{\text{disk}} \epsilon\,|\mathbf{E}(\mathbf{r})|^2\,\mathrm{d}^3 r}\right)$$
$$\approx \frac{\omega_c}{2}\frac{\partial}{\partial z}\left(\frac{\nu_{\text{SiN}}^2 - 1}{\nu_{\text{SiO}_2}^2}\left|\frac{\mathbf{E}_{\text{max}}^{\text{beam}}}{\mathbf{E}_{\text{max}}^{\text{disk}}}\right|^2\frac{V_{\text{beam,opt}}}{V_{\text{disk,opt}}},\right)$$

where the approximation parametrises the coupling rate in terms of an effective optical volume of the disk (beam), $V_{\text{disk(beam),opt}}$, given by the optical energy in the disk (beam) normalised to the maximum electric field intensity in the disk (beam), i.e.,

$$V_{\text{disk(beam),opt}}\left|\mathbf{E}_{\text{max}}^{\text{disk}}\right|^2 = \int_{\text{disk(beam)}} |\mathbf{E}|^2\,\mathrm{d}^3 r.$$

Secondly, for a beam whose transverse dimensions are much less than the vertical evanescent decay length [21, 22],

$$\ell_{\text{ev}} \approx \frac{\lambda_c/2\pi}{\sqrt{\nu_{\text{SiO}_2}^2 - 1}} \approx \frac{\lambda_c}{12} \sim 100\,\text{nm},$$

of the cavity mode, its effective optical volume may be simplified into the form, $V_{\text{beam,opt}} = \mathscr{A}_{\text{beam}}\ell_{\text{eff}}$, where $\mathscr{A}_{\text{beam}}$ is the geometric cross-sectional area and ℓ_{eff} is an effective sampling length. Similarly for the disk, its optical volume may be approximated by, $V_{\text{disk,opt}} = 2\pi r_{\text{disk}}\mathscr{A}_{\text{disk}}$, where r_{disk} is the geometric radius, and $\mathscr{A}_{\text{disk}}$ is the effective transverse area of the whispering-gallery mode. Finally, assuming that the vertical evanescence leads to an exponential decay in the maximum field intensity between the disk and the beam, i.e. $\left|\mathbf{E}_{\text{max}}^{\text{beam}}\right|^2 = \left|\mathbf{E}_{\text{max}}^{\text{disk}}\right|^2 \cdot \alpha\exp(-z/\ell_{\text{ev}})$, the near-field coupling strength may be approximated by [13],

$$G \approx \frac{\omega_c}{2\ell_{\text{ev}}}\frac{\nu_{\text{SiN}}^2 - 1}{\nu_{\text{SiO}_2}^2}\frac{\mathscr{A}_{\text{beam}}\ell_{\text{eff}}}{2\pi r_{\text{disk}}\mathscr{A}_{\text{disk}}}\alpha e^{-z/\ell_{\text{ev}}}. \tag{5.1.4}$$

In practice, the evanescent length ℓ_{ev}, the mode cross-section $\mathscr{A}_{\mathrm{disk}}$, and the geometric pre-factor α need to be determined from a numerical solution of Maxwell's equations [13]. A simple estimate may however be made by replacing these parameters using known approximations for the case of a toroidal cavity [14]: the evanescent length $\ell_{\mathrm{ev}} \approx \lambda_c/12$, while the mode cross-section $\mathscr{A}_{\mathrm{disk}} \approx 0.15 \cdot r_{\mathrm{disk}}^{7/12} t_{\mathrm{disk}}^{1/4} \lambda_c^{7/6}$ and the geometric pre-factor, $\alpha \approx 1.1(\lambda_c/r_{\mathrm{disk}})^{1/3}$. Using the known refractive index, $\nu_{\mathrm{SiN}} = 2$, and the typical geometric parameters of the disk and beam, Eq. (5.1.4) suggests a coupling strength of $G \approx 2\pi \cdot 1\,\mathrm{GHz/nm}$—within 10% of the prediction from a full finite-element simulation.

In order to arrive at the vacuum optomechanical coupling rate, g_0, the zero-point motion of the effective point mass equivalent of the beam must be known. In principle, this again requires detailed knowledge of the overlaps between the extended elastic mode profile, $\tilde{\mathbf{u}}$, of the beam, and the optical cavity field, $\mathbf{E}$, of the disk [23]. The simple approximation, valid for the fundamental mode with an anti-node at the optical mode, involves assuming, $x_{\mathrm{zp}}^2 \approx \hbar/(2\rho\mathscr{A}_{\mathrm{disk}}\ell_{\mathrm{eff}}\Omega_{\mathrm{m}}) \approx (33\,\mathrm{fm})^2$, giving, $g_0 = G x_{\mathrm{zp}} \approx 2\pi \cdot 33\,\mathrm{kHz}$—within 15% of the measured value.

5.1.2 Mechanical Properties of Stressed Radio-Frequency Beams

In the context of linear elastodynamics, in Sect. 3.1, it was mentioned that elastic media with large aspect ratios in one or two dimensions, like a beam or membrane, acquires additional contributions to its equations of motion due to bending of the medium. Following the arguments outlined in Sect. 3.1, the motion of the transverse displacement, $u_i(z,t)$, along either of the two transverse directions $i = x, y$, of a one-dimensional prismatic (i.e. with a uniform transverse profile) beam, described by a coordinate z along its length, is determined by the Euler-Bernoulli equation,[2]

$$\rho\mathscr{A}\frac{\partial^2 u_i(z,t)}{\partial t^2} - T\frac{\partial^2 u_i(z,t)}{\partial z^2} + K M_i \frac{\partial^4 u_i(z,t)}{\partial z^4} = 0, \qquad (i = x, y). \tag{5.1.5}$$

Here, the first term is due to the inertia of the beam, and is characterized by the total moving mass; ρ is the mass density ($\rho_{\mathrm{SiN}} \approx 3100\,\mathrm{kg/m^3}$), and $\mathscr{A} = \ell_x \ell_y$ is the transverse area. The second term is due to stress along the beam; $T > 0$ ($T < 0$) is the tensile (compressive) force. The third term, characteristic of long aspect ratio beams, is due to bending energy; K is the bulk modulus ($K_{\mathrm{SiN}} \approx 100 - 300\,\mathrm{GPa}$),

[2]We here ignore the two-dimensional extension of the beam, justified by the fact that in the cases of interest, the characteristic extension along this dimension is much smaller than the length of the beam. In doing so, we treat the two possible transverse displacements, along the coordinate $x-$ and $y-$directions, using independent Euler-Bernoulli equations; the small coupling between these two mode families, arising from a finite Poisson's ratio ($\varsigma_{\mathrm{SiN}} \approx 0.24$), is therefore ignored. Within this approximation, effects arising from rotation and shear in the transverse direction are also ignored [24, 25].

and M_i is the moment of inertia about the axis perpendicular to the displacement u_i (for example, $M_x = \ell_x \ell_y^3/12$, while $M_y = \ell_y \ell_x^3/12$). The motion of the beam is fixed by the boundary conditions,

$$u_i(0,t) = u_i(\ell_z,t) = 0, \quad \text{and,} \quad \partial_z u_i(0,t) = \partial_z u_i(\ell_z,t) = 0, \tag{5.1.6}$$

describing the doubly-clamped configuration adopted in the devices of interest.

For each of the two transverse mode families $i = x, y$, Eqs. (5.1.5) and (5.1.6) have a series of oscillating solutions, characterized by the mode index $n > 0$, described by the ansatz,

$$u_{i,n}(z,t) = \tilde{u}_{i,n}(z) e^{i\Omega_{i,n} t}. \tag{5.1.7}$$

We also introduce dimensionless variables, viz.

$$\zeta := z/\ell_z, \quad \tilde{v}_{i,n}(\zeta) := \tilde{u}_{i,n}(z)/\ell_i. \tag{5.1.8}$$

Inserting Eqs. (5.1.7) and (5.1.8) in Eq. (5.1.5) gives,

$$\epsilon_i \frac{\partial^4 \tilde{v}_{i,n}}{\partial \zeta^4} - \frac{\partial^2 \tilde{v}_{i,n}}{\partial \zeta^2} = \left(\frac{\Omega_{i,n}}{\Omega_0}\right)^2 \tilde{v}_{i,n}. \tag{5.1.9}$$

Here we have introduced a dimensionless parameter,

$$\epsilon_i := \frac{K M_i}{T \ell_z^2} = \frac{1}{12}\left(\frac{K}{T/\mathscr{A}}\right)\left(\frac{\ell_{i'}}{\ell_z}\right)^2,$$

which quantifies the ratio of bending energy to tension (here, the notation i' means the complement of i, so $x' = y$, $y' = x$ etc.); and a characteristic frequency,

$$\Omega_0 := \left(\frac{T/\ell_z}{\rho \mathscr{A} \ell_z}\right)^{1/2},$$

which characterizes the relative contribution of tension to inertia. Indeed, for the typical nanobeams used in this thesis [13] ($\ell_x \approx 1\,\mu\text{m}$, $\ell_y \approx 100\,\text{nm}$, $\ell_z \approx 100\,\mu\text{m}$), the large tension ($T/\mathscr{A} \approx 1\,\text{GPa}$) and slender geometry ($\ell_{i'}/\ell_z < 10^{-2}$), implies that tensile energy dominates bending throughout the bulk of the nanobeam; in fact, $\epsilon_i < 10^{-5}$. Similarly, a large tension, and a small mass ($\rho \mathscr{A} \ell_z$) imply characteristic radio-frequency vibrations, i.e. $\Omega_0 \approx 2\pi \cdot 1\,\text{MHz}$.

The frequencies of the two mode families ($i = x, y$) can in principle be determined from an exact solution of Eq. (5.1.9) with clamped boundary conditions Eq. (5.1.6). However, since $\epsilon \ll 1$, an approximate solution for small ϵ suffices, giving (see solution in Appendix B, and Eq. (B.47)),

$$\Omega_{i,n} \approx n\pi\Omega_0 \sqrt{1 + (n\pi)^2 \epsilon_i}. \tag{5.1.10}$$

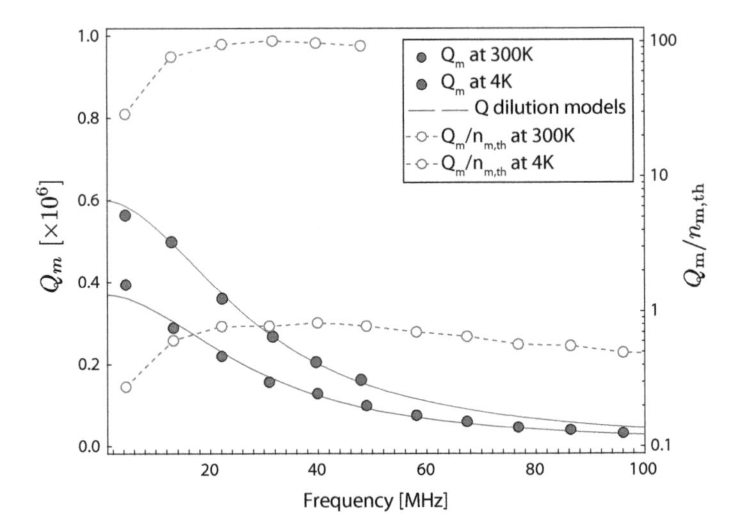

Fig. 5.2 Mechanical quality factor. The blue (red) solid circles show measured mechanical quality factor, Q_m, of the odd ordered out-of-plane modes of the nanobeam, as function of the mode frequency, at 4 K (room temperature, 300 K). By design, even-order modes do not have an appreciable optomechanical coupling; while in the cryostat, measurements of higher order modes, beyond $\Omega_\mathrm{m} \approx 50\,\mathrm{MHz}$, is precluded by a lack of thermal noise signal. The solid blue/red lines show fit to the measured quality factor using the stress-dilution model. Open blue (red) circles show the number of oscillations executed in the decoherence time of the oscillator, i.e. $\Omega_\mathrm{m}/n_\mathrm{m,th}\Gamma_\mathrm{m} = Q_\mathrm{m}/n_\mathrm{m,th}$. When $Q_\mathrm{m}/n_\mathrm{m,th} > 1$, the oscillator is "quantum enabled"

The experimentally observed mode frequencies are in good (within 10%) agreement with this prediction. Indeed, the frequency offset between the fundamental in-plane ($i = x$) and out-of-plane ($i = y$) modes is well described by the aspect ratio of the beam, as predicted. Henceforth, focusing on the out-of-plane mode family, we drop the index i.

Mechanical losses, owing to a variety of factors, determine the linewidth, Γ_n, of each of the modes n, and thence the respective quality factor, $Q_n = \Omega_n/\Gamma_n$. For amorphous materials, such as SiN, supporting radio-frequency modes, the intrinsic quality factor is presumed to be largely determined by losses due to the scattering of phonons off two-level defects in the material bulk [26, 27], or scattering off of geometric defects at the surface [28]. However, the universal characteristics of these mechanisms predict a room-temperature quality factor that is a few orders of magnitude smaller than what is observed in both beam [10, 14] and membrane [29, 30] geometries.

Figure 5.2 shows the measured quality factor, Q_n, as a function of the mode frequencies, $\Omega_n \approx 2\pi n\Omega_0$. Briefly, thermal noise spectrum of the various nanobeam modes is measured (see Sect. 5.2 ahead for details), from which the intrinsic linewidth Γ_n is extracted from Lorentzian fits. For the fundamental mode, with the smallest linewidth, ringdown measurements confirm the values measured from thermal noise spectra. In both cases, care is taken to ensure that changes in Γ_n due to dynamical back-action is negligible.

The increased quality factor, Q_n, of the nth mode is understood as arising from "stress-dilution" [11, 12, 31, 32]: the intrinsic quality factor, Q_n^0, is diluted by the tensile stress applied on the material, to give,

$$Q_n = Q_n^0 \left(1 + \frac{1}{2\sqrt{\epsilon} + (n\pi\sqrt{\epsilon})^2}\right). \tag{5.1.11}$$

Qualitatively, for mechanical oscillators where most of the elastic energy is stored in the tensile stress ($\epsilon \ll 1$), the kinetic energy lost each cycle can be effectively "diluted". Quantitatively, each of the two distinct factors in the denominator, scaling as $\epsilon^{1/2}$ and ϵ, can be shown to arise from the influence of the $\sqrt{\epsilon}$ – scale deviation of the elastic mode profile from a sinusoid at the boundary (see Eq. (B.4.8)), and the bending at the anti-nodes of the sinusoid in the bulk of the elastic medium, respectively [32]. The data shown in Fig. 5.2 agrees well with the stress-dilution model in Eq. (5.1.11); the the value of ϵ used for the model in the plot is inferred from the observed frequency dispersion of the modes in conjunction with its theoretical prediction given in Eq. (5.1.10).

5.2 Measurement and Calibration of Thermomechanical Motion

Since the motion of the mechanical oscillator imparts phase fluctuations commensurate with the motion in the intracavity field, a phase discriminator (i.e. a phase-to-amplitude conversion) is required to infer the mechanical motion. In one approach, the cavity may be used as a phase discriminator, for which detuned operation (i.e. $\Delta \neq 0$) is necessary—so-called *side-of-line* detection. In another approach, the cavity may be embedded in one arm of an interferometer, in which case resonant probing (i.e. $\Delta = 0$) is possible.

The technical ease of side-of-line detection is considerably offset by two deficiencies: (a) the modification of mechanical susceptibility due to dynamic back-action at any finite detuning $\Delta \neq 0$ with high probe power, and, (b) the difficulty of achieving shot-noise-limited detection when the probe is considerably reduced to avoid dynamic back-action. Indeed, from Eq. (3.2.23), the transmitted signal power required to be shot-noise-limited in direct photodetection for a detector with finite NEP, is

$$P_{\text{shot}} = \frac{\mathscr{R}}{2\eta q_e} \bar{S}_P^{\text{NE}}[\Omega].$$

For typical trans-impedance-amplified silicon ($\eta \approx 0.85$) photodetectors (for example, NewFocus 1801 characterised in Fig. 3.2), $P_{\text{shot,PD}} \approx 100\,\mu\text{W}$. Avalanche photodetectors (APD) can have much lower $P_{\text{shot,APD}} \approx 0.5 - 1\,\mu\text{W}$, however, their high-sensitivity avalanche stage tends to saturate around $2P_{\text{shot,APD}}$.

For these reasons, it has proven useful to employ phase discrimination using an external interferometer, i.e. homodyning or heterodyning the cavity transmission with a strong LO.[3] In this case, the LO sets the shot noise background for detection, so that large powers $P_{\text{LO}} \approx 1 - 5\,\text{mW}$ may be employed, limited only by the damage threshold of the silicon photodiodes of the balanced detector (for unbalanced detectors, saturation of the amplifying electronics sets a lower limit on the usable LO power).

Figure 5.3a shows the standard experimental setup to detect thermal motion of the nanobeam coupled to the optical microdisk cavity. When the cavity is probed on resonance, and the homodyne interferometer is locked to the phase quadrature, the voltage spectrum of its trans-impedance-amplified (gain H_{VI}) photocurrent takes the form (see Eq. 3.2.32),

$$\bar{S}_V^{\text{hom}}[\Omega] = |H_{VI}[\Omega]|^2 \left(\mathscr{R}^2 \bar{S}_P^{\text{NE}} + 2\eta q_e \mathscr{R} P_{\text{LO}} + 4\eta^2 \mathscr{R}^2 P_{\text{LO}} P_{\text{sig}} \bar{S}_\phi^{\text{cav}}[\Omega]\right),$$

where $\bar{S}_\phi^{\text{cav}}$ is the total phase noise contribution from the cavity. In the vicinity of mechanical modes, $\bar{S}_\phi^{\text{cav}}[\Omega] = (G/\Omega)^2 \bar{S}_x[\Omega]$, where G is the frequency pull parameter of the relevant mechanical mode. Figure 5.3d shows such a voltage noise spectrum measured using a spectrum analyser at the output of the homodyne interferometer.

In order to calibrate such a spectrum in physically relevant units (example, apparent mechanical motion), it appears to be necessary to have detailed knowledge of the transfer function of each element in the measurement chain. However this is not necessary. For example, if it is known that the mechanical oscillator is in equilibrium at a certain temperature T, then the voltage noise spectrum, expressed in the form (irrespective of the nature of the linear detection scheme),

$$\bar{S}_V[\Omega] = |H_{Vx}[\Omega]|^2 \left(\bar{S}_x^{\text{imp}}[\Omega] + \bar{S}_x[\Omega]\right) =: \bar{S}_V^{\text{imp}}[\Omega] + \underbrace{|H_{Vx}[\Omega]|^2\, \bar{S}_x[\Omega]}_{\bar{S}_V^{\text{mech}}}, \quad (5.2.1)$$

may be calibrated by using the equipartition identity (see Eq. (3.1.27)),

$$\text{Var}\left[\hat{x}\right] = (2n_{\text{m,th}} + 1)x_{\text{zp}}^2 \approx 2x_{\text{zp}}^2 \frac{k_B T}{\hbar\Omega_{\text{m}}}.$$

In practice, implementing this also requires the assumption that $|H_{Vx}[\Omega]|$ is constant in the vicinity of the thermal noise; for high-Q oscillators, this assumption is almost always true. Essentially, calibration involves fixing the number $|H_{Vx}[\Omega_{\text{m}}]|$. The equipartition identity gives,

---

[3] A common-path interferometer (for example using the frequency modulation spectroscopy technique described in Sect. 3.2.4), is technically easier to implement. However, it is not preferred in a cryogenic environment due to the large (mW range) LO powers that might scatter in the cold environment.

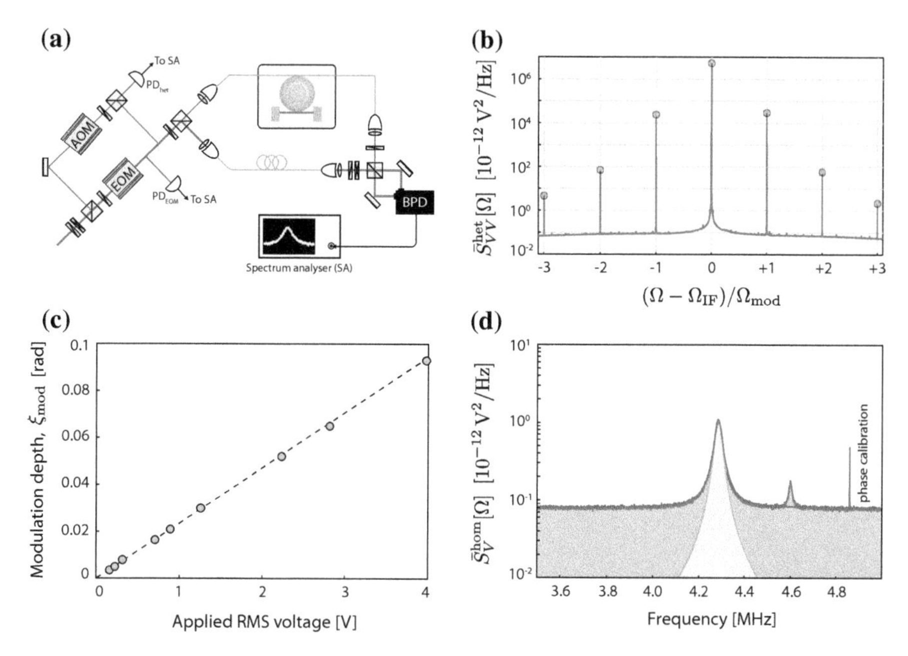

Fig. 5.3 **Frequency noise calibration. a** Schematic of the experimental setup used to perform direct calibration of the frequency noise imparted on the transmitted optical field due to thermomechanical motion. An EOM is used to impart a known phase modulation at a pre-determined frequency; an AOM is used to derive a heterodyne local oscillator. **b** Output of the heterodyne detector, visualising the phase modulation sidebands imparted by the EOM. Here the modulation frequency, $\Omega_{\text{mod}} \approx 2\pi \cdot 4.85\,\text{MHz}$, while the heterodyne intermediate frequency, $\Omega_{\text{IF}} = 2\pi \cdot 78\,\text{MHz}$. **c** The modulation depth ξ_{mod} is verified to be linear in the voltage applied to the EOM over a decade in both. **d** Voltage noise spectrum at the output of the homodyne detector, showing thermal motion of the fundamental out-of-plane mode at $\Omega_{\text{m}} \approx 2\pi \cdot 4.315\,\text{MHz}$, together with the phase modulation calibration tone. Here the oscillator is in contact with buffer gas so that to a good degree it may be assumed to be in thermal equilibrium; however, the mechanical Q is deteriorated by gas-damping (see Sect. 5.3)

$$|H_{Vx}[\Omega_{\text{m}}]| \approx \frac{\text{Var}\,[V_{\text{mech}}]}{2x_{\text{zp}}^2 n_{\text{m,th}}} = \frac{1}{2x_{\text{zp}}^2 n_{\text{m,th}}} \int \left(\bar{S}_V[\Omega] - \bar{S}_V^{\text{imp}}[\Omega] \right) \frac{\text{d}\Omega}{2\pi},$$

where the last equality follows from the general relation between the variance of a process and its spectral density given in Eq. (2.1.7). If the voltage noise spectrum is further only required to be calibrated in units relative to the zero-point spectral density $\bar{S}_x^{\text{zp}}$, knowledge of x_{zp} is rendered irrelevant.

Ultimately, the equipartition identity may be used to determine the frequency pull parameter G, or the vacuum optomechanical coupling rate g_0. However, in order to do this, a reference phase/frequency noise is required that is transduced identical to the mechanical motion through the measurement chain. For side-of-line detection and homodyne detection, it is known that laser frequency fluctuation imparted on the probe laser at the cavity input, as shown in Fig. 5.3a, is transduced identical to cavity frequency fluctuation [33]. Assuming that the detected voltage noise spectrum is related to cavity frequency fluctuation linearly, i.e.

$$\bar{S}_V[\Omega] = \bar{S}_V^{\text{imp}}[\Omega] + |H_{V\omega}[\Omega]|^2\, \bar{S}_\omega[\Omega],$$

the problem is to infer the function $|H_{V\omega}[\Omega]|$. Using an EOM to imprint a known phase modulation of depth ξ_{mod} (see Eq. (3.2.55)) at frequency Ω_{mod}, the voltage noise spectrum consists of two terms,

$$\bar{S}_V[\Omega] = \bar{S}_V^{\text{imp}}[\Omega] + |H_{V\omega}[\Omega]|^2 \left(\bar{S}_\omega^{\text{mod}}[\Omega_{\text{mod}}] + \bar{S}_\omega^{\text{cav}}[\Omega]\right), \tag{5.2.2}$$

where the frequency noise spectrum due to the injected modulation is given by,

$$\bar{S}_\omega^{\text{mod}}[\Omega_{\text{mod}}] = \frac{\xi_{\text{mod}}^2}{2} \Omega_{\text{mod}}^2\, \delta[\Omega - \Omega_{\text{mod}}].$$

Assuming that the cavity frequency noise arises from the thermal motion of a high-Q oscillator at frequency Ω_{m}, i.e. $\bar{S}_\omega^{\text{cav}}[\Omega] = G^2 \bar{S}_x[\Omega] = g_0^2 \bar{S}_{x/x_{\text{zp}}}[\Omega]$, Eq. (5.2.2) takes the form,

$$\bar{S}_V[\Omega] \approx \bar{S}_V^{\text{imp}}[\Omega] + \underbrace{|H_{V\omega}[\Omega_{\text{mod}}]|^2\, \frac{\xi_{\text{mod}}^2 \Omega_{\text{mod}}^2}{2}\, \delta[\Omega - \Omega_{\text{mod}}]}_{\bar{S}_V^{\text{mod}}} + \underbrace{|H_{V\omega}[\Omega_{\text{m}}]|^2\, g_0^2 \bar{S}_{x/x_{\text{zp}}}[\Omega]}_{\bar{S}_V^{\text{mech}}}.$$

Again invoking the equipartition identity, $\text{Var}\left[\hat{x}/x_{\text{zp}}\right] = 2n_{\text{m,th}} + 1$, and relating the integral of the voltage noise spectrum to its variance, gives,

$$\frac{\text{Var}\left[V_{\text{mech}}\right]}{\text{Var}\left[V_{\text{cal}}\right]} \approx \left|\frac{H_{V\omega}[\Omega_{\text{m}}]}{H_{V\omega}[\Omega_{\text{mod}}]}\right|^2 \frac{4 g_0^2 n_{\text{m,th}}}{\xi_{\text{mod}}^2 \Omega_{\text{mod}}^2}.$$

If the modulation frequency is sufficiently close to the mechanical frequency such that the first factor on the right-hand side can be justified to be unity, then this equation may be used to calibrate the vacuum optomechanical coupling g_0 to the injected frequency modulation. Employing this on the example spectrum shown in Fig. 5.3d, gives $g_0 \approx 2\pi \cdot 19\,\text{kHz}$.

This frequency noise calibration technique crucially relies on the ability to produce a pure known frequency modulation. Residual amplitude modulation (RAM) in the EOM is a well known technical challenge to achieving a pure frequency modulation [34, 35]. The effect of RAM in our experiments is two-fold: excessive RAM leads to large uncertainties in the phase modulation power in the reference tone used for calibration; secondly, since the same EOM is used for producing phase modulation tones for resonantly locking the laser (as described in Fig. 3.8), RAM leads to DC offsets in the lock error signal. In the experiment, the presence of RAM is verified and corrected in two ways. A direct photodetector placed immediately in the transmission

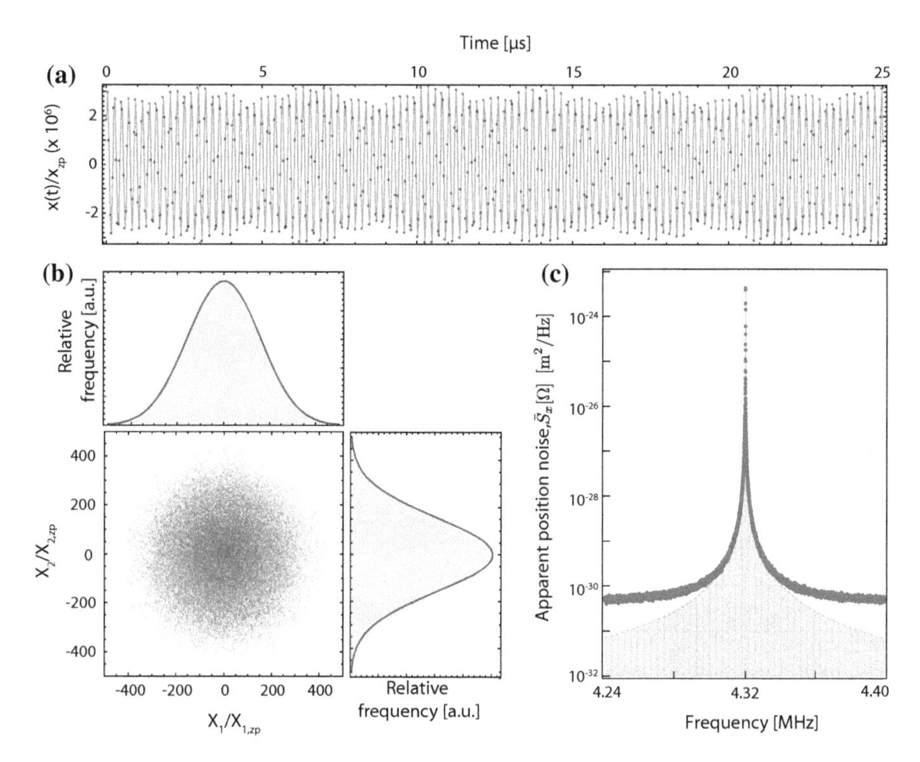

Fig. 5.4 Incarnations of thermal motion. Plots show various representations of thermal fluctuations of the position of the fundamental mode of the nanobeam, inferred from a calibrated homodyne photocurrent record. **a** Time domain trace (bandpass filtered around the mechanical frequency) showing the random amplitude fluctuations caused due to the thermal Langevin force. **b** Quadratures $X_{1,2}(t)$ of the apparent motion $x(t)$, and their marginal distributions (see text for details). **c** Power spectral density computed from the full photocurrent record in **a**

of the EOM is used to nullify any RAM by adjusting the input polarisation to the EOM; we have achieved relative RAM suppression at the level of 20 dB like this, stable over the course of typical experiments. In addition, heterodyne detection of the EOM transmission, for sufficiently large modulation depth $\xi_{\mathrm{mod}} \gtrsim 0.01$, resulting in cascaded phase modulation sidebands can be used to diagnose the presence of RAM a posteriori. In the presence of both AM and PM, the first order sidebands exhibit an asymmetry due to the interference of the PM and AM, whereas the higher order sidebands are purely due to PM. Figure 5.3b shows a typical heterodyne voltage noise spectrum showing cascaded sidebands, with the blue points showing the prediction from a pure PM model. It has also proven useful, especially for stable operation exceeding an hour, to actively stabilise the RAM by using the direct photodetection signal after the EOM to feedback on the DC bias of the EOM in a slow feedback loop [35, 36] (see experiment layout in Fig. 5.9).

Figure 5.4 shows several forms of thermal motion of the fundamental out-of-plane mode of the nanobeam. The data is taken using a resonant probe optically demodulated in a homodyne interferometer. Panel (a) shows a part of the photocurrent record, bandpass filtered around the mechanical frequency, referred to mechanical position. The amplitude scale is calibrated assuming thermal equilibrium at a temperature of 4 K (details of cryogenic operation follow in Sect. 5.3). Panel (c) shows a periodogram power spectral density estimate [37] of the full photocurrent record. Here, the spectrum is calibrated in terms of an absolute position noise through a knowledge of the effective mass of a doubly-clamped beam, $m_{\text{eff}} \approx m_{\text{phys}}/2 \approx 85\,\text{pg}$. Panel (b) shows the apparent motion $x(t)$ decomposed into its quadratures, $X_{1,2}(t)$ [38]:

$$
\begin{aligned}
x(t) &= X_1(t)\cos\Omega_{\text{m}}t + X_2(t)\sin\Omega_{\text{m}}t \\
\text{i.e.,}\quad X_1[\Omega] &= x[\Omega+\Omega_m] + x[\Omega-\Omega_{\text{m}}] \\
\text{and,}\quad X_2[\Omega] &= -i\left(x[\Omega+\Omega_{\text{m}}] - x[\Omega-\Omega_{\text{m}}]\right).
\end{aligned}
$$

These quadratures are obtained from the calibrated apparent position $x(t)$ by demodulating it at the known mechanical frequency $\Omega_{\text{m}} \approx 2\pi \cdot 4.315\,\text{MHz}$ digitally. Due to the large signal-to-noise of the measurement (seen in panel (c)), the phase-space distribution shown in panel (b) may be assumed to be only negligibly ($\lesssim$ 1 ppm) contaminated by the shot-noise of the optical field used for the measurement. The marginal distributions, also shown in panel (b), have variances, $\text{Var}\left[X_{1,2}\right] = \frac{1}{2}\text{Var}\left[x\right] = (n_{\text{m,th}} + \frac{1}{2})x_{\text{zp}}^2$ consistent with the known thermal occupation.

5.3 Cryogenic Operation

The relatively low mechanical frequency (fundamental mode, $\Omega_{\text{m}} \approx 2\pi \cdot 4\,\text{MHz}$) means that the ambient phonon occupation at room temperature is, $n_{\text{m,th}} \approx 10^6$, while the decoherence rate, $n_{\text{m,th}}\Gamma_{\text{m}} \approx 2\pi \cdot 10\,\text{MHz}$, is larger than a single mechanical period. The necessary condition to observe quantum coherent mechanical oscillations, i.e. $\Omega_{\text{m}} \gtrsim n_{\text{m,th}}\Gamma_{\text{m}}$, is thus not met at room temperature for the systems used in this thesis. It is therefore necessary to employ passive cryogenic cooling to decrease the thermal decoherence rate to a level where this condition is met. For example, by operating at 10 K it would be possible to achieve $\Omega_{\text{m}} \gtrsim 5 \cdot n_{\text{m,th}}\Gamma_{\text{m}}$.

Figure 5.5 shows the schematic of the ^{3}He buffer gas cryostat (Oxford Instruments, HelioxTL) employed in this thesis, allowing access to temperatures as low as 0.3 K. The cryostat consists of successive concentric layers of thermal isolation: a vacuum chamber whose outer wall is in contact with room temperature pumped down to a pressure of $<10^{-5}$ mbar; followed by a shield filled with liquid nitrogen, maintained at a temperature of 77 K, further cryo-pumping the outer vacuum chamber to pressures well below 10^{-6} mbar; followed by a shield of liquid ^{4}He maintained at a temperature of 4 K. These shields isolate an inner cylindrical bore (diameter $\approx$ 10 cm) where the sample is inserted—the sample volume.

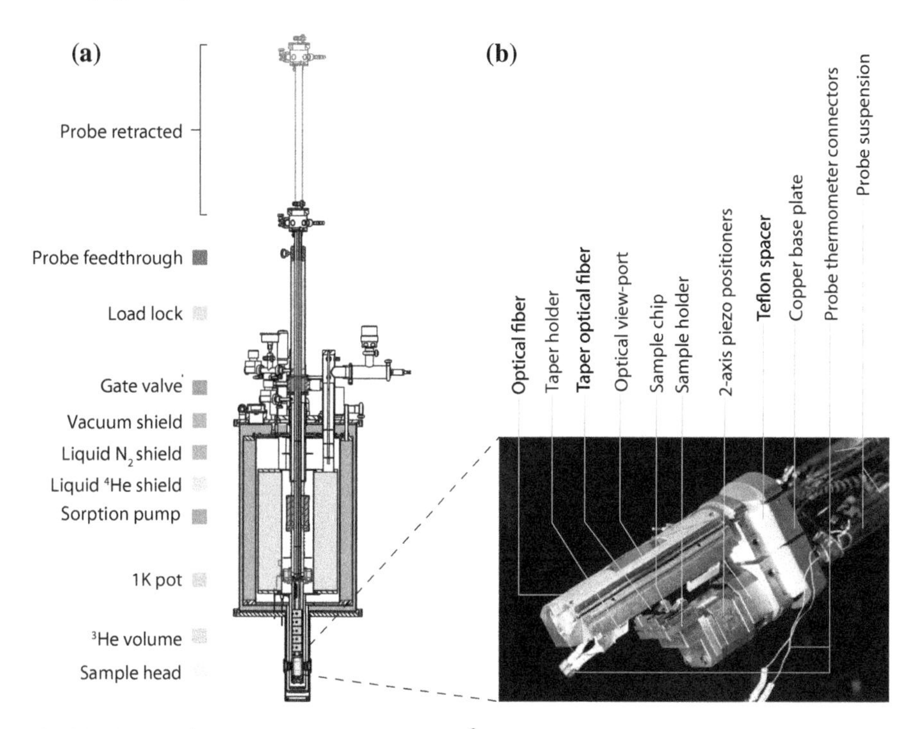

Fig. 5.5 Cryogenic apparatus. a Drawing of the ^{3}He buffer gas cryostat used in experiments. The sample chip, containing multiple optomechanical devices is mounted on the sample head (yellow) embedded in the heart of the cryostat. **b** Photograph of the sample head. The chip containing the samples is mounted on a 2-axis piezo-positioner, allowing for coupling into the optical cavity using tapered optical fiber

The sample—several ($\approx$10) optomechanical devices on a rectangular silicon chip on which they are fabricated—is mounted on the sample head shown in Fig. 5.5b. The head is mounted on the end of a retractable probe which can be attached to the top of the sample volume. Samples can be changed by retracting the probe—it slides on two O-rings between a pumped load-lock volume—from the cryostat bore, closing the gate valve to isolate the cryostat volume from the retracted probe, and then detaching the probe from the cryostat. In this manner, samples may be changed while keeping the cryostat cold.

The sample head (Fig. 5.5b), at the end of the probe, is suspended from hollow steel tubes damped and thermally isolated using Teflon bluffs. The sample chip is mounted on a removable mount and attached onto its face using four clamps. This mount is screwed onto the top of a two-stack piezo-positioner (AttoCube ANPx51/LT). The tapered optical fiber used to optically probe the microcavity is glued onto a custom-made glass holder, which is secured onto the head via a clamp. Care is taken during the assembly of the head to ensure that the desired sample(s) can be coupled using the piezo-positioners—the devices themselves are fabricated on tall ($\approx$100 μm) mesas on the chip so as to ensure that geometric parallelism between the taper and chip

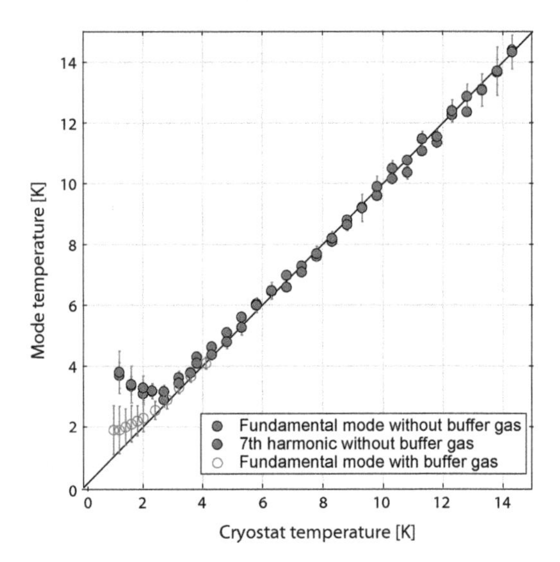

Fig. 5.6 Thermalisation with and without buffer gas. Temperature of the fundamental and 7th harmonic of the nanobeam as a function of the cryostat temperature. Above 4 K, the mechanical modes thermalise to the cryogenic environment despite the absence of buffer gas. In the presence of buffer gas, thermalisation can be achieved down to 2 K. Error bars denote standard errors derived from the statistical dispersion among multiple data points taken at each cryostat temperature

surface is not as stringent. Once the head is assembled, it takes on a copper hood that encloses it on all sides, except for a view-port and inlets for the buffer gas (see Fig. 5.5b).

The sample head, once prepared, is inserted into the cryostat. During insertion, the sample volume is pumped using an internal sorption pump (essentially a clean surface of activated charcoal cooled by a regulated flow of ^{4}He behind it); at the same time, it has proven more reliable to hold all the ^{3}He buffer gas in its external reservoir. The probe is then inserted in a continuous motion into the sample volume; finally, it is rotated so that the view-port on the sample head is aligned to the view-port at the bottom of the cryostat. This view-port is used to peer into the head using a microscope so as to align the sample and tapered fiber for optical coupling.

In order to cool the sample, ^{3}He is introduced back into the sample volume at a very slow rate—the 15 L of the buffer gas leaks in over $\approx$2 h. After an initial phase of radiative cooling from 300 K to about 100 K, the sorption pump is heated up ($\approx$25 K) to eject out the buffer gas, which then thermalises the sample holder to the internal walls of the sample volume at 4 K, in contact with the liquid ^{4}He shield.

To verify thermalisation of the mechanical modes of the nanobeam, we perform calibrated thermal noise measurements on the modes of the beam. A weak laser ($P_{\text{in}} \lesssim 5$ nW) is far detuned on the red side of the cavity (typically $\Delta < -2\kappa$) so as to prevent any dynamical back-action modification of the mechanical oscillator state. The laser, passively stable at this detuning, introduces $n_c \approx 2$, intracavity photons on cavity resonance. Figure 5.6 shows the result of this investigation. With all buffer gas

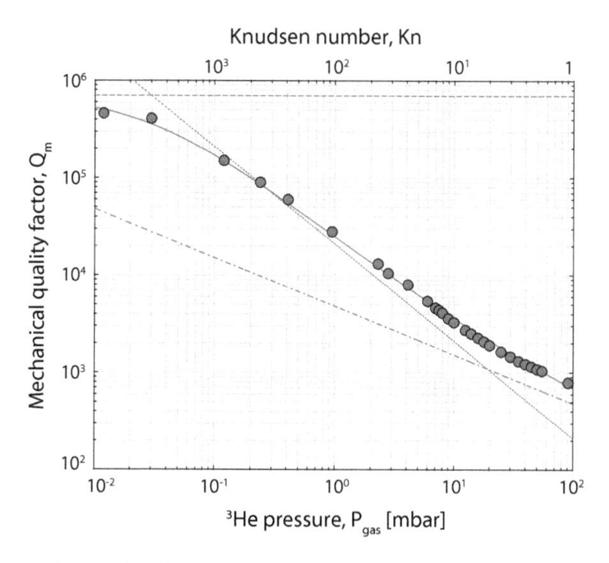

Fig. 5.7 Mechanical damping due to buffer gas. Data points show the mechanical quality factor of the fundamental mode ($\Omega_m = 2\pi \cdot 4.3\,\mathrm{MHz}$) as a function of the buffer gas pressure in the sample volume. The observed data is understood as arising from two contributions to the quality factor: an internal quality factor, $Q_{m,int} \approx 7 \cdot 10^5$, shown in the broken line; and a gas damping contribution shown in chained and dotted lines (see Eq. (5.3.3)). The solid line shows the total model in Eq. (5.3.1)). The Knudsen number in the top axis is estimated from Eq. (5.3.2), assuming $T_{gas} = 4\,\mathrm{K}$, $\ell_{gas} \approx 30\,\mathrm{pm}$ (Bohr radius of He), and using $\ell_b = 1\,\mu\mathrm{m}$

evacuated,[4] all modes of the nanobeam thermalise well up to a temperature of 4 K; Fig. 5.6 filled red and blue points show the mode temperatures of the fundamental and its 7th harmonic. In the presence of buffer gas however, modes thermalise well up to about 2 K. We conjecture that the lack of thermalisation in the absence of buffer gas is due to the lack of thermal conductivity along the beam into the substrate at these very low temperatures. This is consistent with the observation of a universal drop in thermal conductivity of amorphous materials at low temperature [26].

For the fundamental mode of the nanobeam, gas damping prevents operating with buffer gas. In fact, the mechanical quality factor observed as a function of pressure, shown in Fig. 5.7, may be understood as a sum of two contributions,

$$\begin{aligned} Q_m^{-1}(P_{gas}) &= Q_{m,int}^{-1} + Q_{m,gas}^{-1}(P_{gas}) \\ &= Q_{m,int}^{-1} + Q_{m,gas,visc}^{-1}(P_{gas}) + Q_{m,gas,sque}^{-1}(P_{gas}) \end{aligned} \tag{5.3.1}$$

[4]By design, this cryostat is not meant for cold operation without buffer gas. In order to achieve low temperatures without having buffer gas in the sample volume, we first condense ^{3}He: this is done by filling the 1 K pot with liquid ^{4}He and pumping on it to cool it down to <2 K; finally the sorption pump is heated to 25 K to eject out all ^{3}He gas and pressurise it. Once condensed, droplets of liquid ^{3}He accumulate at the bottom of the cryostat (the so-called "^{3}He tail"); pumping on the condensed liquid using the sorption pump evaporatively cools the liquid to as low as 0.3 K. Once all the condensed ^{3}He is evaporated, the sample volume is in vacuum and the sample temperature slowly rises.

where $Q_{\mathrm{m,int}}$ is the intrinsic quality factor (diluted by stress) when all gas is evacuated, and $Q_{\mathrm{m,gas}}(P_{\mathrm{gas}})$ is the quality factor due to losses in the presence of buffer gas [2]. The latter is mainly due to two physical processes: viscous-type retardation due to the gas [6, 39], and, energy lost by the beam into compressing trapped gas—squeeze-film damping [40].

The nature of viscous damping depends on the nature of the fluid flow. The Knudsen number,

$$\mathrm{Kn} := \frac{\ell_{\mathrm{MF}}}{\ell_b} \approx \frac{1}{\sqrt{2}\pi} \cdot \frac{k_B T_{\mathrm{gas}}}{\ell_{\mathrm{gas}}^3 P_{\mathrm{gas}}} \cdot \frac{\ell_{\mathrm{gas}}}{\ell_b}, \tag{5.3.2}$$

which is the ratio between the mean-free path of the gas (ℓ_{MF}) and the characteristic length of the beam (ℓ_b), roughly dictates the damping regime. The second equality above estimates the mean-free path based on a statistical model of an ideal gas; here, T_{gas} is the temperature of the gas, while ℓ_{gas} is the semi-classical radius of the gas atom. Note that Kn essentially parametrises the capacity of the surrounding gas to perform work on the beam via thermal forces, or via compressive forces. In the kinetic regime, characterised by $\mathrm{Kn} \gtrsim 10$, damping is essentially due to recoil from independent incoherent scattering events,[5] while at higher pressure, in the hydrodynamic regime, characterised by $\mathrm{Kn} \gtrsim 1$, damping is due to the inertia or viscosity of the fluid continuum.[6] For the small beams with relatively small thermal velocities, the fluid flow around the beam is dominated by inertia.[7] Under these conditions, the gas damping contribution is given by [2, 39, 40],

$$Q_{\mathrm{m,gas,visc}}(P_{\mathrm{gas}}) = \begin{cases} \frac{\rho_b}{P_{\mathrm{gas}}} \frac{\ell_t}{4} \frac{\Omega_{\mathrm{m}}}{2\pi} \left(\frac{RT_{\mathrm{gas}}}{M_{\mathrm{gas}}}\right)^{1/2}; & \mathrm{Kn} \gtrsim 10 \\ \frac{\rho_b}{\sqrt{P_{\mathrm{gas}}}} \frac{\ell_{\mathrm{BL}}}{6\pi} \left(\frac{\Omega_{\mathrm{m}}}{2\pi}\right)^{1/2} \left(\frac{2RT_{\mathrm{gas}}}{\mu_{\mathrm{gas}}}\right)^{1/2}; & \mathrm{Kn} \approx 1 \end{cases} \tag{5.3.3}$$

where ρ_b is the mass density of the beam, ℓ_t the thickness of the beam in the direction of motion, ℓ_{BL} is a phenomenological boundary layer thickness, M_{gas} the molar mass of the surrounding gas, R is the gas constant, and μ_{gas} is its dynamic viscosity. Importantly, the scaling with pressure characterises the transition from the kinetic to the hydrodynamic regimes. The models shown in Fig. 5.7 (dotted and chained lines) are fits employing this pressure scaling.

Squeeze-film damping, in the regime where at least one transverse dimension of the beam (ℓ_t) is comparable to the thickness of the squeezed gas layer ($\ell_{\mathrm{gas,t}}$), leads to a contribution [40, 44],

---

[5] In confined spaces, even in the regime of $\mathrm{Kn} > 10$, recoil events may not be independent, leading to an excess thermal force [41] with a characteristic time-scale inversely related to the dimensions of the constriction—this effect is not observed here.

[6] It is interesting to note that in the extreme regime, $\mathrm{Kn} \ll 0.1$, the spectrum of gas damping samples the inter-particle collisions of the gas atoms [42, 43] and necessitates a non-Newtonian fluid model—however this regime is not relevant here.

[7] The Reynolds number, Re := (beam velocity)(transverse length)/(kinematic viscosity of He), determines this. For the beam undergoing thermal motion, its root mean square velocity on resonance is $\sqrt{2 n_{\mathrm{m,th}}} x_{\mathrm{zp}} \frac{\Omega_{\mathrm{m}}}{2\pi} \approx 10^{-4}$ m/s. For beam transverse dimension of 0.5 μm, it follows that, $\mathrm{Re} < 10^{-4}$.

$$Q_{\rm m,gas,sque}^{-1} = (\ell_{\rm gas,t}/\ell_t) Q_{\rm m,gas,visc}^{-1},$$

which is not seen to be a relevant to the observed data.[8]

Ultimately, the combination of gas damping and inefficient thermalisation forces experiments to be performed at cryogenic temperatures $T \gtrsim 4\,\mathrm{K}$ with all buffer gas evacuated. In principle, this allows us to achieve, $\Omega_{\rm m} \gtrsim 10 \cdot n_{\rm m,th}\Gamma_{\rm m}$, corresponding to $\Gamma_{\rm m} \approx 2\pi \cdot 6\,\mathrm{Hz}$. The latter value is independently verified using ring-down measurements.

5.3.1 *Nature of Elastic Force: Radiation Pressure Versus Thermoelasticity*

Given that the beam does not necessarily thermalise below 4 K, it needs to be investigated whether the thermal gradients established in the beam produce additional mechanical forces—*thermoelastic forces*—and how they compare against the desired radiation pressure force.

In order to investigate the contribution of these competing forces, we measured the response of the cavity frequency ω_c to modulation of the injected power $P_{\rm in}$, i.e. $\partial\omega_c/\partial P_{\rm in}$. This is done by using a probe laser locked on the resonance of a cavity mode at 780 nm, while a pump laser is locked on resonance to an independent cavity mode at 850 nm. Care is taken to ensure that there is no optical and/or electronic cross-talk between the two; both lasers are attenuated to (each $\approx$100 nW) ensure no spurious static thermal shifts of cavity resonance. The amplitude of the incident pump is modulated by an intensity modulator driven by a network analyser, while the frequency fluctuations in the probe laser are detected using a homodyne detector whose output voltage is demodulated by the network analyser.

Figure 5.8a shows two examples of such a response measurement, taken at 10 K (red) and 4 K (pink). We understand the measured cavity frequency fluctuation $\delta\omega_c$ as arising from three different sources:

$$\delta\omega_c[\Omega] = \delta\omega_c^{\rm th}[\Omega] + \delta\omega_c^{\rm mech,the}[\Omega] + \delta\omega_c^{\rm mech,rad}[\Omega] \qquad (5.3.4)$$

where $\delta\omega_c^{\rm th}$ is the cavity frequency shift due to material and geometric thermal deformation of the cavity volume, $\delta\omega_c^{\rm mech,the}$ is due to mechanical motion driven by thermoelastic forces [46, 47] and $\delta\omega_c^{\rm rad}$ is due to mechanical motion driven by radiation pressure force.

The cavity thermal shift can be understood as arising from change in the cavity length via a finite thermal expansion α_V, or due to material property (refractive index

[8]Due to the nature of the problem, it seems reasonable to imagine that for a narrow longitudinal (along the beam) constriction with a transverse opening, squeeze-films get evacuated much more quickly than diffusion would suggest. In fact molecules in a squeeze film execute Lévy walks [45], lending plausibility to their absence after even a short pumping time.

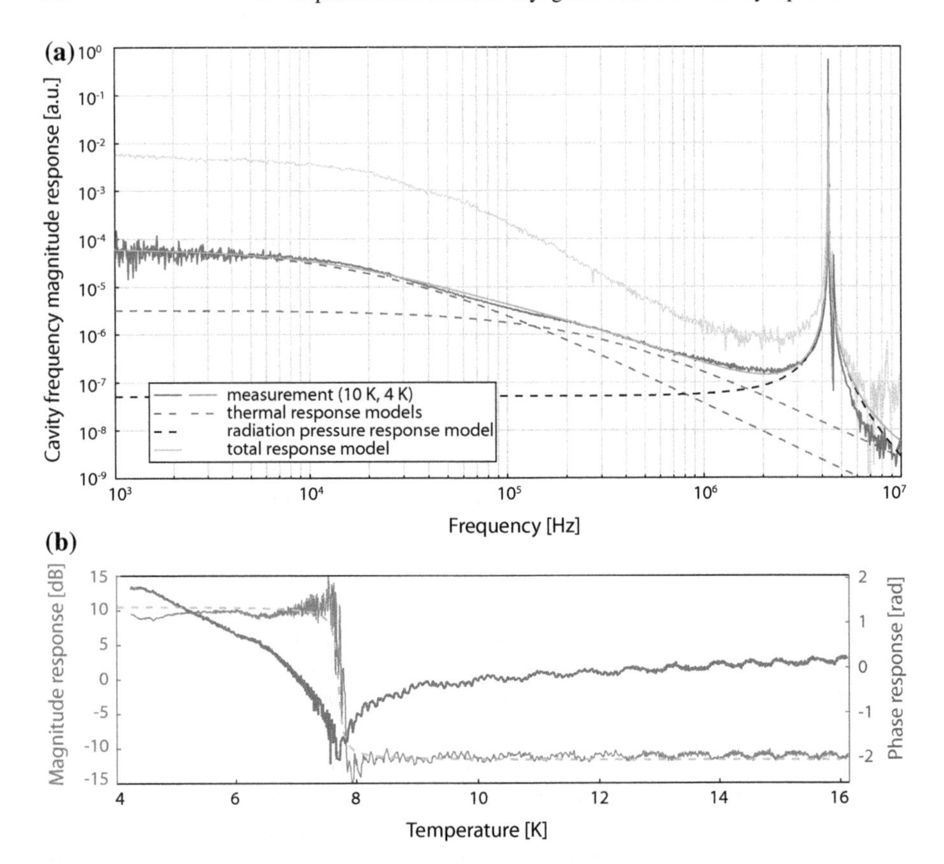

Fig. 5.8 Cavity frequency response to intensity modulation. a Frequency response measured by modulating an independent pump beam and demodulating the effect on a resonant probe using a homodyne detector. Red (pink) trace shows data taken at 10 K (4 K). Black dashed is a fit to expected radiation pressure force, which is prompt with respect to the low mechanical frequency $\Omega_{\mathrm{m}} \approx 2\pi \cdot 4.3\,\mathrm{MHz}$. The blue and purple dashed lines show cavity frequency shifts expected from models of delayed thermal conduction in the disk and/or beam. **b** Low frequency ($\Omega = 2\pi \cdot 2\,\mathrm{kHz}$) response as a function of cryostat temperature. Red (blue) points show the magnitude (phase) response

ν) changes, i.e.

$$\delta\omega_c^{\mathrm{th}}[\Omega] = -\omega_c \left(\alpha_V + \frac{1}{\nu}\frac{\partial \nu}{\partial T}\right)\delta T_c[\Omega] = -\omega_c \left(\alpha_V + \frac{1}{\nu}\frac{\partial \nu}{\partial T}\right)\frac{G_c^{\mathrm{th}}}{1 + i\Omega/\Omega_c^{\mathrm{th}}}\delta n_c, \tag{5.3.5}$$

where the cavity temperature change δT_c is assumed to be driven by pump photon number modulation δn_c, and the temperature relaxes via diffusion, modelled as a single-pole response with cut-off Ω_c^{th} and gain G_c^{th}. In Fig. 5.8a, the low frequency cut-off at 10 kHz may be identified with Ω_c^{th}, consistent with measurements on toroidal microcavities [48]. Further, the low frequency response G_c^{th} as a function of cryostat temperature, shown in Fig. 5.8b, is consistent with the known $\partial\nu/\partial T = 0$ point of silica microcavities at 8 K [49].

The remaining two contributions to $\delta\omega_c$ in Eq. (5.3.4) arise from frequency shifts due to mechanical motion, i.e.

$$\begin{aligned}\delta\omega_c^{\text{mech}} &:= \delta\omega_c^{\text{mech,the}} + \delta\omega_c^{\text{mech,rad}} \\ \delta\omega_c^{\text{mech,the}}[\Omega] &= G_{\text{the}}\chi_x[\Omega]\,\frac{(\kappa_{\text{abs}}/c)\delta n_c}{1+i\Omega/\Omega_{\text{the}}} \\ \delta\omega_c^{\text{mech,rad}}[\Omega] &= G_{\text{rad}}\chi_x[\Omega]\,\frac{(\kappa/c)\delta n_c}{1+i\Omega/\Omega_{\text{rad}}}.\end{aligned} \tag{5.3.6}$$

Here, G_{rad} (G_{the}) is the optomechanical coupling due to radiation pressure (thermoelastic) motion, χ_x is the mechanical oscillator response, $(\kappa/c)\delta n_c$ ($(\kappa_{\text{abs}}/c)\delta n_c$) is the recoil force due to pump photon number modulation, and $\Omega_{\text{rad}} = \kappa/2$ (Ω_{the}) is the characteristic response frequency of the radiation pressure (thermoelastic) force. Note that since $\kappa \gg \Omega_{\text{m}}$, the radiation pressure force is effectively instantaneous, while the thermoelastic force has a finite delay due to the absorption and diffusion of temperature in the nanobeam. It follows that the response measured near the mechanical frequency (where $\delta\omega_c^{\text{th}}$ is negligible),

$$\left|\frac{\delta\omega_c^{\text{mech}}[\Omega]}{\delta n_c[\Omega]}\right| \approx \left|\delta\omega_c^{\text{mech,rad}}[\Omega]\right| \cdot \left|1 + \frac{G_{\text{the}}}{G_{\text{rad}}}\frac{\kappa_{\text{abs}}}{\kappa}\frac{1}{1+i\Omega/\Omega_{\text{the}}}\right|,$$

provides information regarding the fractional contribution of the thermoelastic force compared to radiation pressure (second factor on the left-hand side). From Fig. 5.8a, and other similar measurements, we have determined that radiation pressure dominates at temperatures above 6 K whereas for colder temperatures, thermoelastic back-action is observed.

5.4 Experimental Schematic

Figure 5.9 shows the schematic of the essential optical and electronic layout of the entire experiment. At the heart of the experiment is a ^{3}He cryostat (Oxford Instruments, HelioxTL), in which is embedded a chip containing multiple optomechanical devices. A desired device is probed by coupling to its optical cavity using a tapered optical fiber. The tapered optical fiber can be driven by one of three laser sources: a tunable 780 nm external cavity diode laser (NewFocus, Velocity), an 850 nm ECDL, or a Ti:Sa laser (MSquared, Solstis). Typically, the 780 nm laser (ECDL or Ti:Sa) is used for probing the cavity, while the 850 nm laser is used for actuating the cavity and/or the mechanical oscillator.

Both lasers can be locked to cavity resonance; they are frequency modulated using an EOM, and the modulation sideband is demodulated after the cavity transmission

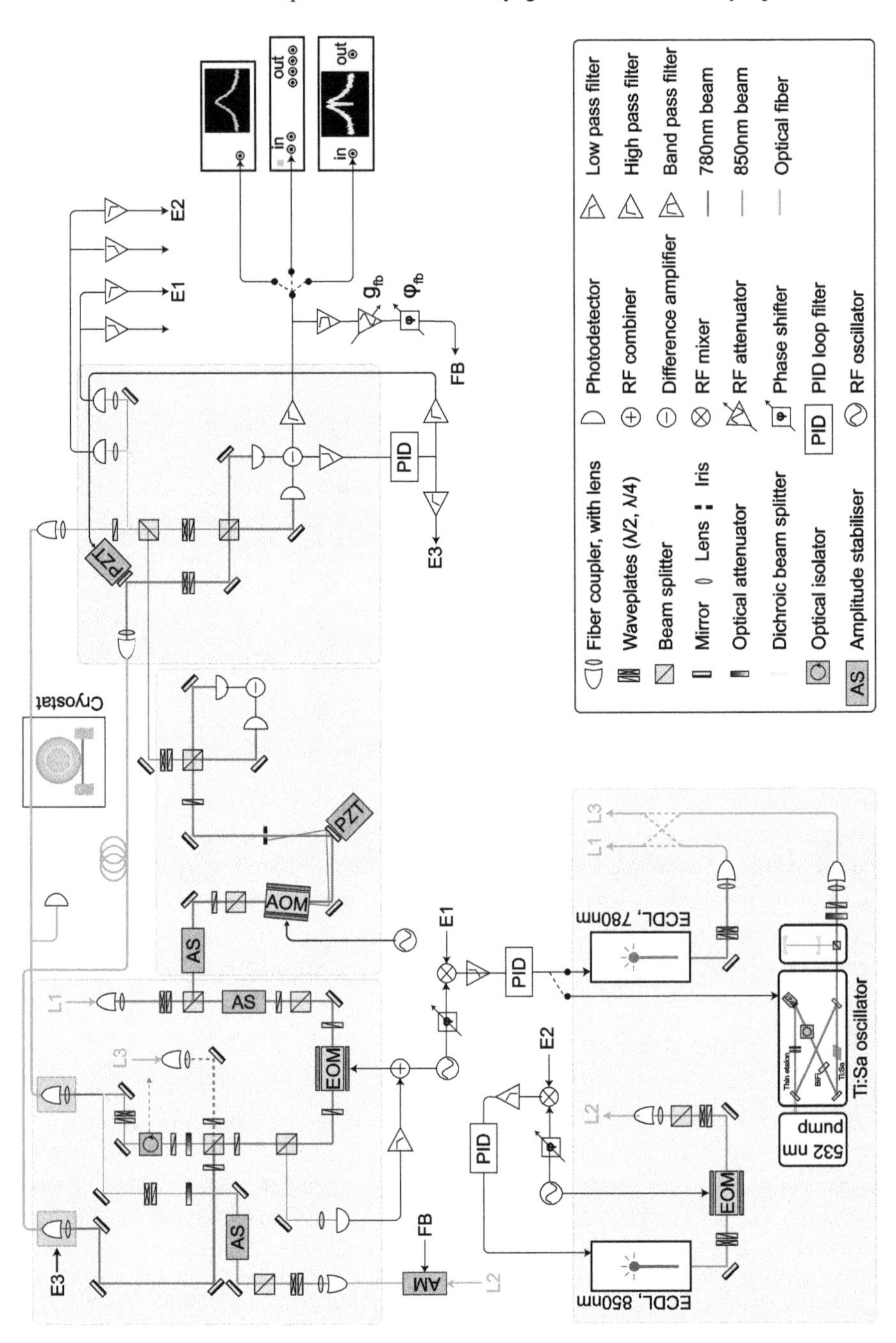

Fig. 5.9 Layout of the experiment

is detected on avalanche photodiodes. The cavity is placed in one arm of a balanced Mach-Zehnder interferometer and is directed onto either a balanced homodyne detector, or a balanced heterodyne detector, depending on what quadrature(s) of the light need to be observed.

In Chap. 6, the homodyne configuration, operated using the 780 nm ECDL is used to measure mechanical motion, while the 850 nm ECDL is used to perform feedback control. In Chap. 7, the same configuration is employed, with the addition that a part of the 780 nm transmission is directed onto the balanced heterodyne detector. The homodyne detector may also be locked onto the amplitude quadrature, as in Chap. 7, to probe for optical squeezing.

During measurements reported in Chap. 7, to study quantum correlations in the transmitted optical beam, it was found convenient to synchronise the internal clocks of all RF sources and receivers: this was done by using the 10 MHz signal derived from an atomic clock (not shown in Fig. 5.9).

References

1. M. Aspelmeyer, T.J. Kippenberg, F. Marquardt, Rev. Mod. Phys. **86**, 1391 (2014)
2. W.E. Newell, Science **161**, 1320 (1968)
3. H.G. Craighead, J. Sci. **290**, 1532 (2000)
4. K.L. Ekinci, M.L. Roukes, Rev. Sci. Instr. **76**, 061101 (2005)
5. K.C. Schwab, M.L. Roukes, Phys. Today **58**, 36 (2005)
6. K.Y. Yasumura, T.D. Stowe, E.M. Chow, T. Pfafman, T.W. Kenny, B.C. Stipe, D. Rugar, J. Microelectromech. Syst. **9**, 117 (2000)
7. P. Mohanty, D.A. Harrington, K.L. Ekinci, Y.T. Yang, M.J. Murphy, M.L. Roukes, Phys. Rev. B **66**, 085416 (2002)
8. M. Imboden, P. Mohanty, Phys. Rep. **534**, 89 (2014)
9. S.S.Verbridge, J.M. Parpia, R.B. Reichenbach, L.M. Bellan, H.G. Craighead, J. Appl. Phys. **99** (2006)
10. S.S. Verbridge, H.G. Craighead, J.M. Parpia, Appl. Phys. Lett. **92**, 013112 (2008a)
11. Q.P. Unterreithmeier, T. Faust, J.P. Kotthaus, Phys. Rev. Lett. **105**, 027205 (2010)
12. S. Schmid, K.D. Jensen, K.H. Nielsen, A. Boisen, Phys. Rev. B **84**, 165307 (2011)
13. R. Schilling, H. Schütz, A. Ghadimi, V. Sudhir, D. Wilson, T. Kippenberg, Phys. Rev. Applied **5**, 054019 (2016)
14. G. Anetsberger, O. Arcizet, Q.P. Unterreithmeier, R. Rivière, A. Schliesser, E.M. Weig, J.P. Kotthaus, T.J. Kippenberg, Nat. Phys. **5**, 909 (2009)
15. G. Anetsberger, Novel cavity optomechanical systems at the micro- and nanoscale and quantum measurements of nanomechanical oscillators. Ph.D. thesis, LMU Munich, 2010
16. E. Gavartin, Optonanomechanical systems for measurement applications. Ph.D. thesis, EPFL, 2013
17. M. Nieto-Vesperinas, P.C. Chaumet, A. Rahmani, Phil. Trans. Roy. Soc. A **362**, 719 (2004)
18. Q.P. Unterreithmeier, E.M. Weig, J.P. Kotthaus, Nature **458**, 1001 (2009)
19. M. Eichenfield, R. Camacho, J. Chan, K.J. Vahala, O. Painter, Nature **459**, 550 (2009)
20. D. van Thourhout, J. Roels, Nat. Phot. **4**, 211 (2010)
21. B.E. Littleand, H.A. Haus, J. Lightwave Tech. **17**, 704 (1999)
22. M.L. Gorodetsky, A.E. Fomin, IEEE J. Selec. Top. Quant. Elec. **12**, 33 (2006)
23. M. Pinard, Y. Hadjar, A. Heidmann, Eur. Phys. J. D **7**, 107 (1999)
24. S. Timoshenko, *Vibration Problems in Engineering*, 2nd edn. (D.Van Nostrand Company, 1937)

25. S.M. Han, H. Benaroya, T. Wei, J. Sound Vib. **225**, 935 (1999)
26. R.O. Pohl, X. Liu, E. Thompson, Rev. Mod. Phys. **74**, 991 (2002)
27. T. Faust, J. Rieger, M.J. Seitner, J.P. Kotthaus, E.M. Weig, Phys. Rev. B **89**, 100102 (2014)
28. L.G. Villanueva, S. Schmid, Phys. Rev. Lett. **113**, 227201 (2014)
29. B.M. Zwickl, W.E. Shanks, A.M. Jayich, C. Yang, A.C. Bleszynski, J.D. Thompson, J.G.E. Harris, Appl. Phys. Lett. **92**, 103125 (2008)
30. D.J. Wilson, C.A. Regal, S.B. Papp, H.J. Kimble, Phys. Rev. Lett. **103**, 207204 (2009)
31. G.I. González, P.R. Saulson, J. Acoust. Soc. Am. **96**, 207 (1994)
32. P.-L. Yu, T.P. Purdy, C.A. Regal, Phys. Rev. Lett. **108**, 083603 (2012)
33. M.L. Gorodetsky, A. Schliesser, G. Anetsberger, S. Deleglise, T.J. Kippenberg, Opt. Exp. **18**, 23236 (2010)
34. E.A. Whittaker, M. Gehrtz, G.C. Bjorklund, J. Opt. Soc. Am. B **2**, 1320 (1985)
35. N.C. Wong, J.L. Hall, J. Opt. Soc. Am. B **2**, 1527 (1985)
36. W. Zhang, M.J. Martin, C. Benko, J.L. Hall, J. Ye, C. Hagemann, T. Legero, U. Sterr, F. Riehle, G.D. Cole, M. Aspelmeyer, Opt. Lett. **39**, 1980 (2014)
37. R.B. Blackman, J.W. Tuckey, *The Measurement of Power Spectra* (Dover, 1959)
38. T. Briant, P.F. Cohadon, M. Pinard, A. Heidmann, Eur. Phys. J. D **22**, 131 (2003)
39. F.R. Blom, S. Bouwstra, M. Elwenspoek, J.H.J. Fluitman, J. Vac. Sci. Tech. B **10**, 19 (1992)
40. M. Bao, H. Yang, Sens. Act. A **136**, 3 (2007)
41. A. Cavalleri, G. Ciani, R. Dolesi, A. Heptonstall, M. Hueller, D. Nicolodi, S. Rowan, D. Tombolato, S. Vitale, P.J. Wass, W.J. Weber, Phys. Rev. Lett. **103**, 140601 (2009)
42. V. Yakhot, C. Colosqui, J. Fluid Mech. **586**, 249 (2007)
43. D.M. Karabacak, V. Yakhot, K.L. Ekinci, Phys. Rev. Lett. **98**, 254505 (2007)
44. S.S. Verbridge, R. Ilic, H.G. Craighead, J.M. Parpia, Appl. Phys. Lett. **93**, 013101 (2008b)
45. S. Schlamminger, C.A. Hagedorn, J.H. Gundlach, Phys. Rev. D **81**, 123008 (2010)
46. C.H. Metzger, K. Karrai, Nature **432**, 1002 (2004)
47. S. De Liberato, N. Lambert, F. Nori, Phys. Rev. A **83**, 033809 (2011)
48. A. Schliesser, Cavity optomechanics and optical frequency comb generation with silica whispering-gallery-mode microresonators. Ph.D. thesis, LMU Munich, 2010
49. O. Arcizet, R. Rivière, A. Schliesser, G. Anetsberger, T. Kippenberg, Phys. Rev. A **80**, 021803(R) (2009)

Chapter 6
Observation and Feedback-Suppression of Measurement Back-Action

Moby Dick seeks thee not. It is thou, thou, that madly seekest him!

Herman Melville

Realizing a quantum-noise-limited measurement of the position of a harmonic oscillator has been a 30 year old white whale of experimental physics, stretching back to the conception of interferometric gravitational wave antennae [1–3]. Meeting this requirement is tantamount to two conditions: a quantum-noise-limited measurement imprecision, and quantum-noise-limited measurement back-action. Together they imply that the Heisenberg uncertainty product (Eq. 4.3.13),

$$\bar{S}_{FF}^{\text{tot}}[\Omega]\bar{S}_{xx}^{\text{imp}}[\Omega] \geq \frac{\hbar^2}{4},$$

is saturated. Here, $\bar{S}_{FF}^{\text{tot}}$ is the total force acting on the oscillator, including, but not typically limited to, quantum measurement back-action; $\bar{S}_{xx}^{\text{imp}}$ is the total measurement imprecision, including, but not typically limited to, optical quantum noise. A basic technical difficulty in performing a Heisenberg-uncertainty-limited position measurement in large scale interferometers, despite their record low quantum-noise-limited measurement imprecision, is that at the optical powers required to overwhelm the thermal force with back-action, dynamical instabilities set in [4, 5]. At the other end of the spectrum in scale, single atoms in optical traps are easily driven by back-action [6]; however, unavailability of high-efficiency large dynamic range transducers of their motion precludes a sufficiently low measurement imprecision. About 10 years ago, mesoscopic-scale cavity opto-/electro-mechanical systems were thus foreseen as an avenue to explore Heisenberg-limited measurements [7]. Section 6.1 details our experiment that realizes the closest approach to the Heisenberg uncer-

V. Sudhir, *Quantum Limits on Measurement and Control of a Mechanical Oscillator*, Springer Theses, https://doi.org/10.1007/978-3-319-69431-3_6

tainty product to date, corresponding to an uncertainty product of $5 \cdot \hbar/2$. This is achieved by simultaneously realizing a measurement imprecision that is 43 dB below that at the standard quantum limit, and observing quantum measurement back-action comparable to thermal motion [8].

Vis-a-vis measurement back-action, the standard quantum limit ($\bar{S}_{xx}^{\mathrm{SQL}} = 2\bar{S}_{xx}^{\mathrm{zp}}$), prevents the observation of the zero-point motion of the mechanical oscillator, despite the exquisite measurement sensitivity realized in our experiment. This is a specific example of the fundamental impediment posed by measurement back-action in the ability to measure physical quantities [9]. Various techniques have been proposed to mitigate this problem. Section 6.2 experimentally explores one such solution—feedback control to suppress back-action [10, 11]. In fact, it should not be surprising that if the record of the measurement of the oscillator position contains traces of the back-action-induced motion, a real-time feedback controller can anticipate and cancel the disturbance [12]. We have been able to use measurement-based feedback to suppress more than 40 dB of the oscillator's motion—including thermal motion and back-action motion—and prepare the oscillator with an average phonon occupation of 5.3 [8].

6.1 Quantum-Noise-Limited Position Measurement

We here study a specific example of the type of optomechanical system described in detail in Chap. 5. Specifically, the system consists of a 65 μm × 400 nm × 70 nm (effective mass $m \approx 2.9$ pg) nanobeam placed ~50 nm from the surface of a 30 μm diameter microdisk. The microdisk is optically probed using a low-loss (≈6%) fiber-taper and light supplied by a tunable diode laser. Mechanical motion is observed in the phase of the transmitted cavity field using a balanced homodyne interferometer. We interrogate two optical modes: a *meter* mode (used for homodyne readout) at $\lambda_c \approx 775$ nm that exhibits an intrinsic photon decay rate of $\kappa_0 \approx 2\pi \cdot 0.44$ GHz and a *feedback* mode (used for radiation pressure actuation) at $\lambda_c \approx 843$ nm that exhibits a decay rate of $\kappa_0 \approx 2\pi \cdot 1$ GHz. For the mechanical oscillator, we use the $\Omega_m \approx 2\pi \cdot 4.3$ MHz fundamental out-of-plane mode of the nanobeam. The optomechanical coupling strength between the oscillator and the sensor mode is $g_0 \approx 2\pi \cdot 20$ kHz (see discussion surrounding Fig. 6.3), corresponding to a frequency pulling factor of $G \approx 2\pi \cdot 0.70$ GHz/nm for the estimated zero-point amplitude of $x_{\mathrm{zp}} = \sqrt{\hbar/2m\Omega_m} \approx 29$ fm. The experiments were conducted in a ^{3}He buffer gas cryostat at an operating temperature of $T \approx 4.4$ K ($n_{\mathrm{m,th}} = k_B T/\hbar\Omega_m \approx 2.1 \cdot 10^4$) and at gas pressures below 10^{-3} mbar. Ring-down measurements here reveal a mechanical damping rate of $\Gamma_m \approx 2\pi \cdot 5.7$ Hz ($Q_m \approx 7.6 \cdot 10^5$). Our system is thus able to operate with a near-unity single-photon cooperativity $C_0 = 4g_0^2/\kappa\Gamma_m \approx 0.64$.

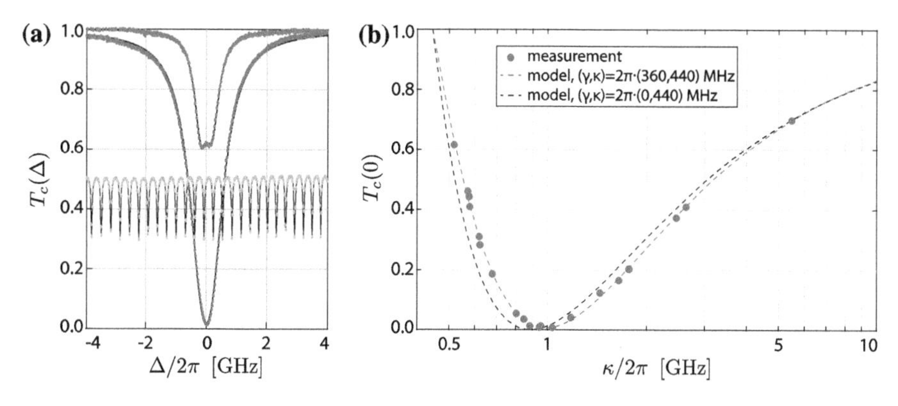

Fig. 6.1 Steady state spectroscopy of a cavity mode split by internal scattering. a Red traces show the cavity transmission, $T_c(\Delta)$, recorded as the laser-cavity detuning Δ is changed. The traces show the cavity under-coupled, where the effect of the splitting γ is obvious, and over-coupled where the effect is masked by the loaded linewidth $\kappa = \kappa_0 + \kappa_{\text{ex}}$. The gray trace shows the transmission through a calibrated fiber-loop cavity, recorded simultaneously; its free spectral range provides frequency markers to calibrate the frequency sweep. **b** Transmission on resonance, $T_c(0)$, plotted as a function of the external coupling to the cavity

6.1.1 Measurement Imprecision and Back-Action in a Split-Mode Cavity

Spectroscopy of the optical cavity, shown in Fig. 6.1, indicates that the optical resonance is split. In whispering gallery optical micro-cavities, such splitting is an indication of scattering centers that couple light circulating along the conventional direction of the injected field ("clockwise") into the mode against this direction ("counter-clockwise") [13, 14]. In the following we develop a theoretical model that allows to understand the spectroscopic data in Fig. 6.1, as well as predict the modification to measurement imprecision and back-action when employing such a split-mode cavity. The following discussion thus takes into account non-idealities arising from mode splitting which results in deviations from the ideal considerations given in Sect. 4.3.

We adopt the following set of coupled Langevin equations to model the dynamics of the cavity mode (characterised by the slowly varying amplitude of the intracavity field, $\hat{a}$) and the mechanical mode (characterised by its displacement $\hat{x}$):

$$\begin{aligned}\dot{\hat{a}}_+ &= \left(i\Delta_0 - \frac{\kappa}{2}\right)\hat{a}_+ + \frac{i\gamma}{2}\hat{a}_- + ig_0\hat{z}\hat{a}_+ + \sqrt{\eta_c\kappa}\,\hat{a}_{\text{in}}^+ + \sqrt{(1-\eta_c)\kappa}\,\delta\hat{a}_0^+ \\ \dot{\hat{a}}_- &= \left(i\Delta_0 - \frac{\kappa}{2}\right)\hat{a}_- + \frac{i\gamma}{2}\hat{a}_+ + ig_0\hat{z}\hat{a}_- + \sqrt{\eta_c\kappa}\,\hat{a}_{\text{in}}^- + \sqrt{(1-\eta_c)\kappa}\,\delta\hat{a}_0^-\end{aligned} \tag{6.1.1a}$$

$$m\left(\ddot{\hat{x}} + \Gamma_{\text{m}}\dot{\hat{x}} + \Omega_{\text{m}}^2\hat{x}\right) = \delta\hat{F}_{\text{th}} + \hat{F}_{\text{BA}}. \tag{6.1.1b}$$

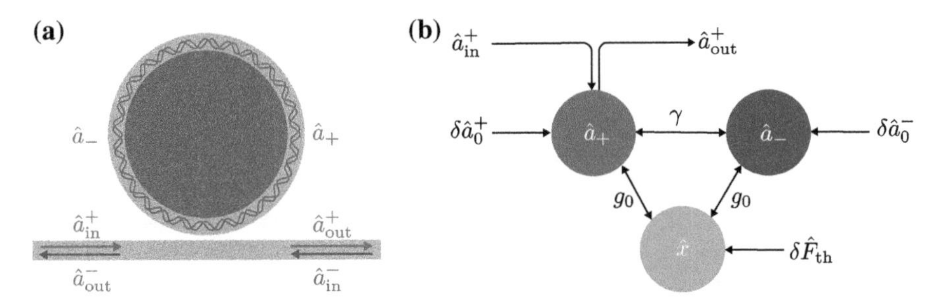

Fig. 6.2 Schematic of optomechanics using a split-mode cavity. a Physical model of the split-mode cavity: red shows the conventional mode ("clockwise") that is coupled to by a laser injected in the optical fiber, while blue shows the mode ("counter-clockwise") that is established by scattering centers in the cavity. **b** Schematic of the mutual coupling between the cavity modes ($\hat{a}_\pm$) via the scattering interaction at rate γ, and their common coupling to the mechanical mode ($\hat{x}$) at rate g_0

Notably, in Eq. (6.1.1a) we use a two-mode model to describe the microdisk cavity. Subscripts $+$ and $-$ refer to whispering gallery modes propagating along ('clockwise') and against ('counter-clockwise') the conventional direction ($+$) of the injected field, respectively. The two modes are coupled at a rate γ by scattering centres [14]. Since the split modes are (nearly) spatially orthogonal [13], the optomechanical coupling of either mode to the oscillator can be assumed to be independent, while the geometry of evanescent coupling means that they may be assumed to share a common vacuum optomechanical coupling rate, g_0. The resulting radiation-pressure back-action force is,

$$\hat{F}_{\rm BA} = \hbar g_0 \left(\hat{a}_+^\dagger \hat{a}_+ + \hat{a}_-^\dagger \hat{a}_-\right) \frac{\hat{x}}{x_{\rm zp}}. \tag{6.1.2}$$

Light is physically coupled to the microdisk cavity using an optical fiber as discussed in Sect. 3.2.4. As depicted in Fig. 6.2 and Eq. (6.1.1a), we model this coupler as a two port waveguide. Fields entering(exiting) the 'clockwise' port, $\hat{a}^+_{\rm in(out)}$, couple directly to the clockwise cavity mode. Fields entering(exiting) the 'counter-clockwise' port, $\hat{a}^-_{\rm in(out)}$, couple directly to the counter-clockwise mode. The cavity-waveguide coupling rate is $\kappa_{\rm ex} = \eta_c \kappa$, where $\kappa = \kappa_{\rm ex} + \kappa_0$ is the total cavity decay rate and κ_0 is the intrinsic cavity decay rate. In addition, each cavity mode is driven through its intrinsic decay channel by a vacuum state with amplitude $\delta\hat{a}_0^\pm$. Input field amplitudes are here normalised in the conventional manner, so that $P_{\rm in}^\pm = \hbar\omega_\ell^\pm |\bar{a}_{\rm in}^\pm|^2$ is the injected power. $\Delta_0 = \omega_{\rm c} - \omega_\ell^\pm$ denotes the detuning of the drive field carrier frequency, $\omega_\ell^\pm$, from the centre frequency of the optical mode doublet, ω_c.

For notational simplicity, in the following, we express the equation of motion of the mechanical oscillator (Eq. 6.1.1b) in terms of the normalized position, $\hat{z} := \hat{x}/x_{\rm zp}$, i.e.,

$$\ddot{\hat{z}} + \Gamma_{\rm m}\dot{\hat{z}} + \Omega_{\rm m}^2 \hat{z} = \delta\hat{f}_{\rm th} + \hat{f}_{\rm BA}. \tag{6.1.3}$$

Here, the forces are normalized such they have dimensions of (time)$^{-2}$; the actual forces (as used in Sect. 6.2.1), in units of Newtons, are given by $\hat{F}_i = m x_{\mathrm{zp}} \hat{f}_i$. Using this convention, the thermal Langevin force is given by

$$\delta \hat{f}_{\mathrm{th}} = \Omega_{\mathrm{m}} \Gamma_{\mathrm{m}} \sqrt{2(2n_{\mathrm{m,th}} + 1)}\, \delta \hat{\xi}_{\mathrm{th}}, \tag{6.1.4}$$

where $\delta \hat{\xi}_{\mathrm{th}}$ is a unit variance white noise process modelling the thermal fluctuations. The radiation pressure force, in Eq. (6.1.2), similarly normalised, becomes,

$$\hat{f}_{\mathrm{BA}} = \Omega_{\mathrm{m}} g_0 (\hat{a}_+^\dagger \hat{a}_+ + \hat{a}_-^\dagger \hat{a}_-). \tag{6.1.5}$$

In the following treatment, both optical modes are driven by optical fields entering the clockwise port of the optical fiber. The field driving mode doublet $\hat{a}_\pm$ is identified as the meter field. The counter-clockwise port of the optical fiber is used to monitor the transmitted sensor field, but is otherwise left open.

Steady State

When the cavity is excited by the meter field, the static component of the ensuing radiation pressure force displaces the oscillator to a new steady-state position, $\bar{z}$, and leads to a renormalisation of the laser-cavity detuning to $\Delta = \Delta_0 + g_0 \bar{z}$. In practice the frequency of the sensor field is stabilised so that $\Delta = 0$. In this case the steady state intracavity field amplitude ($\bar{a}$) and oscillator position are given by

$$\begin{aligned} &\bar{a}_+ = \sqrt{n_+}, \quad \bar{a}_- = i\sqrt{n_-} \quad \text{and} \quad \bar{z} = \frac{g_0}{\Omega_{\mathrm{m}}}(n_+ + n_-), \\ &\text{where,} \quad n_+ = \frac{4\eta_{\mathrm{c}}}{\kappa} \frac{P_{\mathrm{in}}^+/\hbar\omega_{\mathrm{c}}}{(1+\gamma^2/\kappa^2)^2} \quad \text{and} \quad n_- = \left(\frac{\gamma}{\kappa}\right)^2 n_+. \end{aligned} \tag{6.1.6}$$

denote the mean intracavity photon number of the clockwise and counter-clockwise modes, respectively. Note that henceforth, we shall denote, $n_{\mathrm{c}} = n_+$, i.e. the intracavity photon number established in the direction of the injected power is that of the clockwise mode.

Splitting of the cavity resonance can be observed spectroscopically in the normalised steady state transmission. Using the input-output relation $\bar{a}_{\mathrm{out}}^+ = \bar{a}_{\mathrm{in}}^+ - \sqrt{\eta_{\mathrm{c}}\kappa}\, \bar{a}_+$ gives the steady-state cavity transmission,

$$T_c(\Delta) := \frac{P_{\mathrm{out}}^+}{P_{\mathrm{in}}^+} = \left|\frac{\bar{a}_{\mathrm{out}}^+}{\bar{a}_{\mathrm{in}}^+}\right|^2 = 1 - \eta_{\mathrm{c}}\kappa^2 \frac{\left(\Delta^2 + (\gamma/2)^2 + (\kappa/2)^2\right) - \eta_{\mathrm{c}}\left(\Delta^2 + (\kappa/2)^2\right)}{\left(\Delta^2 - (\kappa/2)^2 - (\gamma/2)^2\right)^2}, \tag{6.1.7}$$

consisting of a Lorentzian-like dip, but with the peaks split by γ.

The experimentally observed steady state transmission, shown in Fig. 6.1, is well described by the above expression. Figure 6.1a depicts the steady-state cavity transmission, $T_c(\Delta)$, and shows the effect of cavity mode splitting when under-coupled. Figure 6.1b plots the resonant transmission, $T_c(0)$, as the cavity coupling efficiency is varied (by varying the physical taper-cavity coupling point). By proper choice of the taper-cavity coupling point, it is possible to achieve over-coupled operation $\eta_c \approx 0.9$ necessary for high-quantum-efficiency measurement of the mechanical motion.

Fluctuations

Fluctuations of the cavity field, $\delta\hat{a} = \hat{a} - \bar{a}$, and the mechanical position, $\delta\hat{z} = \hat{z} - \bar{z}$, are coupled according to Eq. (6.1.1). To first order:

$$\delta\dot{\hat{a}}_{\pm} = \left(i\Delta - \frac{\kappa}{2}\right)\delta\hat{a}_{\pm} + \frac{i\gamma}{2}\delta\hat{a}_{\mp} + ig_0\bar{a}_{\pm}\,\delta\hat{z} + \sqrt{\eta_c\kappa}\,\delta\hat{a}_{\text{in}}^{\pm} + \sqrt{(1-\eta_c)\kappa}\,\delta\hat{a}_0^{\pm} \tag{6.1.8a}$$

$$\delta\ddot{\hat{z}} + \Gamma_{\text{m}}\delta\dot{\hat{z}} + \Omega_{\text{m}}^2\delta\hat{z} = \delta\hat{f}_{\text{th}} + g_0\Omega_{\text{m}}\sum_{j=\pm}(\bar{a}_j\delta\hat{a}_j^{\dagger} + \bar{a}_j^{*}\delta\hat{a}_j). \tag{6.1.8b}$$

The ensuing radiation pressure force fluctuations

$$\delta\hat{f}_{\text{BA}} = g_0\Omega_{\text{m}}\sum_{j=\pm}(\bar{a}_j\delta a_j^{\dagger} + \bar{a}_j^{*}\delta a_j) \tag{6.1.9}$$

contain both a dynamic and stochastic component, as detailed in Sects. 6.1.1.1 and 6.1.1.2, respectively.

Taking the Fourier transforms of Eq. (6.1.8) recasts the optomechanical interaction in terms of optical(mechanical) susceptibilities, $\chi_{a(z)}$:

$$\begin{aligned}\chi_a^{(\gamma)}[\Omega]^{-1}\,\delta\hat{a}_{\pm}[\Omega] = & \,ig_0\left(\bar{a}_{\pm} + \frac{i\gamma}{2}\bar{a}_{\mp}\,\chi_a^{(0)}[\Omega]\right)\delta\hat{z}[\Omega] \\ & + \sqrt{(1-\eta_c)\kappa}\left(\delta\hat{a}_{\text{vac}}^{\pm}[\Omega] + \frac{i\gamma}{2}\chi_a^{(0)}[\Omega]\,\delta\hat{a}_{\text{vac}}^{\mp}[\Omega]\right) \\ & + \sqrt{\eta_c\kappa}\left(\delta\hat{a}_{\text{in}}^{\pm}[\Omega] + \frac{i\gamma}{2}\chi_a^{(0)}[\Omega]\,\delta\hat{a}_{\text{in}}^{\mp}[\Omega]\right)\end{aligned} \tag{6.1.10a}$$

$$\left(\chi_z[\Omega]^{-1} + \chi_{\text{BA}}[\Omega]^{-1}\right)\delta\hat{z} = \delta\hat{f}_{\text{th}} + \delta\hat{f}_{\text{BA}}. \tag{6.1.10b}$$

Here χ_{BA} is the modification to the intrinsic mechanical susceptibility due to dynamic back-action, and $f_{\text{BA,th}}$ represents the stochastic measurement back-action force. Before elaborating, we emphasise the following simplifications in the experimentally relevant 'bad-cavity' limit, $\kappa \gg \Omega_{\text{m}}$, assuming a resonantly driven cavity ($\Delta = 0$):

$$\begin{aligned}\chi_a^{(0)}[\Omega]^{-1} &:= -i(\Omega + \Delta) + \frac{\kappa}{2} \approx \frac{\kappa}{2} \\ \chi_a^{(\gamma)}[\Omega]^{-1} &:= \frac{\chi_a^{(0)}[\Omega]^{-1}}{\chi_a^{(0)}[\Omega]^{-2} + (\gamma/2)^2} \approx \frac{\kappa}{2}\left(1 + \frac{\gamma^2}{\kappa^2}\right) \\ \chi_z[\Omega]^{-1} &:= \Omega_{\text{m}}^2 - \Omega^2 - i\Omega\Gamma_{\text{m}} \\ \chi_{\text{BA}}[\Omega]^{-1} &:= \Omega_{\text{BA}}^2(\Omega) - i\Omega\Gamma_{\text{BA}}(\Omega) \approx 0.\end{aligned} \tag{6.1.11}$$

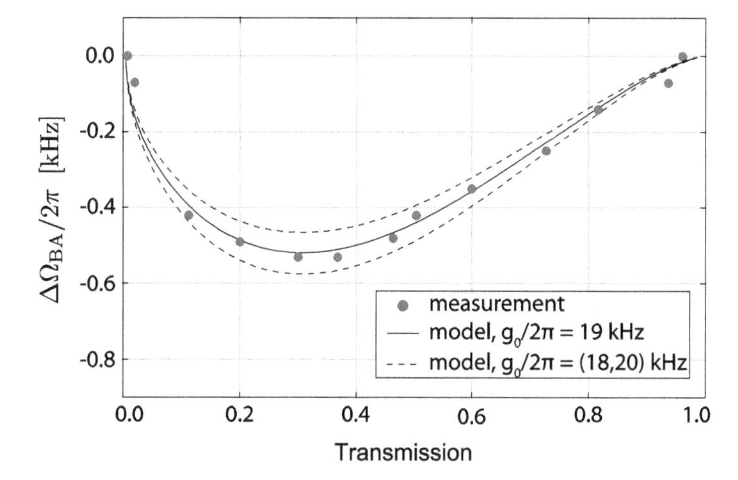

Fig. 6.3 Mechanical frequency shift due to dynamic back-action. Plot shows mechanical frequency shift due to dynamic back-action at various laser-cavity detunings. Model curves are derived from Eqs. (6.1.7) and (6.1.12a)

6.1.1.1 Effect of Dynamic Back-Action

When the cavity is driven away from resonance ($\Delta \neq 0$), classical correlations between the radiation pressure back-action force and the mechanical position give rise to dynamic radiation pressure back-action [15, 16] (see Sect. 4.2.2). In the high-Q ($\Omega_\mathrm{m} \gg \Gamma_\mathrm{m}$), bad-cavity ($\kappa \gg \Omega_\mathrm{m}$) limit relevant to our experiment, dynamic back-action manifests as a displaced mechanical frequency and passive cold-damping [16]. Accounting for cavity mode splitting, the optically-induced frequency shift ($\Delta\Omega_\mathrm{BA}$) and damping rate (Γ_BA) are given by:

$$\Delta\Omega_\mathrm{BA} := \Omega_\mathrm{BA}(\Omega_\mathrm{m}) - \Omega_\mathrm{m} \approx \frac{2g_0^2}{\kappa}\frac{4\eta_\mathrm{c}P_\mathrm{in}^+}{\kappa\hbar\omega_\mathrm{c}}\sum_{j=\pm}\frac{(\kappa/2)^3(\Delta + j\gamma/2)}{\left[(\Delta + j\gamma/2)^2 + (\kappa/2)^2\right]^2} \tag{6.1.12a}$$

$$\Gamma_\mathrm{BA}(\Omega_\mathrm{m}) \approx \frac{\Omega_\mathrm{m}}{4\kappa}\cdot\frac{2g_0^2}{\kappa}\frac{4\eta_\mathrm{c}P_\mathrm{in}^+}{\kappa\hbar\omega_\mathrm{c}}\sum_{j=\pm}\frac{\kappa^5(\Delta - j\gamma/2)}{\left[(\Delta + j\gamma/2)^2 + (\kappa/2)^2\right]^3}. \tag{6.1.12b}$$

Note that both terms vanish for resonant probing.

Figure 6.3 depicts the measured mechanical frequency shift with an input power, $P_\mathrm{in}^+ \approx 1\,\mu\mathrm{W}$, at various detunings; data is plotted against the fraction of the transmitted power, $T_c(\Delta)$ in Eq. (6.1.7). In order to make these measurements, the sample is operated with buffer gas evacuated from the cryostat, so as to eliminate deleterious effects from gas damping (see Eq. 5.3.3). The measured dynamic back-action effect provides an independent check of the vacuum optomechanical coupling rate g_0: Fig. 6.3 shows the observed data with model curves derived from Eqs. (6.1.7) and (6.1.12a) for the case of a split cavity resonance. The value of the coupling rate obtained thus, $g_0 = 2\pi \cdot 19\,\text{kHz}$, is consistent with a direct calibration (giving $g_0 = 2\pi \cdot 21\,\text{kHz}$). In fact, the $\approx$10% discrepancy provides error estimates for the single-photon cooperativity C_0, and the ideal measurement imprecision, n_{imp}.

6.1.1.2 Measurement Back-Action

When the cavity is driven on resonance ($\Delta = 0$), the quantum component of the radiation pressure back-action force takes the form

$$\delta \hat{f}_{\text{BA}} = \frac{8 g_0 \Omega_{\text{m}}}{\sqrt{\kappa}\left(1+\gamma^2/\kappa^2\right)} \Big\{ \Big(\sqrt{n_+} + \frac{\gamma}{\kappa}\sqrt{n_-}\Big)\sqrt{\eta_{\text{c}}}\delta\hat{q}^+_{\text{in}} + \Big(\sqrt{n_+} + \frac{\gamma}{\kappa}\sqrt{n_-}\Big)\sqrt{1-\eta_{\text{c}}}\delta\hat{q}^+_{\text{vac}}$$
$$- \Big(\frac{\gamma}{\kappa}\sqrt{n_+} - \sqrt{n_-}\Big)\sqrt{\eta_{\text{c}}}\delta\hat{p}^-_{\text{in}} - \Big(\frac{\gamma}{\kappa}\sqrt{n_+} - \sqrt{n_-}\Big)\sqrt{1-\eta_{\text{c}}}\delta\hat{p}^-_{\text{vac}} \Big\}, \tag{6.1.13}$$

where $\hat{q}\,(\hat{p})$ denote the amplitude(phase) quadrature of each field. In Eq. (6.1.13), we have retained the explicit dependence on $n_\pm$ in order to emphasise their role in weighting the various noise components. We note that as a consequence of the scattering process, (amplitude)phase fluctuations entering the (clockwise)counter-clockwise mode are converted to force fluctuations by two pathways.

Assuming that the drive field is shot-noise limited in its amplitude quadrature ($\bar{S}^{\text{in}}_{qq} = \frac{1}{2}$) and that the cavity is otherwise interacting with a zero temperature bath ($\bar{S}^{\text{vac}}_{qq} = \frac{1}{2} = \bar{S}^{\text{vac}}_{pp}$), we find that the effective thermal occupation due to measurement back-action is given by

$$n_{\text{m,BA}} = C_0 \frac{1}{1+\gamma^2/\kappa^2}(n_+ + n_-) = C_0 n_+, \tag{6.1.14}$$

which is exactly the same as the expression for the case where the cavity modes are unsplit.

6.1.1.3 Modification to Measurement Imprecision

The cavity transmission, $\delta\hat{a}^+_{\text{out}} = \delta\hat{a}^+_{\text{in}} - \sqrt{\eta_{\text{c}}\kappa}\,\delta\hat{a}_+$, at $\Delta = 0$ is given by,

$$\delta\hat{a}^+_{\text{out}} = -i\sqrt{\eta_{\text{c}}}\frac{2 g_0\sqrt{n_+}}{\sqrt{\kappa}}\left(\frac{1-\gamma^2/\kappa^2}{1+\gamma^2/\kappa^2}\right)\delta\hat{z}$$
$$+ \left(1 - \frac{2\eta_{\text{c}}}{1+\gamma^2/\kappa^2}\right)\delta\hat{a}^+_{\text{in}} - \frac{\gamma}{\kappa}\left(\frac{2\eta_{\text{c}}}{1+\gamma^2/\kappa^2}\right)\delta\hat{a}^-_{\text{in}} \tag{6.1.15}$$
$$- \frac{2\sqrt{\eta_{\text{c}}(1-\eta_{\text{c}})}}{1+\gamma^2/\kappa^2}\left(\delta\hat{a}^+_{\text{vac}} + i\frac{\gamma}{\kappa}\delta\hat{a}^-_{\text{vac}}\right).$$

The transmitted field is amplified in a balanced homodyne receiver with a coherent local oscillator (LO) $\hat{a}_{\mathrm{LO}}$. Following the discussion of homodyne detection in Sect. 3.2.3.2, the operator corresponding to the homodyne detector photocurrent is,

$$\delta \hat{I}_{\mathrm{hom}} \propto |\bar{a}_{\mathrm{LO}}| \left(\delta \hat{q}_{\mathrm{out}}^{+} \cos \theta_{\mathrm{hom}} + \delta \hat{p}_{\mathrm{out}}^{+} \sin \theta_{\mathrm{hom}} \right),$$

where $|\bar{a}|_{\mathrm{LO}}$ is the amplitude of the large coherent LO field, and θ_{hom} the relative mean phase between the LO and the cavity transmission. The path length of the LO arm is electronically locked to maintain $\theta_{\mathrm{hom}} \approx \pi/2$, so that the homodyne signal picks out the phase quadrature of the cavity transmission containing the position fluctuations $\delta\hat{z}$. The resulting shot-noise-normalised spectrum of photocurrent fluctuations is given by,

$$\bar{S}_{I}^{\mathrm{hom}}[\Omega] \propto 1 + \eta \frac{16 g_0^2 n_+}{\kappa} \left(\frac{1 - \gamma^2/\kappa^2}{1 + \gamma^2/\kappa^2} \right)^2 \bar{S}_z[\Omega],$$

where η is the total detection efficiency. Thus, the imprecision in the estimation of $\delta\hat{z}$ from the homodyne photocurrent is,

$$\bar{S}_{z}^{\mathrm{imp}}[\Omega] = \frac{\kappa}{16 \eta g_0^2 n_+} \left(\frac{1 + \gamma^2/\kappa^2}{1 - \gamma^2/\kappa^2} \right)^2.$$

Expressed as an equivalent phonon occupation,

$$n_{\mathrm{imp}} = \left(\frac{1}{16 \eta C_0 n_+} \right) \left(\frac{1 + \gamma^2/\kappa^2}{1 - \gamma^2/\kappa^2} \right)^2. \tag{6.1.16}$$

Note that mode splitting causes the optical susceptibility (see Eq. (6.1.8)) to flatten near resonance, leading to divergence of n_{imp} when $\gamma = \kappa$.

Thus the effect of mode-splitting is an additional efficiency penalty. Indeed, for a given injected power, the quantum back-action arises due to photons populating both the cavity modes, while transmission measurements only collect photons from the clockwise mode.

6.1.2 *Measurement Imprecision*

In all position sensors, extraneous thermal fluctuations pose a fundamental limit to the achievable imprecision. In cavity-optomechanical sensors, the main sources of extraneous imprecision arise from thermomechanical [17, 18] and thermodynamic fluctuations of the cavity substrate [19–23]. These result in excess cavity frequency noise, $\bar{S}_{\omega}^{\mathrm{imp,ex}}$. Figure 6.4 shows the extraneous noise floor of our sensor over a broad range of frequencies surrounding the oscillator resonance. We obtained this spectrum by subtracting shot noise from a measurement made with a large intracavity photon

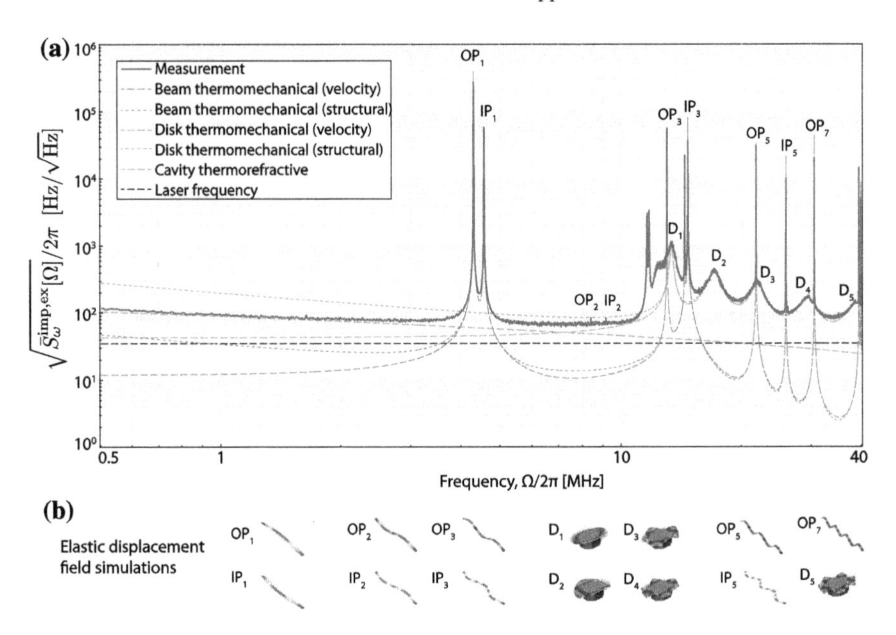

Fig. 6.4 Extraneous measurement imprecision. a Red trace shows the shot-noise subtracted homodyne photocurrent calibrated as an apparent frequency noise. The various high-Q peaks are understood to arise from the in-plane (marked IP_n) and out-of-plane (marked OP_n) modes of the nanobeam (blue), while the low-Q peaks are from the mechanical modes of the disk (marked D_n) (green). In the vicinity of the fundamental out-of-plane mode, at $\Omega_\mathrm{m} = 2\pi \cdot 4.3$ MHz, measurement imprecision is dominated by a combination of cavity thermorefractive noise (black), and a small contribution from estimated laser frequency noise (orange). **b** Finite-element model simulations of the various mechanical modes seen in the measurement; red shows maximum displacement, and blue shows no displacement, on an arbitrary scale. The simulated displacement field frequencies have excellent agreement with the observed frequencies and dispersion

number, $n_\mathrm{c} > 10^5$. (To mitigate thermo-optic and optomechanical instabilities, the measurement was in this case conducted using $\approx$10 mbar of gas pressure at an elevated temperature of 15.7 K.) The relevant noise peak, at $\Omega_\mathrm{m} \approx 2\pi \cdot 4.3$ MHz, due to the thermal motion of the fundamental out-of-plane mode, is measured against an extraneous imprecision, $\bar{S}_\omega^{\mathrm{imp,ex}}$, which we understand as arising from three sources:

$$\bar{S}_\omega^{\mathrm{imp,ex}}[\Omega_\mathrm{m}] = \bar{S}_\omega^{\mathrm{imp,ex,mech}}[\Omega_\mathrm{m}] + \bar{S}_\omega^{\mathrm{imp,ex,cav}}[\Omega_\mathrm{m}] + \bar{S}_\omega^{\mathrm{imp,ex,laser}}[\Omega_\mathrm{m}];$$

noise due to extraneous thermomechanical motion of other modes of the beam and the cavity, extraneous frequency fluctuations due to the cavity substrate, and extraneous frequency noise from the probe laser.

High- and low-Q noise peaks correspond to thermal motion of the nanobeam and the microdisk, respectively (see Sect. 6.1.2.1 below). In the vicinity of the fundamental noise peak, we observe an extraneous frequency noise background of $\bar{S}_\omega^{\mathrm{imp,ex}} \approx (2\pi \cdot 30\mathrm{Hz}/\sqrt{\mathrm{Hz}})^2$, corresponding to an extraneous position imprecision of $\bar{S}_x^{\mathrm{imp,ex}} \approx (4.3 \cdot 10^{-17}\mathrm{m}/\sqrt{\mathrm{Hz}})^2$. We identify this noise as a combination of of microdisk thermorefractive noise [24] (see Sect. 6.1.2.2), diode laser frequency noise [25], and off-resonant thermal motion of the neighbouring beam mode at 4.6 MHz. Owing to the large zero-point motion of the nanobeam, $\bar{S}_\omega^{\mathrm{zp}} = 4g_0^2/\Gamma_\mathrm{m} = (2\pi \cdot 6.7\ \mathrm{kHz}/\sqrt{\mathrm{Hz}})^2$ ($\bar{S}_x^{\mathrm{zp}} = (0.95 \cdot 10^{-14}\mathrm{m}/\sqrt{\mathrm{Hz}})^2$), the equivalent bath occupancy of this noise has an exceptionally low value of $n_{\mathrm{imp}}^{\mathrm{ex}} := \bar{S}_\omega^{\mathrm{imp,ex}}/2\bar{S}_\omega^{\mathrm{zp}} \approx 1.0 \cdot 10^{-5}$, nearly 44 dB below the value at the SQL.

In the following, we go through the various sources of extraneous classical noise that constitutes the budget shown in Fig. 6.4. Briefly, these fall into three categories: noise due to thermal motion of extraneous mechanical modes in the optomechanical system, noise due to thermodynamic fluctuations in the cavity, and noise on the laser as it is conveyed to the system.

6.1.2.1 Imprecision Due to Thermomechanical Noise

High- and low-Q noise peaks in Fig. 6.4 correspond to the thermal motion of the extraneous modes of the nanobeam and the microdisk (radial breathing, and flexural, modes that have significant optomechanical coupling), respectively. Assuming that all these modes, with position fluctuation $\hat{x}_i$, are equilibrated at temperature T, and have an optomechanical coupling G_i, their contribution to the frequency noise background under the fundamental mode is,

$$\bar{S}_\omega^{\mathrm{imp,ex,mech}}[\Omega_\mathrm{m}] = \sum_i G_i^2 \bar{S}_{x_i}[\Omega_\mathrm{m}] = \sum_i G_i^2 \left|\chi_{x_i}[\Omega_\mathrm{m}]\right|^2 \cdot 4m_i\Gamma_i k_B T. \quad (6.1.17)$$

Here, we have approximated the extraneous modes as independent oscillators, each of effective mass m_i and damping rate Γ_i, driven by a thermal force noise given by the fluctuation-dissipation theorem (see Eq. 2.2.10) in the limit that their mean thermal phonon occupation, $n_{i,\mathrm{th}} := \frac{k_B T}{\hbar\Omega_i} \gg 1$.

Structural resonances like these, typical of bulk resonators, are known to exhibit damping that is not proportional to velocity [26]. The so-called *structural damping* model posits a frequency dependent damping rate such that the different modes have approximately uniform mechanical Q; i.e.,

$$\Gamma_i[\Omega] = \begin{cases} \Gamma_i & \text{velocity damping} \\ \frac{\Gamma_i \Omega_i}{\Omega} & \text{structural damping} \end{cases}$$

leading to the replacement, $\Gamma_i \to \Gamma_i[\Omega]$ in the susceptibility and the force noise in Eq. (6.1.17). Assuming that the fundamental mode frequency is smaller than the frequencies of the extraneous modes,

$$\bar{S}_\omega^{\mathrm{imp,ex,mech}}[\Omega_\mathrm{m} \ll \Omega_i] \approx \sum_i \frac{g_i^2}{\Omega_i} \times \begin{cases} n_{i,\mathrm{th}} \frac{\Gamma_i}{\Omega_i} & \text{velocity damping} \\ \frac{n_{i,\mathrm{th}}}{Q_i} \frac{\Omega_i}{\Omega_\mathrm{m}} & \text{structural damping} \end{cases}$$

implying that if the extraneous modes are structurally damped, a very low frequency for the fundamental mode would be susceptible to larger imprecision due to extraneous thermomechanical noise.

In Fig. 6.4, the pair of blue (green) traces show models of thermomechanical noise arising from extraneous modes of the nanobeam (disk). The dashed (dotted) curves assume a velocity (structural) damped model for the motion. Despite the low-Q ($\approx$10) of the disk modes, it is seen that the data is incompatible with a structural damping mechanism for these modes. For the beam modes, measurements at the frequencies presented in the figure, do not allow discrimination between either model.

6.1.2.2 Imprecision Due to Cavity Substrate Noise

Macroscopic optical cavities, like the whispering-gallery cavities we use, equilibrated at some temperature T, experience fundamental thermodynamic fluctuations in its resonance frequency ω_c. Within the electrodynamic description of cavity frequency fluctuations (in Chap. 4), the two possible causes are fluctuations in volume V and fluctuations in the dielectric constants of the cavity substrate. For an optical cavity, the latter is equivalent to fluctuations in the refractive index ν. The cavity frequency, $\omega_c(\nu, V)$, therefore undergoes fluctuations,

$$\delta\omega_c = \frac{\partial\omega_c}{\partial\nu}\delta\nu + \frac{\partial\omega_c}{\partial V}\delta V =: \delta\omega_c^{\text{TRN}} + \delta\omega_c^{\text{TEN}}.$$

When refractive index fluctuations and volume fluctuations are caused by underlying thermodynamic causes, these two contributions lead to thermorefractive (TRN) [21] and thermoelastic (TEN) [20] frequency noise.

Note that the underlying thermodynamic fluctuations are transduced via the coefficients $\frac{\partial\omega_c}{\partial\nu}, \frac{\partial\omega_c}{\partial V}$ measured in *equilibrium*, i.e. at constant temperature T. Although temperature itself does not fluctuate in equilibrium, an apparent temperature fluctuation may be ascribed to the fluctuations in the total energy; the variance of this apparent temperature fluctuation is,[1]

[1]This is derived as follows. Assume the body is in thermal equilibrium at temperature T, so that it is described by the canonical thermal state $\hat{\rho} = e^{-\beta\hat{H}}/Z$, with $Z := \text{Tr}\, e^{-\beta\hat{H}}$, and $\beta := (k_B T)^{-1}$. Then, the average energy is given by, $\langle\hat{H}\rangle := \text{Tr}\,\hat{H}\hat{\rho} = -\frac{1}{Z}\partial_\beta Z$, while its second moment is, $\langle\hat{H}\rangle^2 := \text{Tr}\,\hat{H}^2\hat{\rho} = \frac{1}{Z}\partial_\beta^2 Z = -\partial_\beta\langle\hat{H}\rangle + \langle\hat{H}\rangle^2$. Subtracting these two expressions give the variance in the energy:

$$\text{Var}\left[\hat{H}\right] := \langle\hat{H}^2\rangle - \langle\hat{H}\rangle^2 = -\partial_\beta\langle\hat{H}\rangle = k_B T^2\,\partial_T\langle\hat{H}\rangle = k_B T^2 C_V.$$

Here, $C_V := \partial_T\langle\hat{H}\rangle$ is the specific heat at constant volume as defined conventionally. To refer the above variance in energy to an apparent variance in temperature, we again use the definition of the specific heat, $\delta T = \delta E/C_V$, to arrive at, $\text{Var}\,[T] = \text{Var}\,[E]\,/C_V^2 = k_B T^2/C_V$.

$$\mathrm{Var}\,[T] = \frac{k_B T^2}{C_V} = \frac{k_B T^2}{\rho V c_V}, \tag{6.1.18}$$

where the first equality is expressed in terms of the heat capacity at constant volume C_V, while the second is expressed in terms of the specific heat at constant volume, $c_V = C_V/(\rho V)$, an intensive material property. The implied variance in frequency due to TRN and TEN is therefore,

$$\begin{aligned}\mathrm{Var}\left[\delta\omega_c^{\mathrm{TRN}}\right] &= \left(\frac{\partial\omega_c}{\partial\nu}\frac{\partial\nu}{\partial T}\right)^2 \mathrm{Var}\,[T] = \left(\frac{\omega_c}{\nu}\frac{\partial\nu}{\partial T}\right)^2 \frac{k_B T^2}{\rho V c_V} \\ \mathrm{Var}\left[\delta\omega_c^{\mathrm{TEN}}\right] &= \left(\frac{\partial\omega_c}{\partial V}\frac{\partial V}{\partial T}\right)^2 \mathrm{Var}\,[T] = (\omega_c\alpha_V)^2 \frac{k_B T^2}{\rho V c_V},\end{aligned} \tag{6.1.19}$$

where $\alpha_V := (1/V)\partial V/\partial T$ is the isobaric thermal expansion coefficient. For macroscopic ultra-stable cavities, TRN has been observed to be a limitation on frequency stability at room temperature [23], and demonstrated to be suppressed at cryogenic temperatures [27]. For small mode-volume microcavities, TRN poses a much larger problem, and has been observed to limit frequency imprecision at the level of $10^3\,\mathrm{Hz}^2/\mathrm{Hz}$ at Fourier frequencies of about 1 MHz, at room temperature [24].

For $\mathrm{SiO_2}$, the material constituting our cavity, the coefficient of transduction for TRN is roughly 100 times larger than the coefficient of transduction for TEN in a wide range of temperatures down to about $T = 1\,\mathrm{K}$ [28, 29]; we therefore focus on TRN here. In order to understand the distribution of the variance, $\mathrm{Var}\left[\delta\omega_c^{\mathrm{TRN}}\right]$, given in Eq. (6.1.19), in frequency, it is necessary to subscribe to a dynamic model of temperature in the cavity [28]. Assuming diffusive thermal transport, it can be shown that for low-order optical modes, at Fourier frequencies high compared to the inverse thermal diffusion time, $\tau_T^{-1} := D_T/(2\pi r_{\mathrm{disk}}^2) \approx (10-50)\,\mathrm{kHz}$ (here D_T is the thermal diffusivity of silica, and r_{disk} the radius of the microdisk cavity), the power spectral density of TRN is approximated by [28],

$$\bar{S}_{\omega}^{\mathrm{TRN}}[\Omega] \approx \left(\frac{\omega_c}{\nu}\frac{\partial\nu}{\partial T}\right)^2 \cdot \frac{k_B T^2}{\rho V c_V} \cdot \frac{(16\pi)^{1/3}\tau_T}{(\Omega\tau_T)^{1/2}(1+(\Omega\tau_T)^{3/4})^2}. \tag{6.1.20}$$

The thermal diffusivity, $D_T = K_T/\rho c_V$, by being strongly temperature dependent through parameters like the thermal conductivity K_T and specific heat c_V [30], is only known to within 50% uncertainty for silica at cryogenic temperatures, $T \lesssim 10\,\mathrm{K}$, and depends weakly on the presence of impurities [31]. Using a thermal time constant, $\tau_T^{-1} \approx 5\,\mathrm{MHz}$, consistent with the known material constants, and estimates of the whispering-gallery mode cross-section, Eq. (6.1.20) gives a qualitatively correct scaling of the measured low frequency imprecision noise in Fig. 6.4, deviating by about 50% in absolute magnitude.

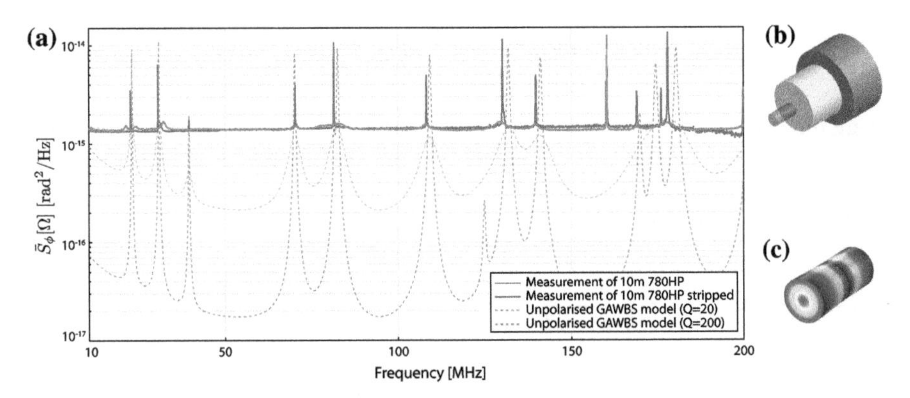

Fig. 6.5 Extraneous phase noise from optical fiber. a Blue shows measured phase noise spectrum of a 10 m long segment of a standard 780 HP optical fiber. The observed low-Q peaks are well described by a theoretical model (blue dashed, $Q = 20$) of unpolarised GAWBS. Red shows the same segment of fiber, but measured after its cladding is etched using buffered hydro-flouric acid. Removal of the surrounding cladding increases the Q of the modes, leading to localisation of GAWBS-induced phase noise in narrow spectral intervals. Red dashed shows a theoretical model assuming $Q = 200$. **b** Schematic of a typical optical fiber (not to scale), showing the core (blue), which carries the optical field, the cladding surrounding the core (yellow) that provides the refractive index contrast to constrain the propagating optical field transversally, and the coating (brown) that provides for mechanical rigidity. **c** Theoretical elastic displacement field of the core of a 4.5 μm radius (typical of 780 HP) glass core of the fiber. Red shows compression and blue shows elongation, the scale is arbitrary.

6.1.2.3 Imprecision Due to Noise in Optical Fiber

In addition to the above two sources of frequency noise that arise from the optomechanical system, our experiment is also sensitive to frequency noise arising from the optical path in the arm of the interferometer containing the optomechanical system. As described in Sect. 5.3, this path, passing through the cryostat, predominantly consists of a ≈10 m long (single-mode) optical fiber (780 HP). Roughly half this length passes through the cryostat (at 4 K), while the other half is at room temperature. Transverse elastic modes of the fiber core, undergoing thermal motion, can inelastically scatter photons off of the longitudinally propagating optical field via the strain-optical effect; this process—*guided acoustic-wave Brillouin scattering* (GAWBS)—is known to cause excess frequency noise in the field exiting the fiber[2] [32, 33].

GAWBS-induced excess phase noise is measured by inserting a 10 m long optical fiber in the signal arm of our homodyne interferometer. Figure 6.5a blue trace is the result of such a measurement, showing excess frequency noise $\bar{S}_\omega^{\mathrm{GAWBS}}[\Omega] = \Omega^2 \bar{S}_\phi^{\mathrm{GAWBS}}[\Omega] \approx (2\pi \cdot 1\,\mathrm{Hz}/\sqrt{\mathrm{Hz}})^2$ at Fourier frequencies of $\Omega \approx 2\pi \cdot 20\,\mathrm{MHz}$, and increasing quadratically with Fourier frequency.

[2] Longitudinal elastic modes have a similar effect, but their frequency being larger, the resulting phase noise isn't relevant in our experiment.

The observed spectrum (blue in Fig. 6.5) can be understood using a simple model [32]. The transverse axis-symmetric elastic field of a cylinder of radius r_c—general solution of the Navier equation (Eq. 3.1.18) in cylindrical coordinates with free boundary conditions—is given by Eq. (3.1.21), viz.

$$\mathbf{u}(\mathbf{r}, t) = \sum_n x_n(t)\mathbf{u}_n(\mathbf{r}),$$

where $\mathbf{u}_n$ are the (orthogonal) spatial mode functions of the elastic cylinder, given by (here J_k is the Bessel function of order k and $\mathbf{e}_r$ is the unit radial vector),

$$\mathbf{u}_n(\mathbf{r}) = J_1\left(\alpha_n \frac{r}{r_c}\right)\mathbf{e}_r,$$

and $x_n(t)$ are standard harmonic oscillator amplitudes driven by thermal noise. The frequency of the nth mode, $\Omega_n = \alpha_n(c_T/r_c)$, is determined by the transverse elastic velocity c_T (see Appendix B.2), and α_n, fixed by the boundary conditions, is the nth root of the characteristic equation (here c_L is the longitudinal velocity, see Appendix B.2),

$$\frac{J_0(\alpha)}{J_2(\alpha)} = \frac{(c_T/c_L)^2}{1-(c_T/c_L)^2},$$

and describes the dispersion of the elastic modes. The frequencies observed in the measured data in Fig. 6.5a, agree well with the frequencies of these transverse elastic modes, shown as the broad resonances with a phenomenological quality factor $Q \approx 20$ in the blue dashed curve.

The phase shift caused by the elastic mode is determined by a combination of two factors: the amplitude of the thermally driven elastic mode x_n, and, the forward scattering cross-section. The amplitude of the thermally driven elastic motion, x_n, is fixed by the equipartition principle [34]:

$$\frac{1}{2}k_B T = \int_0^L \mathrm{d}z \int_0^{2\pi} \mathrm{d}\theta \int_0^{r_c} r\,\mathrm{d}r\, \frac{1}{2}\rho\Omega_n^2 \langle \mathbf{u}_n^*(\mathbf{r})\cdot\mathbf{u}_n(\mathbf{r})\rangle$$
$$\Rightarrow \quad \mathrm{Var}[x_n] = \frac{k_B T}{m_n \Omega_n^2}, \quad \text{where,} \quad m_n = \rho\cdot\pi r_c^2 L \int_0^{r_c} J_1^2(\alpha_n r/r_c)\, r\,\mathrm{d}r, \tag{6.1.21}$$

is the effective mass of the elastic mode. Such thermal motion of the fiber core leads to refractive index fluctuations, $\delta\nu$, via the strain-optic effect [32, 35], viz.

$$\delta\nu(\mathbf{r}, t) = \frac{\nu^3}{2}(p_{11}+p_{12})\frac{1}{r}\frac{\partial}{\partial r}[r\mathbf{u}(\mathbf{r}, t)]\cdot\mathbf{e}_r,$$

where p_{ij} are the elements of the strain-optic tensor for silica, $p_{11} \approx 0.12$, $p_{12} \approx 0.27$. These fluctuations induce fluctuations in the phase, $\delta\phi$, of the longitudinally propagating electric field; it can be approximated as an average of the transverse field profile, $E(r) \approx (\pi w_c^2)^{-1} e^{-r^2/w_c^2}$, over the refractive index fluctuation profile, viz.,

$$\begin{aligned} \delta\phi_{\text{GAWBS}}(t) &= \frac{2\pi}{\lambda} \int_0^L \mathrm{d}z \int_0^{2\pi} \mathrm{d}\theta \int_0^\infty r\mathrm{d}r\, \delta\nu(\mathbf{r}, t) E(r) \\ &= \frac{\pi \nu^3}{\lambda} (p_{11} + p_{22}) \frac{L}{r_c} \sum_n \alpha_n e^{-\alpha_n^2 w_c^2 / 4 r_c^2} x_n(t) \end{aligned} \tag{6.1.22}$$

The spectrum of phase fluctuations due to GAWBS in the optical fiber is thus,

$$\bar{S}_\phi^{\text{GAWBS}}[\Omega] = \left(\frac{\pi \nu^3}{\lambda} (p_{11} + p_{22}) \frac{L}{r_c} \right)^2 \sum_n \alpha_n^2 e^{-\alpha_n^2 w_c^2 / 2 r_c^2} \bar{S}_x^n[\Omega], \tag{6.1.23}$$

where $\bar{S}_x^n$ is the thermal motion consistent with the equipartition principle given in Eq. (6.1.21), viz.,

$$\bar{S}_x^n[\Omega] \approx \frac{\text{Var}\,[x_n]}{\Gamma_n} \frac{\Gamma_n^2}{(\Omega - \Omega_n)^2 + (\Gamma_n/2)^2}. \tag{6.1.24}$$

Here m_n is the effective mass given in Eq. (6.1.21) and $\Omega_n = \alpha_n (c_T / r_c)$ the elastic resonance frequency. The decay rate of the modes, $\Gamma_n \approx \Omega_n / Q$ is dominated by clamping losses due to the fiber cladding. Model curves in Fig. 6.5a are plots of Eqs. (6.1.23) and (6.1.24).

Frequency noise imprecision, due to GAWBS, around the resonance of the nanobeam mode at $\Omega_{\text{m}} \approx 2\pi \cdot 4.3\,\text{MHz}$ is solely due to the low frequency part of $\bar{S}_\phi^{\text{GAWBS}}$, given by,

$$\bar{S}_\omega^{\text{imp,ex,fiber}}[\Omega] := \Omega_{\text{m}}^2 \bar{S}_\phi^{\text{GAWBS}}[\Omega_{\text{m}} \ll \Omega_n] \approx \left(\frac{\pi \nu^3}{\lambda} (p_{11} + p_{22}) \frac{L}{r_c} \right)^2 \sum_n \frac{\alpha_n^2}{Q} \frac{k_B T}{m_n \Omega_n}. \tag{6.1.25}$$

Clearly, a large mechanical quality factor for the transverse elastic modes of the optical fiber significantly reduces classical extraneous imprecision due to GAWBS. Figure 6.5a red trace shows engineering of the quality factor of the GAWBS modes, and consequent reduction of frequency noise. The higher Q is achieved by etching the 10 m long fiber in buffered HF (40% solution, for an hour), which reduces the cladding diameter from 125 µm to about 90 µm. The resulting increase in Q by a factor of 10, results in $\bar{S}_\omega^{\text{imp,ex,fiber}}[\Omega_{\text{m}}] \lesssim (2\pi \cdot 10^{-3}\,\text{Hz}/\sqrt{\text{Hz}})$.

6.1.3 *Measurement Back-Action*

In principle, given an extraneous imprecision of $n_{\text{imp}}^{\text{ex}} \approx 10^{-5}$, it must be possible to perform a quantum-noise-limited measurement of the oscillator position by injecting sufficient optical power. The performance of our sensor is limited in practice by constraints on the usable optical power, including photon collection efficiency, photothermal and radiation pressure instabilities, and extraneous sources of measurement back-action, such as heating due to optical absorption. We investigate these constraints by recording $n_{\rm imp}$ and $n_{\rm m}$ as a function of intracavity photon number, comparing their product to the uncertainty-limited value, $4\sqrt{n_{\rm imp}n_{\rm m}} > 1$ (Fig. 6.6). Two considerations are crucial to this investigation. First, in order to efficiently collect photons from the cavity, it is necessary to increase the taper-cavity coupling rate to $\kappa_{\rm ex} \gtrsim \kappa_0$, thereby increasing the total cavity decay rate to $\kappa = \kappa_0 + \kappa_{\rm ex}$. We operate at a near-critically coupled ($\kappa_{\rm ex} \approx \kappa_0$) value of $\kappa \approx 2\pi \cdot 0.91$ GHz, thus reducing the single photon cooperativity to $C_0 \approx 0.31$ in exchange for a higher output coupling efficiency of $\eta_c = (\kappa - \kappa_0)/\kappa \approx 0.52$. Second, in order to minimise $\bar{S}_x^{\rm imp}$, it is necessary to maximise intracavity photon number while mitigating associated dynamic instabilities. We accomplish this by actively damping the oscillator using radiation pressure feedback. Feedback was performed by modulating the drive intensity, and therefore the intracavity photon number, of the secondary feedback mode using an electronically amplified and delayed (by $\tau \approx 3\pi/2\Omega_{\rm m}$) copy of the homodyne photocurrent as an error signal [36]. The resulting viscous radiation pressure reduces the phonon occupancy of the mechanical mode to a mean value of $n_{\rm m}' \approx n_{\rm m}\Gamma_{\rm m}/(\Gamma_{\rm m}+\Gamma_{\rm fb})$, where $\Gamma_{\rm fb}$ is the optically-induced damping rate. It should be noted that added damping leads to an apparent imprecision $n_{\rm imp}' = n_{\rm imp}(\Gamma_{\rm m} + \Gamma_{\rm fb})/\Gamma_{\rm m}$ that differs from the intrinsic value ($\Gamma_{\rm fb} = 0$). We here restrict our attention to the latter, noting that the associated cooling preserves the apparent imprecision-back-action product: $n_{\rm m}'n_{\rm imp}' = n_{\rm m}n_{\rm imp}$.

Representative measurements of the oscillator's thermal motion are shown in Fig. 6.6b. We determine $n_{\rm m}$ and $n_{\rm imp}$ by fitting each noise peak to a Lorentzian with a linewidth of $\Gamma_{\rm eff} = \Gamma_{\rm m} + \Gamma_{\rm fb} + \Gamma_{\rm BA}$ (including a minor contribution from dynamic back-action, $\Gamma_{\rm BA}$), a peak amplitude of $\bar{S}_\omega[\Omega_{\rm m}] \approx 2n_{\rm m}(\Gamma_{\rm m}/\Gamma_{\rm eff})^2\bar{S}_\omega^{\rm zp}$, and an offset of $\bar{S}_\omega^{\rm imp} = 2n_{\rm imp}\bar{S}_\omega^{\rm zp}$. For low intracavity photon number, $n_c \ll n_{\rm m,th}/C_0$, we observe that the effective bath occupation is dominated by the cryostat, $n_{\rm m} \approx n_{\rm m,th}$, and that imprecision scales as $n_{\rm imp} = (16\eta C_0 n_c)^{-1}$, where $\eta \approx 0.23$. η represents the ideality of the measurement, and includes both optical losses and reduction in the cavity transfer function due to mode splitting (see Sect. 6.1.1.3). Operating with higher input power—ultimately limited by the onset of parametric instability in higher-order beam modes—the lowest imprecision we have observed is $n_{\rm imp} \approx 2.7(\pm 0.2) \cdot 10^{-5}$, corresponding to an imprecision 39.7 ± 0.3 dB below that at the SQL.

For large measurement strengths, quantum measurement back-action [37, 38] should in principle exceed the ambient thermal force. As shown in Fig. 6.6, our system deviates from this ideal behaviour due to extraneous back-action, manifesting as an apparent excess cooperativity, $C_0^{\rm ex}$, and limiting the fractional contribution of quantum back-action to $C_0/(C_0 + C_0^{\rm ex}) \approx 35\%$. Combining this extraneous back-action with non-ideal measurement transduction/efficiency, we model the apparent imprecision-back-action product of our measurement (green curve in Fig. 6.6a) as

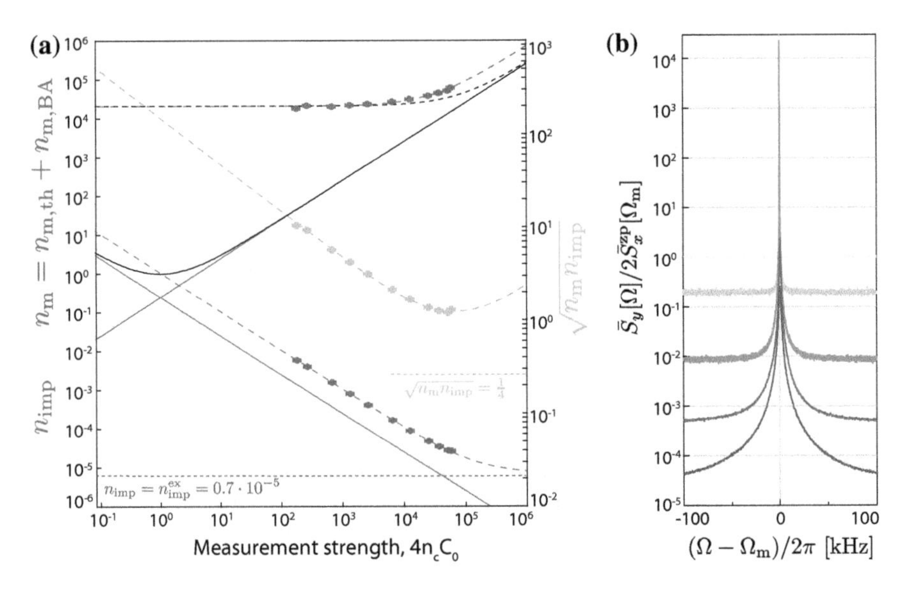

Fig. 6.6 **Measurement imprecision and back-action versus intracavity photon number. a** Red, blue, and green points correspond to measurements of total effective bath occupation, $n_{\mathrm{m}} = n_{\mathrm{m,th}} + n_{\mathrm{m,BA}}$, measurement imprecision referred to an equivalent bath occupation, n_{imp}, and the apparent imprecision-back-action product, $4\sqrt{n_{\mathrm{m}} n_{\mathrm{imp}}}$, respectively. Solid black line depicts the ideal SQL model, for a zero temperature oscillator, consisting of quantum-limited imprecision (solid blue) $n_{\mathrm{imp}} = (16 C_0 n_c)^{-1}$, and quantum back-action (solid red) $n_{\mathrm{m,BA}} = C_0 n_c$. Dashed black represents the SQL curve for the case of finite thermal occupation of the oscillator, described by $n_{\mathrm{m}} = n_{\mathrm{m,th}} + C_0 n_c$. Dashed red and blue lines highlight excursion from their counterparts due to extraneous back-action, $C_0^{\mathrm{ex}} = 0.56$, extraneous imprecision, $n_{\mathrm{imp}}^{\mathrm{ex}} = 0.70 \cdot 10^5$, and imperfect detection efficiency, $\eta = 0.23$, as described in the text. Green line models the apparent force-imprecision product using the Eq. (6.1.26). **b** Spectra of the position fluctuations of the oscillator at various measurement strengths. Yellow line marks the peak spectral density at the SQL

$$4\sqrt{n_{\mathrm{imp}} n_{\mathrm{m}}} = \sqrt{\frac{1}{\eta}\left(1 + \frac{n_{\mathrm{m,th}}}{n_{\mathrm{m,BA}}} + \frac{C_0^{\mathrm{ex}}}{C_0}\right)\left(1 + \frac{n_{\mathrm{c}}}{n_{\mathrm{c}}^{\mathrm{ex}}}\right)}, \tag{6.1.26}$$

where $n_c^{\mathrm{ex}} \equiv (16\eta C_0 n_{\mathrm{imp}}^{\mathrm{ex}})^{-1}$ is the photon number at which extraneous and shot-noise imprecision are equal. Operating at $n_{\mathrm{c}} \approx 5 \cdot 10^4 \ll n_{\mathrm{c}}^{\mathrm{ex}}$, we observe a minimum imprecision-back-action product of $4\sqrt{n_{\mathrm{imp}} n_{\mathrm{m}}} \approx 5.0$, i.e. a factor 5 away from the ideal value predicted by the uncertainty principle.

6.1.3.1 Back-Action Due to Ohmic Heating

The origin of the excess back-action cooperativity, C_0^{ex}, remains not fully understood. However, heating due to laser noise can be fully ruled out. Preliminary measurements of the back-action heating of multiple mechanical modes of the nanobeam are shown

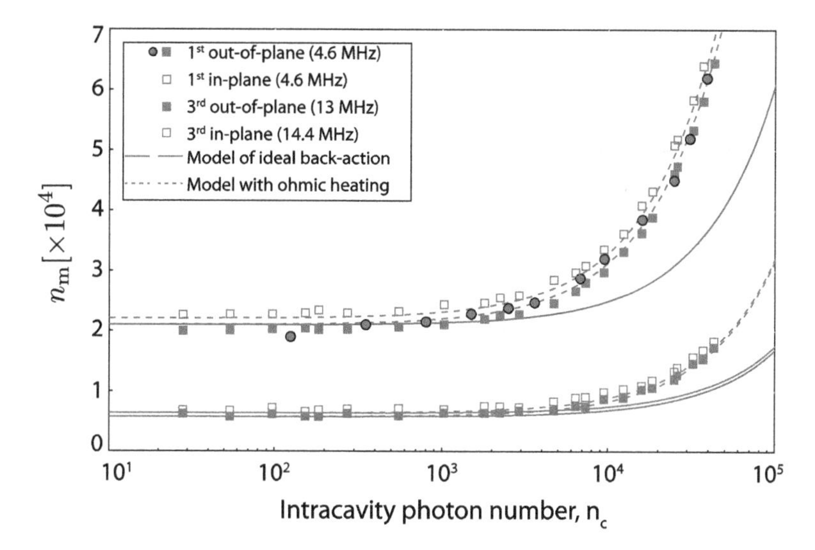

Fig. 6.7 Excess back-action due to ohmic heating. Plot shows a compilation of several measurements of the back-action heating of the four lowest order modes of the nanobeam. Black-circled-red points are data from Fig. 6.6a, for the fundamental out-of-plane mode. The square data report the back-action heating of the fundamental out-of-plane (solid red), fundamental in-plane (red), third harmonic out-of-plane (solid blue) and third harmonic in-plane (blue) modes. The solid lines are predictions from a model based on quantum back-action heating alone

in Fig. 6.7. The fact that the ratio of observed back-action among the different modes do not scale with their known cooperativites strongly suggest that laser noise heating can be ruled out.

Ohmic heating via absorption of laser light in the beam remains a strong candidate. However, attempts to model this scenario using a simple heat transfer model: assuming a point heat source at the centre of the beam taken to be in equilibrium at its clamping points, implies a mode-dependent heating that is lower than what is observed. Given that all modes equilibrate at sufficiently high pressure, where we do not observe any appreciable back-action, we conjecture that the modification of thermal transport (both in terms of a drop in thermal conductivity [30], and maybe even transport mechanism [39]) along the slender beam may be responsible for deviations from simple heat transfer.

6.2 Feedback Suppression of Back-Action

The ability to exert control over physical systems to a degree where their quantum mechanical behaviour can be probed and manipulated has been the focus of a second quantum revolution [40]. The past 40 years have witnessed an exquisite level of control being achieved on atomic systems, mostly fuelled by the availability of the laser as a highly coherent element that synthesises forces to steer atoms towards some desired state. This approach, termed *autonomous control*, essentially relies on engineering a hamiltonian [41, 42] (or a dissipative reservoir [43–45]) such that the associated open-loop evolution achieves some desired target. In this scenario the essential requirement is to be able to perform control operations within the decoherence time of the desired state.

The conventional paradigm of *feedback* control—wherein the output of an auxiliary system is fed-back to steer a plant towards a desired target, has two inequivalent manifestations in quantum mechanics. This has to do with whether the auxiliary system performs measurement [46, 47], or whether it coherently cascades its output into the input of the plant [48]. In the former case—*measurement-based feedback control*—the burden of timescale is shifted to the measurement: the measurement must be strong enough to track the quantum fluctuations of the system under control, while simultaneously being weak enough to not impart excess back-action [12]. Spectacular applications of such measurement-based feedback, for example to stabilize microwave Fock states [49, 50] and persistent Rabi oscillations of an artificial atom [51], have been limited to well isolated quantum systems.

For mechanical oscillators, the advent of quantum-noise-limited measurements, as described in the previous section, provides for measurement strong enough to resolve the oscillator's quantum state in the timescale of its decoherence. The rate at which the oscillator's zero-point motion is measured [52],

$$\Gamma_{\text{meas}} := \frac{x_{\text{zp}}^2}{\bar{S}_{xx}^{\text{imp}}[\Omega_{\text{m}}]} = \frac{\Gamma_{\text{m}}}{4n_{\text{imp}}},$$

equals the total decoherence rate, $\Gamma_{\text{decoh}} := (n_{\text{m,th}} + n_{\text{m,BA}})\Gamma_{\text{m}}$ (i.e. $\Gamma_{\text{meas}} = \Gamma_{\text{decoh}}$) when,

$$(n_{\text{m,th}} + n_{\text{m,BA}})n_{\text{imp}} = \frac{1}{4}. \tag{6.2.1}$$

That is, a quantum-noise-limited measurement is precisely the one that resolves the zero-point motion of the oscillator in its decoherence time. Remarkably, despite the fact that a quantum-noise-limited measurement perturbs the thermally driven oscillator by measurement back-action, feedback of the measurement record can suppress this back-action [10, 53].

Practically therefore, the ability to perform measurement-based feedback control of a mechanical oscillator necessitates that its motion be measured with an imprecision that is a factor $n_{\text{m,th}}$ below that at the standard quantum limit [53], or that its thermal motion is resolved with a signal-to-noise ratio of $n_{\text{m,th}}^2$. The measurement reported in the previous section, summarised in Fig. 6.6, features an imprecision that meets this condition at a bath temperature below 10 K. The sufficient condition, of being quantum-noise-limited, i.e. satisfy Eq. (6.2.1), is approached by within an order of magnitude, specifically, $\Gamma_{\text{meas}} \approx 0.2 \cdot \Gamma_{\text{decoh}}$.

In the rest of this section, we describe the design of a feedback controller, and its experimental implementation, that enables the suppression of measurement back-action. It turns out that the optimal controller, given phase-quadrature homodyne measurements, is the so-called cold-damping strategy [36, 54, 55]. Finally we present data where the oscillator is feedback cooled to a final phonon occupation of 5.3, suppressing more than $2 \cdot 10^4$ quanta of measurement back-action.

6.2.1 Synthesis of a Linear Quadratic Gaussian Controller

In the following, we precisely formulate, and solve, the following control problem: synthesise a feedback force based on the record of a continuous linear measurement of the position of a harmonic oscillator driven by Gaussian forces, such that it prepares the oscillator close to its ground state.

It turns out that in the quantum mechanical setting, as long as all elements that measure and apply feedback are linear and susceptible to Gaussian noises, there exists a formal analogy between this problem and the class of problems studied in the classical paradigm of LQG (linear quadratic Gaussian) control; here quadratic refers to the cost function whose optimisation is the control objective. In the classical setting, powerful techniques have been developed by Wiener [56] (for one dimensional systems) and Kalman (for multi-dimensional systems) that solves the LQG synthesis problem [57]. These techniques have been extended to the quantum LQG setting, with results analogous to the classical Kalman filter [58–61]. We adopt a frequency domain approach where an appropriate modification of Wiener's approach (i.e. without adopting a multi-dimensional state-space model) gives the required control strategy. A complementary treatment, shadowing the state-space approach via quantum trajectories, is given in [62].

6.2.1.1 Dynamics with Linear Measurement and Feedback

Concretely, we consider a harmonic oscillator with position fluctuations $\delta\hat{x}^{(0)}$ when it is in equilibrium with two baths: the thermal bath modelled by the force $(\delta\hat{F}_{\text{th}})$ associated with the ambient environment (as in Sect. 3.1.3), and the measurement bath due to the meter (nominally to an optical cavity as in Sect. 4.3) modelled by a back-action force $(\delta\hat{F}_{\text{BA}})$. In this open-loop case—the oscillator is coupled to a meter but the output of the meter is not fed-back—its dynamics is described by the equation,

$$\begin{aligned} m\left(\delta\ddot{\hat{x}}^{(0)} + \Gamma_{\text{m}}\delta\dot{\hat{x}}^{(0)} + \Omega_{\text{m}}^2\delta\hat{x}^{(0)}\right) &= \delta\hat{F}_{\text{th}}(t) + \delta\hat{F}_{\text{BA,th}}(t) \\ \text{i.e.,}\quad \delta\hat{x}^{(0)}[\Omega] &= \chi_x^{(0)}[\Omega]\left(\delta\hat{F}_{\text{th}}[\Omega] + \delta\hat{F}_{\text{BA,th}}[\Omega]\right). \end{aligned} \tag{6.2.2}$$

Here the susceptibility, $\chi_x^{(0)}[\Omega] = m^{-1}\left(\Omega_{\mathrm{m}}^2 - \Omega^2 - i\Omega\Gamma_{\mathrm{m}}\right)^{-1}$, encapsulates the effect of dynamical back-action from the meter, while $\delta\hat{F}_{\mathrm{BA,th}}$ models the fluctuating quantum back-action force.

Control is affected by using the output of a linear detector, which gives the instantaneous apparent position,

$$\delta\hat{y}^{(\mathrm{fb})} := \delta\hat{x}^{(\mathrm{fb})} + \delta\hat{x}_{\mathrm{imp}}, \tag{6.2.3}$$

and synthesising a feedback force linearly proportional to this position estimate,

$$\delta\hat{F}_{\mathrm{fb}}[\Omega] := -\chi_{\mathrm{fb}}^{-1}[\Omega]\delta\hat{y}^{(\mathrm{fb})} + \delta\hat{F}_{\mathrm{fb,th}}[\Omega]. \tag{6.2.4}$$

Here $\delta\hat{x}^{(\mathrm{fb})}$ is the physical motion modified by the application of feedback, $\chi_{\mathrm{fb}}^{-1}[\Omega]$ is a *causal and stable*[3] function that describes the response of the feedback controller, and $\delta\hat{F}_{\mathrm{fb,th}}$ models noise added by the feedback loop.

The physical motion $\delta\hat{x}^{(\mathrm{fb})}$ in the presence of feedback is determined by the equation,

[3]A function $f(t)$ is (asymptotically) stable if $|f(t\to\infty)| < \infty$; it is causal if $f(t<0)=0$. These properties can be expressed in terms of the Fourier transform,

$$f[\Omega] = \int_{-\infty}^{\infty} f(t)e^{i\Omega t}\,\mathrm{d}t,$$

extended to the complex plane (i.e. $\Omega \mapsto \Omega + i\Gamma$). Firstly the bound (an instance of the triangle inequality),

$$|f[\Omega+i\Gamma]| = \left|\int f(t)e^{(i\Omega-\Gamma)t}\,\mathrm{d}t\right| \leq \int |f(t)|\,e^{-\Gamma t}\,\mathrm{d}t,$$

together with stability implies that, $|f[\Omega+i\Gamma]| < \infty$ for $\Gamma > 0$; thus, all singularities of $f[\Omega+i\Gamma]$ are in the lower-half plane (real axis included). Secondly, causality in time domain can be expressed as the identity $f(t) = \Theta(t)f(t)$, where Θ is the Heaviside step function; taking its Fourier transform gives,

$$f[\Omega] = \frac{i}{\pi}\int_{-\infty}^{\infty}\frac{f[\Omega']}{\Omega'-\Omega}\,\mathrm{d}\Omega',$$

a constraint imposed by casuality for real frequencies (separating out the real and imaginary parts of this equation gives the Kramers-Kronig relation). We now consider the integral of $f[\Omega']/(\Omega'-\Omega)$ along a contour C in the complex plane that consists of the real line and an arc at infinity enclosing the upper-half plane; by stability, the latter integral is zero; by causality, the former satisfies the above constraint; thus, we can show that,

$$\oint_C \frac{f[\Omega']}{\Omega'-\Omega}\mathrm{d}\Omega' = 0.$$

Therefore (by Cauchy's theorem) $f[\Omega]$ is analytic in the upper-half plane [63–66]. (A stronger version of this line of inference goes by the name of Titchmarsh's theorem, which identifies causal and stable functions as boundary values of analytic functions [67, Section 17.8].)

$$m\left(\delta\ddot{\hat{x}}^{(\text{fb})}+\Gamma_\text{m}\delta\dot{\hat{x}}^{(\text{fb})}+\Omega_\text{m}^2\delta\hat{x}^{(\text{fb})}\right)=\delta\hat{F}_\text{th}(t)+\delta\hat{F}_{\text{BA,th}}(t)+\delta\hat{F}_\text{fb}(t)$$

$$\text{i.e.,}\quad \delta\hat{x}^{(\text{fb})}[\Omega]=\chi_x^{(0)}[\Omega]\left(\delta\hat{F}_\text{th}[\Omega]+\delta\hat{F}_{\text{BA,th}}[\Omega]+\delta\hat{F}_\text{fb}[\Omega]\right)=\delta\hat{x}^{(0)}[\Omega]+\chi_x^{(0)}[\Omega]\delta\hat{F}_\text{fb}[\Omega]. \tag{6.2.5}$$

Inserting Eqs. (6.2.3) and (6.2.4) into Eq. (6.2.5), and solving for $\delta\hat{x}^{(\text{fb})}$ gives,

$$\delta\hat{x}^{(\text{fb})}[\Omega]=\frac{\delta\hat{x}^{(0)}[\Omega]}{1+\chi_x^{(0)}[\Omega]\chi_\text{fb}^{-1}[\Omega]}-\frac{\delta\hat{x}_\text{imp}[\Omega]}{1+(\chi_x^{(0)}[\Omega]\chi_\text{fb}^{-1}[\Omega])^{-1}}=:\alpha[\Omega]\delta\hat{y}^{(0)}[\Omega]-\delta\hat{x}_\text{imp}[\Omega]. \tag{6.2.6}$$

Here we have absorbed $\delta\hat{F}_{\text{fb,th}}$ into $\delta\hat{x}_\text{imp}$ as an apparent imprecision via the redefinition, $\delta\hat{x}_\text{imp}\to\delta\hat{x}_\text{imp}+\chi_\text{fb}\delta\hat{F}_{\text{fb,th}}$; further, in passing to the second equality, we have defined the dimensionless response function,

$$\alpha[\Omega]:=\frac{1}{1+\chi_x^{(0)}[\Omega]\chi_\text{fb}^{-1}[\Omega]}, \tag{6.2.7}$$

and the apparent out-of-loop position,

$$\delta\hat{y}^{(0)}:=\delta\hat{x}^{(0)}+\delta\hat{x}_\text{imp}. \tag{6.2.8}$$

The relevance of the second form of $\delta\hat{x}^{(\text{fb})}$ in Eq. (6.2.6) is that it expresses the spectrum of the physical position,

$$\bar{S}_{xx}^{(\text{fb})}[\Omega]=|\alpha[\Omega]|^2\,\bar{S}_{yy}^{(0)}[\Omega]+\bar{S}_{xx}^{\text{imp}}[\Omega]-2\text{Re}\;\alpha[\Omega]\bar{S}_{y^{(0)}x_\text{imp}}[\Omega], \tag{6.2.9}$$

such that the correlation (third) term describes measurement-induced correlations alone, and not those invariably induced by the act of feedback ($\propto\bar{S}_{yx_\text{imp}}$). Expressed in this form, it is apparent that the act of feedback introduces an additional noise, vis-a-vis the feedback of the imprecision noise; in the ideal case, when the measurement imprecision is set by quantum fluctuations in the meter, this noise sets a fundamental limit to the performance of the feedback protocol (as detailed in Sect. 6.2.2.1).

It can be shown from the definition Eq. (6.2.7) that α is causal and stable whenever $\chi_x^{(0)}$ and χ_fb are so. Physically, this can be seen from the expression for the in-loop position,

$$\delta\hat{y}^{(\text{fb})}[\Omega]=\alpha[\Omega]\delta\hat{y}^{(0)}[\Omega].$$

This equation provides an operational meaning to $\alpha[\Omega]$: it describes the difference between an out-of-loop and in-loop probe of the oscillator position. Being therefore a physical quantity, α must be causal and stable.

6.2.1.2 Control Objective

The specific objective of feedback control is to prepare the oscillator in a state where back-action is minimised. Since the oscillator begins in a (Gaussian) thermal state, and all measurements and control are linear, the target state is at best another Gaussian state. We thus consider minimising an appropriate norm of the (symmetrised) covariance matrix of the target state, viz.,

$$\mathbf{\Sigma} := \begin{pmatrix} \frac{\langle \hat{x}^2 \rangle}{x_{\rm zp}^2} & \frac{1}{2}\frac{\langle \hat{x}\hat{p}+\hat{p}\hat{x} \rangle}{x_{\rm zp}p_{\rm zp}} \\ \frac{1}{2}\frac{\langle \hat{x}\hat{p}+\hat{p}\hat{x} \rangle}{x_{\rm zp}p_{\rm zp}} & \frac{\langle \hat{p}^2 \rangle}{p_{\rm zp}^2} \end{pmatrix}.$$

We assume that the target state is represented by a legitimate density operator, so that $\mathrm{Det}\,\mathbf{\Sigma} \geq \frac{1}{4}$, i.e. the generalized uncertainty principle in for the mechanical oscillator. (When the equality condition is met, the Gaussian state is pure.) Under this assumption, a suitable norm that is positive and quadratic in the position and momentum of the oscillator is the trace of the covariance matrix,

$$\mathrm{Tr}\,\mathbf{\Sigma} = \frac{\langle \hat{x}^2 \rangle}{x_{\rm zp}^2} + \frac{\langle \hat{p}^2 \rangle}{p_{\rm zp}^2} = \frac{\langle \hat{H}_{\rm m} \rangle}{\hbar\Omega_{\rm m}/4},$$

where,

$$\langle \hat{H}_{\rm m} \rangle = \frac{\langle \hat{p}^2 \rangle}{2m} + \frac{m\Omega_{\rm m}^2 \langle \hat{x}^2 \rangle}{2},$$

is the average energy of the target state. Thus, the task of realizing a minimum-uncertainty target state is equivalent to the task of minimising the energy of the state, given linear measurements and feedback, and an initial Gaussian state.

We therefore define the control action by the constraint,

$$\min_{\chi_{\rm fb}} \langle \hat{H}_{\rm m} \rangle = \min_{\alpha} \langle \hat{H}_{\rm m} \rangle. \tag{6.2.10}$$

Using the fact that,

$$\langle \hat{x}^2 \rangle = \int \bar{S}_{xx}^{(\rm fb)}[\Omega]\frac{\mathrm{d}\Omega}{2\pi}, \quad \text{and,} \quad \langle \hat{p}^2 \rangle = m^2 \langle \dot{\hat{x}}^2 \rangle = m^2 \int \Omega^2 \bar{S}_{xx}^{(\rm fb)}[\Omega]\frac{\mathrm{d}\Omega}{2\pi},$$

where $\bar{S}_{xx}^{(\rm fb)}$ is the spectral density of the physical position in the presence of feedback (i.e. $\delta\hat{x}^{(\rm fb)}$), the average energy takes the form,

$$\langle \hat{H}_{\rm m} \rangle = \frac{m\Omega_{\rm m}^2}{2} \int \left(1 + \frac{\Omega^2}{\Omega_{\rm m}^2}\right) \bar{S}_{xx}^{(\rm fb)}[\Omega]\frac{\mathrm{d}\Omega}{2\pi}.$$

Thus, the constraint in Eq. (6.2.10) can be expressed purely in terms of the spectral density of the oscillator position. Specifically, it involves the minimisation of a functional $\mathscr{C}$, defined by,

$$\min_{\alpha} \mathscr{C}[\![\alpha]\!] := \min_{\alpha} \int \left(1 + \frac{\Omega^2}{\Omega_\mathrm{m}^2}\right) \bar{S}_{xx}^{(\mathrm{fb})}[\Omega] \frac{\mathrm{d}\Omega}{2\pi}. \tag{6.2.11}$$

This form is advantageous because the measurement record only contains an estimate of the position of the oscillator, while its momentum has to be inferred by taking appropriate derivatives.

6.2.1.3 Solution of the Control Problem

Inserting the expression for the spectral density $\bar{S}_{xx}$ from Eq. (6.2.9) into the constraint in Eq. (6.2.11), we have,

$$\min_{\alpha} \mathscr{C}[\![\alpha]\!] = \min_{\alpha} \int \tau[\Omega] \left(|\alpha[\Omega]|^2 \bar{S}_{yy}^{(0)}[\Omega] + \bar{S}_{xx}^{\mathrm{imp}}[\Omega] - 2\mathrm{Re}\, \alpha[\Omega] \bar{S}_{y^{(0)} x_{\mathrm{imp}}}[\Omega]\right) \frac{\mathrm{d}\Omega}{2\pi},$$

where, $\tau[\Omega] := \left(1 + \Omega^2/\Omega_\mathrm{m}^2\right)$. Using standard results from variational calculus [68], the extremisation is equivalent to the first variation, $\mathrm{D}\mathscr{C} := \mathscr{C}[\![\alpha + \mathrm{D}\alpha]\!] - \mathscr{C}[\![\alpha]\!]$ being equal to zero for all valid variations $\mathrm{D}\alpha$ of the response function. Here validity refers to the requirement that the perturbed function $\alpha + \mathrm{D}\alpha$ be a physical response. In particular, this implies that $\mathrm{D}\alpha$ is itself a physical response functions; thus it has to satisfy two constraints: (a) $\mathrm{D}\alpha[\Omega]^* = \mathrm{D}\alpha[-\Omega]$, since the time-domain response function is real, and, (b) $\mathrm{D}\alpha$ is analytic in the upper-half plane, since the response function is causal and stable (see footnote 3 on page 124). The variation $\mathrm{D}\mathscr{C}$ computed by incorporating the first constraint (and appropriate symmetry properties of the spectra), takes the form,

$$\mathrm{D}\mathscr{C} = 2 \int \mathrm{D}\alpha[\Omega]\, \tau[\Omega] \left(\alpha[\Omega] \bar{S}_{yy}^{(0)}[\Omega] - \bar{S}_{y^{(0)} x_{\mathrm{imp}}}[\Omega]\right) \frac{\mathrm{d}\Omega}{2\pi}.$$

The extremisation condition, $\frac{\mathrm{D}\mathscr{C}}{\mathrm{D}\alpha} = 0$, now reads,

$$\int \tau[\Omega] \left(\alpha[\Omega] \bar{S}_{yy}^{(0)}[\Omega] - \bar{S}_{y^{(0)} x_{\mathrm{imp}}}[\Omega]\right) \frac{\mathrm{d}\Omega}{2\pi} = 0, \quad \text{for,} \quad \alpha \text{ causal and stable.} \tag{6.2.12}$$

The clause of causal stability prevents the naive solution obtained by setting the integrand equal to zero.

An ingenious argument due to Wiener and Hopf [56, 57] gracefully accommodates the causality constraint. In fact, if the integrand in Eq. (6.2.12) is a causal function, then it is guaranteed that α is also causal. Thus, we introduce a function $f_-[\Omega]$ that is arbitrary in the lower-half plane, but zero (and analytic) in the upper-half plane, i.e.

$$\alpha[\Omega] \left(\tau[\Omega] \bar{S}_{yy}^{(0)}[\Omega]\right) - \left(\tau[\Omega] \bar{S}_{y^{(0)} x_{\mathrm{imp}}}[\Omega]\right) = f_-[\Omega]. \tag{6.2.13}$$

Note that this equation is simply a place-holder for the fact that the left-hand side is a causal function. In order to convert this to a concrete solution for a causal response α, we take recourse in a mathematical fact, namely, that $\bar{S}_{yy}^{(0)}$ can be factorised in the form,[4]

$$\bar{S}_{yy}^{(0)}[\Omega] = \left\{\bar{S}_{yy}^{(0)}[\Omega]\right\}_+ \left\{\bar{S}_{yy}^{(0)}[\Omega]\right\}_- ,$$

where the first (second) factor is analytic in the upper(lower)-half plane. More generally, such a factorisation also exists for the product, $\tau \bar{S}_{yy}^{(0)}$. Inserting this in Eq. (6.2.13) allows it to be expressed in the form,

$$\alpha \left\{\tau \bar{S}_{yy}^{(0)}\right\}_+ = \frac{\tau \bar{S}_{y^{(0)} x_{\mathrm{imp}}}}{\left\{\tau \bar{S}_{yy}^{(0)}\right\}_-} + \frac{f_-}{\left\{\tau \bar{S}_{yy}^{(0)}\right\}_-};$$

where the second term on the right-hand side is anti-causal and anti-stable. Since we expect the left-hand side to be causal and stable, it must be that it is equal to the causal stable part of the right-hand side, i.e.,

$$\alpha[\Omega] = \frac{1}{\left\{\tau[\Omega]\bar{S}_{yy}^{(0)}[\Omega]\right\}_+} \left[\frac{\bar{S}_{y^{(0)} x_{\mathrm{imp}}}[\Omega]}{\left\{\tau[\Omega]\bar{S}_{yy}^{(0)}[\Omega]\right\}_-}\right]_+ . \tag{6.2.14}$$

Here, $[\cdots]_+$ represents the additive decomposition of its argument into its stable causal part, analogous to $\{\cdots\}_+$ representing the multiplicative decomposition. This is the formal solution for the physical response function of the feedback network that prepares the oscillator in a minimum-uncertainty Gaussian state.

This formal solution can now be applied to the problem at hand. We consider the case of phase quadrature homodyne measurement, in which case,

$$\begin{aligned}
\bar{S}_{y^{(0)} x_{\mathrm{imp}}}[\Omega] &= \bar{S}_{xx}^{\mathrm{imp}}[\Omega] = n_{\mathrm{imp}} \cdot 2\bar{S}_{xx}^{\mathrm{zp}}[\Omega_{\mathrm{m}}] \\
\bar{S}_{yy}^{(0)}[\Omega] &= \bar{S}_{xx}^{\mathrm{imp}}[\Omega] + \bar{S}_{xx}^{(0)}[\Omega] = \left(n_{\mathrm{imp}} + \frac{(n_{\mathrm{tot}} + \frac{1}{2})(\Omega_{\mathrm{m}}\Gamma_{\mathrm{m}})^2}{(\Omega^2 - \Omega_{\mathrm{m}})^2 + (\Omega\Gamma_{\mathrm{m}})^2}\right) \cdot 2\bar{S}_{xx}^{\mathrm{zp}}[\Omega_{\mathrm{m}}].
\end{aligned}$$

[4]This is generally true of symmetrized spectra of the observables of linear Markovian systems driven by white-noise. Firstly, the spectra of the observables of such systems is a rational function of the frequency, i.e. a ratio of polynomials (Markovianity is essential for this to be true since arbitrarily long time delays preserve linearity but do not have a rational frequency response). Secondly, considering Ω to be extended into the complex plane, it can be shown that the spectrum of an observable, say y, satisfies the relation, $\bar{S}_{yy}[\Omega] = \bar{S}_{yy}^*[-\Omega^*]$ (for real frequencies, this reduces to, $\bar{S}_{yy}[\Omega] = \bar{S}_{yy}[-\Omega]$, given in Eq. 2.1.16). This symmetry implies a characteristic symmetry for the roots of the numerator and denominator polynomials: purely real roots occur in positive/negative pairs, while purely imaginary ones in conjugate pairs, and complex roots occur in conjugate positive/negative quadruplets. Thus, both the numerator and denominator can be factorised such that the factors have zeros symmetric about the real-axis. Collecting those factors with zeros and poles in either half-plane gives the required factorisation.

Since we expect the response α to only have a significant contribution around the mechanical resonance in a bandwidth around its linewidth, we may safely assume $\tau[\Omega]$ is a constant and omit it. Putting the above expressions into Eq. (6.2.14), and assuming $n_{\mathrm{tot}} \gg 1 \gg n_{\mathrm{imp}}$, we can work out, after considerable algebra, the form of the feedback filter, viz.

$$\chi_{\mathrm{fb}}^{-1}[\Omega] = \chi_x^{(0)}[\Omega]^{-1}\left(\alpha[\Omega]^{-1} - 1\right) \approx im\Omega_{\mathrm{m}}\Gamma_{\mathrm{m}} \cdot \sqrt{\frac{n_{\mathrm{tot}}}{n_{\mathrm{imp}}}}. \tag{6.2.15}$$

This feedback filter, coinciding with the so-called cold damping strategy [54, 55], has a very simple physical interpretation: the oscillator is damped by applying a feedback force, $\delta\hat{F}_{\mathrm{fb}} \approx -\chi_{\mathrm{fb}}^{-1}\delta\hat{x} \propto i\Omega_{\mathrm{m}}\delta\hat{x}^{(\mathrm{fb})}$, that is proportional to the instantaneous velocity.

6.2.2 *Feedback by Cold Damping*

The objective of feedback cooling is to nullify the effect of the total ambient thermal force, $\delta\hat{F}_{\mathrm{th}} + \delta\hat{F}_{\mathrm{BA,th}}$, on the oscillator. The cold damping strategy achieves this heuristically by coupling the oscillator to a colder environment at progressively larger coupling rates [54, 55]. With respect to the standard feedback schematic shown in Fig. 6.8, cold damping corresponds to the choice of the feedback controller,

$$\chi_{\mathrm{fb}}^{-1}[\Omega] = -im\Omega\Gamma_{\mathrm{fb}}[\Omega], \tag{6.2.16}$$

ideally with the feedback damping Γ_{fb} given by,

$$\Gamma_{\mathrm{fb}}[\Omega] = g_{\mathrm{fb}}\Gamma_{\mathrm{m}}, \tag{6.2.17}$$

for some dimensionless real *feedback gain* g_{fb}. Note that the feedback gain simply parametrises the optimal strategy computed in the previous section, and given in Eq. (6.2.15).

To see how this damping leads to cooling, we reconsider the physical motion of the oscillator given in Eq. (6.2.6), but now expressed in the form,

$$\delta\hat{x}^{(g_{\mathrm{fb}})} = \chi_x^{(g_{\mathrm{fb}})}\left(\delta\hat{F}_{\mathrm{th}} + \delta\hat{F}_{\mathrm{BA,th}} + \delta\hat{F}_{\mathrm{fb,th}} - \chi_{\mathrm{fb}}^{-1}\delta\hat{x}_{\mathrm{imp}}\right), \tag{6.2.18}$$

where the effective mechanical susceptibility,

$$\chi_x^{(g_{\mathrm{fb}})}[\Omega]^{-1} := \chi_x[\Omega]^{-1} + \chi_{\mathrm{fb}}[\Omega]^{-1} = m\left(-\Omega^2 + \Omega_{\mathrm{m}}^2 - i\Omega\Gamma_{\mathrm{m}}(1 + g_{\mathrm{fb}})\right), \tag{6.2.19}$$

features an increased damping rate due to feedback. The motion of the oscillator is driven by forces from the three physical environments it is coupled to: (1) the thermal

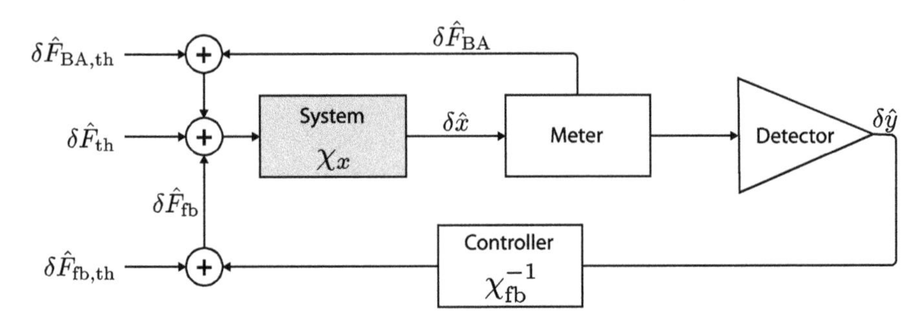

Fig. 6.8 Schematic of measurement-based feedback loop. The oscillator (system) experiences a multitude of forces: the ambient thermal force $\delta\hat{F}_{\mathrm{th}}$, a back-action force $\delta\hat{F}_{\mathrm{BA}}$ arising from its coupling to the optical field (meter), and finally, a feedback force $\delta\hat{F}_{\mathrm{fb}}$ synthesised by the controller

force from the ambient thermal environment, (2) measurement back-action from the meter, and (3) force fluctuations from the feedback network. For a high-Q oscillator, and in the limit where these noises may be assumed to be Gaussian, each of these environments can be assigned a thermal noise equivalent phonon occupation: $n_{\mathrm{m,th}}$, $n_{\mathrm{m,BA}}$ and $n_{\mathrm{m,fb}}$ respectively. Thus the total effective thermal force may be expressed:

$$\begin{aligned} \bar{S}_{FF}^{\mathrm{tot}}[\Omega] &:= \bar{S}_{FF}^{\mathrm{th}}[\Omega] + \bar{S}_{FF}^{\mathrm{BA,th}}[\Omega] + \bar{S}_{FF}^{\mathrm{fb,th}}[\Omega] \\ &= \left(n_{\mathrm{m,th}} + n_{\mathrm{m,BA}} + n_{\mathrm{m,fb}} + \tfrac{1}{2}\right) \cdot |\chi_x^{(0)}[\Omega_{\mathrm{m}}]|^{-2} \cdot 2\bar{S}_{xx}^{\mathrm{zp}}[\Omega_{\mathrm{m}}], \end{aligned} \tag{6.2.20}$$

where we have introduced for convenience the (peak) position spectral density in the ground state (Eq. 3.1.30):

$$\bar{S}_{xx}^{\mathrm{zp}}[\Omega_{\mathrm{m}}] = \frac{2x_{\mathrm{zp}}^2}{\Gamma_{\mathrm{m}}}.$$

We further introduce the imprecision quanta, n_{imp}, as the apparent thermal occupation associated with noise in the measurement:

$$\bar{S}_{xx}^{\mathrm{imp}}[\Omega] = n_{\mathrm{imp}} \cdot 2\bar{S}_{xx}^{\mathrm{zp}}[\Omega_{\mathrm{m}}]. \tag{6.2.21}$$

The apparent position—the position-equivalent output of the detector—is described by the equation,

$$\delta\hat{y}^{(g_{\mathrm{fb}})} = \chi_x^{(g_{\mathrm{fb}})}\left(\delta\hat{F}_{\mathrm{th}} + \delta\hat{F}_{\mathrm{BA,th}} + \delta\hat{F}_{\mathrm{fb,th}} + \chi_x^{(0)-1}\delta\hat{x}_{\mathrm{imp}}\right) \tag{6.2.22}$$

Using the expression for the total force noise spectrum (Eq. (6.2.20)), we get the spectrum of the physical position and the apparent position, viz.

$$\frac{\bar{S}_{xx}^{(g_{\mathrm{fb}})}[\Omega]}{2\bar{S}_{xx}^{\mathrm{zp}}[\Omega_{\mathrm{m}}]} = \frac{(n_{\mathrm{m,th}} + n_{\mathrm{m,BA}} + n_{\mathrm{m,fb}} + \frac{1}{2})\Omega_{\mathrm{m}}^2\Gamma_{\mathrm{m}}^2 + n_{\mathrm{imp}}\, g_{\mathrm{fb}}^2\Omega^2\Gamma_{\mathrm{m}}^2}{(\Omega_{\mathrm{m}}^2 - \Omega^2)^2 + (\Omega\Gamma_{\mathrm{m}}(1+g_{\mathrm{fb}}))^2}$$
$$\frac{\bar{S}_{yy}^{(g_{\mathrm{fb}})}[\Omega]}{2\bar{S}_{xx}^{\mathrm{zp}}[\Omega_{\mathrm{m}}]} = n_{\mathrm{imp}} + \frac{(n_{\mathrm{m,th}} + n_{\mathrm{m,BA}} + n_{\mathrm{m,fb}} + \frac{1}{2})\Omega_{\mathrm{m}}^2\Gamma_{\mathrm{m}}^2 - n_{\mathrm{imp}} g_{\mathrm{fb}}(g_{\mathrm{fb}}+2)\Omega^2\Gamma_{\mathrm{m}}^2}{(\Omega_{\mathrm{m}}^2 - \Omega^2)^2 + (\Omega\Gamma_{\mathrm{m}}(1+g_{\mathrm{fb}}))^2}. \tag{6.2.23}$$

The mean phonon occupancy of the cooled oscillator is then given by,

$$n_{\mathrm{m}} + \frac{1}{2} = \frac{1}{2}\int_{-\infty}^{\infty} \frac{\bar{S}_{xx}[\Omega]}{x_{\mathrm{zp}}^2}\frac{d\Omega}{2\pi} = \frac{(n_{\mathrm{m,th}} + n_{\mathrm{m,BA}} + n_{\mathrm{m,fb}} + \frac{1}{2}) + n_{\mathrm{imp}} g_{\mathrm{fb}}^2}{1+g_{\mathrm{fb}}}. \tag{6.2.24}$$

In the relevant case, $n_{\mathrm{m,th}} \gg \frac{1}{2}$, a minimum of

$$n_{\mathrm{m,min}} \approx 2\sqrt{(n_{\mathrm{m,th}} + n_{\mathrm{m,BA}} + n_{\mathrm{m,fb}})n_{\mathrm{imp}}} - \frac{1}{2}, \tag{6.2.25}$$

is attained at an optimal gain of

$$g_{\mathrm{fb,opt}} \approx \sqrt{\frac{n_{\mathrm{m,th}} + n_{\mathrm{m,ba}} + n_{\mathrm{m,fb}}}{n_{\mathrm{imp}}}}, \tag{6.2.26}$$

consistent with the optimal case given in Eq. (6.2.15). In particular, for the experimentally relevant case of $n_{\mathrm{m,th}} \gg n_{\mathrm{m,fb}}$, the conventional condition for ground state cooling, $n_{\mathrm{m}} < 1$, translates to

$$n_{\mathrm{imp}} < \frac{9}{16}(n_{\mathrm{m,th}} + n_{\mathrm{m,BA}} + n_{\mathrm{m,fb}})^{-1}. \tag{6.2.27}$$

6.2.2.1 Limits Due to Measurement Back-Action

Expressing Eq. (6.2.25) in the form,

$$n_{\mathrm{m,min}} + \frac{1}{2} = \frac{1}{\hbar}\sqrt{\bar{S}_{FF}^{\mathrm{tot}}[\Omega_{\mathrm{m}}]\bar{S}_{xx}^{\mathrm{imp}}[\Omega_{\mathrm{m}}]}, \tag{6.2.28}$$

it is apparent that the minimum phonon occupation attained by cold damping—and indeed, linear feedback of the phase quadrature homodyne measurement record—is limited by the ability to perform a quantum-noise-limited in-loop measurement. In fact, using the imprecision-back-action constraint for an ideal measurement, $n_{\mathrm{imp}}n_{\mathrm{m,BA}} \geq \frac{1}{16}$, the condition for ground state cooling (Eq. (6.2.27)) can be expressed in the form,

$$n_{\mathrm{imp}} < (2n_{\mathrm{m,th}})^{-1}, \tag{6.2.29}$$

equivalently, the measurement imprecision has to satisfy,

$$\bar{S}_{xx}^{\text{imp}} < \frac{\bar{S}_{xx}^{\text{zp}}}{n_{\text{m,th}}} = \frac{2x_{\text{zp}}^2}{n_{\text{m,th}}\Gamma_{\text{m}}} = \frac{4x_{\text{zp}}^2}{\Gamma_{\text{decoh}}}, \tag{6.2.30}$$

where $\Gamma_{\text{decoh}} := \Gamma_{\text{m}} n_{\text{m,th}}$ is the thermal decoherence rate. Notably Eq. (6.2.30) corresponds to an imprecision $n_{\text{th}}/2$ times below that at the standard quantum limit, or equivalently, a measurement rate [52, 69]

$$\Gamma_{\text{meas}} := \frac{x_{\text{zp}}^2}{\bar{S}_{xx}^{\text{imp}}[\Omega_{\text{m}}]} = \frac{\Gamma_{\text{m}}}{4n_{\text{imp}}} > \frac{\Gamma_{\text{decoh}}}{2}, \tag{6.2.31}$$

comparable to the thermal decoherence rate.

6.2.3 Implementation of Feedback

Implementing the ideal cold damping strategy in practise means implementing the ideal filter given in Eq. (6.2.16) – a differentiator ($\propto -i\Omega$) with a variable gain ($\propto g_{\text{fb}}$). Several practical challenges need to be reckoned with when realising a feedback filter: the ability to realise a filter with very low input noise, the ability to accommodate a large dynamic range of about $n_{\text{m,th}}/n_{\text{imp}} \approx 80$ dB, ease to fine-tune its response in the relevant frequency band, and, ease to set its overall gain. After a few iterations, we have found that the simple implementation shown in Fig. 6.9a, meets these criteria.

Figure 6.9a shows the sub-part of the entire experiment (detailed in Fig. 5.9), that realises the feedback controller. Light transmitted from the meter cavity mode (at $\lambda = 780$ nm) is measured in a balanced homodyne detector tuned to the phase quadrature, whose output voltage is (see Sect. 4.3),

$$\begin{aligned}\delta V_{\text{hom}}[\Omega] &= H_{\text{det}}[\Omega]\sqrt{n_{\text{LO}}}\,\delta\hat{p}_{\text{out}} \\ &= H_{\text{det}}[\Omega]\sqrt{2\eta_{\text{m}} C_{\text{m}} n_{\text{LO}} \Gamma_{\text{m}}}\,\frac{\delta\hat{x}[\Omega] + \delta\hat{x}_{\text{imp}}[\Omega]}{x_{\text{zp}}},\end{aligned} \tag{6.2.32}$$

where, H_{det} is the gain of the detector in units of volts/(photons/s), n_{LO} is the homodyne LO photon flux, $\delta\hat{p}_{\text{out}}$ is the phase quadrature of the meter output (red beam in Fig. 6.8a), η_{m} is the detection efficiency of the meter mode, C_{m} is its multiphoton cooperativity, and $\delta\hat{x}_{\text{imp}}$ is the measurement imprecision given in Eq. (6.1.15). A major part of this signal is coupled using a directional coupler (MiniCircuits ZFDC-20-5+) into a single-pole low pass filter[5] at 5 MHz, amplified (Miteq AU-1525), and attenuated using a voltage-controlled attenuator (MiniCircuits ZX73); the resulting voltage,

[5] A theorem due to Bode [63] asserts that for a causal stable filter, a magnitude response that falls off as some polynomial power, Ω^n, has to impart at least a phase change of $n\pi/2$; thus this filter adds a phase of $\pi/2$.

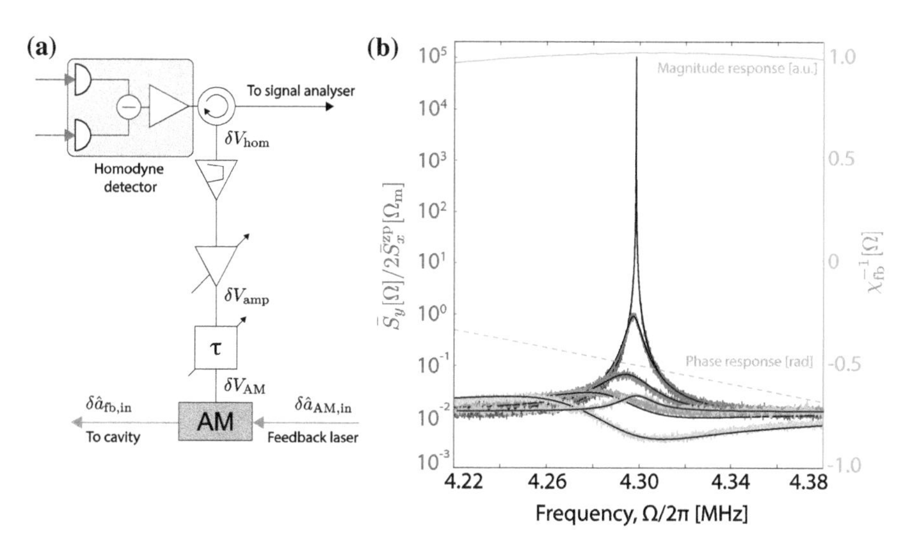

Fig. 6.9 Implementation of cold damping. a Schematic of the experimental implementation of the feedback controller consisting of the homodyne detector output band-passed, amplified, and delayed, before being imprinted onto the amplitude of the feedback laser. **b** In-loop mechanical spectra measured for the case where the feedback filter $\chi_{\rm fb}^{-1}[\Omega]$ has a frequency dependence that is (non-)ideal, in (red) blue. The red traces show measured in-loop spectra as the feedback gain $g_{\rm fb}$ is increased over three orders of magnitude; the green traces show the measured filter response. The black traces show predictions for the in-loop spectra given in Eq. (6.2.23) using the measured filter response as the only free variable

$$\delta V_{\rm amp}[\Omega] = G_{\rm amp}[\Omega](\delta V_{\rm hom}[\Omega] + \delta V_{\rm amp,in}[\Omega]), \tag{6.2.33}$$

features a variable (dimensionless) gain $G_{\rm amp}$ that effectively sets the feedback gain $g_{\rm fb}$, and the input noise of the amplifier $\delta V_{\rm amp,in}$ that sets the dominant classical contribution to the thermal force noise added by the feedback $\delta \hat{F}_{\rm fb,th}$. This signal is delayed using a tunable passive delay line (Stanford Research Systems DB 64), whose output in time domain is $V_{\rm AM}(t) = V_{\rm amp}(t-\tau)$; in the frequency domain,

$$\begin{aligned} V_{\rm AM}[\Omega] &= e^{i\Omega\tau} V_{\rm amp}[\Omega] \approx (1 + i\Omega\tau) V_{\rm amp}[\Omega] \\ \text{i.e.,} \quad \delta V_{\rm AM}[\Omega] &\approx i\Omega\tau\, \delta V_{\rm amp}[\Omega]. \end{aligned} \tag{6.2.34}$$

Finally, this voltage is imprinted on the feedback laser (at $\lambda = 850\,\text{nm}$) as an amplitude modulation using an in-fiber modulator (Photline NIR-MX800-LN10); the field fluctuations of the feedback laser exiting the modulator, acting as the input to the feedback cavity mode, is given by,

$$\delta \hat{a}_{\rm fb,in}[\Omega] = \delta \hat{a}_{\rm AM,in}[\Omega] + \sqrt{n_{\rm fb,in}} \frac{\pi}{V_\pi} \delta V_{\rm AM}[\Omega], \tag{6.2.35}$$

where $\delta\hat{a}_{\mathrm{AM,in}}$ is the fluctuation at the input of the modulator (i.e. the fluctuations of the feedback laser), $n_{\mathrm{fb,in}}$ is the mean photon flux of the feedback laser, and $V_\pi \approx 4\,\mathrm{V}$ is the π−voltage of the modulator. The feedback laser is coupled to an independent mode of the optical cavity, so that the radiation pressure force from it is the actuator in the feedback loop. This radiation pressure force fluctuation is given by,

$$\delta\hat{F}_{\mathrm{fb}}[\Omega] = \hbar G_{\mathrm{fb}}\sqrt{n_{\mathrm{fb}}}\delta\hat{q}_{\mathrm{fb}} = \frac{\hbar}{x_{\mathrm{zp}}}\sqrt{\eta_{\mathrm{fb}}C_{\mathrm{fb}}\Gamma_{\mathrm{m}}}\,\delta\hat{q}_{\mathrm{fb,in}}[\Omega]. \tag{6.2.36}$$

Here, in the first equation, G_{fb} is the frequency pull parameter of the feedback mode, n_{fb} is the mean intracavity photon number in that mode, and, $\delta\hat{q}_{\mathrm{fb}}$ is the intracavity amplitude quadrature; while in the second equation, η_{fb} is the cavity coupling efficiency, C_{fb} is the multiphoton cooperativity, and $\delta\hat{q}_{\mathrm{fb,in}}$ is the amplitude quadrature of the propagating field entering the cavity.

We can now arrive at expressions for the dimensionless feedback gain g_{fb}, and the phonon occupation due to the feedback network $n_{\mathrm{m,fb}}$, in terms of the measurable properties of the various elements in the feedback loop. Using Eqs. (6.2.32)–(6.2.35), the expression for the feedback force in Eq. (6.2.36) takes the form,

$$\begin{aligned}\delta\hat{F}_{\mathrm{fb}}[\Omega] &= i m\Omega\Gamma_{\mathrm{m}} \cdot \underbrace{G_{\mathrm{amp}}[\Omega]\cdot\Omega_{\mathrm{m}}\tau\cdot\frac{\pi H_{\mathrm{det}}[\Omega]}{V_\pi/\sqrt{n_{\mathrm{LO}}n_{\mathrm{fb,in}}}}\sqrt{8\eta_{\mathrm{m}}\eta_{\mathrm{fb}}C_{\mathrm{m}}C_{\mathrm{fb}}}}_{=g_{\mathrm{fb}}}\cdot\delta\hat{x}[\Omega] \\ &\quad + \underbrace{\frac{\hbar}{x_{\mathrm{zp}}}\sqrt{\eta_{\mathrm{fb}}C_{\mathrm{fb}}\Gamma_{\mathrm{m}}}\left(\delta\hat{q}_{\mathrm{AM,in}}[\Omega] + i\pi\Omega\tau\cdot G_{\mathrm{amp}}[\Omega]\cdot\frac{\delta V_{\mathrm{amp,in}}[\Omega]}{V_\pi/\sqrt{n_{\mathrm{fb,in}}}}\right)}_{=\delta\hat{F}_{\mathrm{fb,th}}}.\end{aligned} \tag{6.2.37}$$

Here, we have omitted the term corresponding to feedback of the imprecision noise ($\propto \delta\hat{x}_{\mathrm{imp}}$) since it is explicitly taken into account in our treatment of feedback. Importantly, the feedback gain, $g_{\mathrm{fb}} \propto G_{\mathrm{amp}}$, so that by changing only the electronic gain G_{amp} of the amplifier, while keeping everything else (cavity coupling, optical power, homodyne detector setting etc) fixed, the oscillator can be cold damped. Further, the force noise due to the feedback path, $\delta\hat{F}_{\mathrm{fb,th}}$, leads to an excess phonon occupation,

$$\begin{aligned} n_{\mathrm{m,fb}} &= C_{\mathrm{fb}}\left(1 + (\pi\Omega_{\mathrm{m}}\tau)^2\left|G_{\mathrm{amp}}[\Omega_{\mathrm{m}}]\right|^2\frac{\bar{S}_{VV}^{\mathrm{amp,in}}[\Omega_{\mathrm{m}}]}{V_\pi^2/n_{\mathrm{fb,in}}}\right) \\ &= C_{\mathrm{fb}} + 2g_{\mathrm{fb}}^2\cdot\frac{n_{\mathrm{imp}}}{\eta_{\mathrm{fb}}}\cdot\frac{\bar{S}_{VV}^{\mathrm{amp,in}}[\Omega_{\mathrm{m}}]}{\left|H_{\mathrm{det}}[\Omega_{\mathrm{m}}]\right|^2 n_{\mathrm{LO}}}\end{aligned} \tag{6.2.38}$$

that is due to two sources: quantum back-action from radiation pressure vacuum fluctuations of the feedback beam (first term) which is independent of the feedback gain, and thermal occupation due to classical voltage noise (second term) in the feedback electronics that is proportional to the feedback gain. However, under our typical experimental conditions, the second contribution is $\approx (g_{\text{fb}}/10^9)^2$. Thus, at feedback gains required to reach the ground state ($g_{\text{fb}} \approx 10^5$), the contribution of excess back-action due to classical voltage noise is negligible compared to quantum back-action from the feedback laser. In fact, the latter is also orders of magnitude smaller than quantum back-action from the meter.

6.2.3.1 Effect of Non-ideal Feedback Phase

In the experimentally implemented feedback filter, essentially consisting of a passive delay, the phase has to satisfy

$$\arg\left[\chi_{\text{fb}}[\Omega]^{-1}\right] = -(2k+1)\frac{\pi}{2}, \quad k \in \mathbb{N},$$

at all relevant Fourier frequencies, for efficient cold damping. Note that to satisfy the condition for Markovian feedback [46, 47, 54], the total time delay, $\tau := \arg\left[\chi_{\text{fb}}[\Omega_{\text{m}}]^{-1}\right]/\Omega_{\text{m}}$, has to satisfy, $\tau < 2\pi/\Gamma_{\text{decoh}}$, thus constraining the largest phase wrap tolerable. As detailed above, a single-pole low pass filter adds a phase of $\pi/2$, so that we tune the delay in order to get an addition π phase shift.

Here we consider the effect of any residual dispersion and/or non-ideal phase, which we model by,

$$\begin{aligned}\chi_{\text{fb}}[\Omega]^{-1} &= \exp\left[-i\left((2k+1)\frac{\pi}{2} + \delta\phi_{\text{fb}}[\Omega]\right)\right] m\Omega\, g_{\text{fb}}[\Omega]\Gamma_{\text{m}} \\ &= -im\Omega g_{\text{fb}}[\Omega]\Gamma_{\text{m}}\, e^{-i\delta\phi_{\text{fb}}[\Omega]}.\end{aligned} \tag{6.2.39}$$

The corresponding effective susceptibility,

$$\chi_x^{(g_{\text{fb}})}[\Omega]^{-1} = m\left(\Omega^{(g_{\text{fb}})}[\Omega]^2 - \Omega_{\text{m}}^2 - i\Omega\Gamma^{(g_{\text{fb}})}[\Omega]\right), \tag{6.2.40}$$

features feedback-induced damping, and an additional frequency shift,

$$\begin{aligned}\Gamma^{(g_{\text{fb}})}[\Omega] &= \Gamma_{\text{m}}\left[1 + g_{\text{fb}}[\Omega]\cos\delta\phi_{\text{fb}}[\Omega]\right] \\ \Omega^{(g_{\text{fb}})}[\Omega] &= \Omega_{\text{m}}\left[1 - g_{\text{fb}}[\Omega]\frac{\Omega\Gamma_{\text{m}}}{\Omega_{\text{m}}^2}\sin\delta\phi_{\text{fb}}[\Omega]\right]^{1/2}.\end{aligned} \tag{6.2.41}$$

Thus, deviations from an ideal phase profile can be observed in the feedback-induced frequency shift, while a non-flat gain profile results in a susceptibility that is no longer Lorentzian. Figure 6.9b shows the effect of non-ideal feedback phase in the in-loop signal. As the feedback gain is increased (red traces), the in-loop signal features a distinct red-shift of the oscillator frequency; at the highest feedback gain, an apparent deviation from a Lorentzian spectrum is visible, an effect due to "noise squashing", i.e. classical correlations established in the in-loop signal via feedback of noise and not due to the slight deviation from flatness of the measured feedback gain (green). The black traces show the expected spectrum calculated using the measured feedback gain profiles. By trimming the feedback delay, purely dissipative feedback (corresponding to $\delta\phi_{\mathrm{fb}} \approx 0$) can be realised (blue trace).

6.2.4 *Feedback Cooling to Near the Ground State*

We now consider what temperature can be reached by increasing the strength of the feedback used to damp the oscillator. The effective phonon occupancy of the cooled mechanical mode depends on the balance between coupling to thermal, measurement, and feedback reservoirs at rates Γ_{decoh}, $\Gamma_{\mathrm{m}} n_{\mathrm{m,BA}}$, and $\Gamma_{\mathrm{fb}} n_{\mathrm{imp}}$ respectively, and is given by (Eq. (6.2.24)),

$$n_{\mathrm{m}} + \frac{1}{2} = \frac{1}{1+g_{\mathrm{fb}}}\left(n_{\mathrm{m,th}} + n_{\mathrm{m,BA}}\right) + \frac{g_{\mathrm{fb}}^2}{1+g_{\mathrm{fb}}} n_{\mathrm{imp}} \geq 2\sqrt{n_{\mathrm{imp}}(n_{\mathrm{m,th}} + n_{\mathrm{m,BA}})}. \tag{6.2.42}$$

The minimum value on the RHS of corresponds to suppressing the apparent position noise to the imprecision noise floor (cf. yellow curve in Fig. 6.10, inset). Notably, in the absence of extraneous back-action, $n_{\mathrm{m}} < 1$ requires $n_{\mathrm{imp}} < 1/(2n_{\mathrm{m,th}})$.

Figure 6.10 shows the result of feedback cooling using a measurement with an imprecision far below that at the SQL. For this demonstration, imprecision was deliberately limited to $n_{\mathrm{imp}} = 2.9 \cdot 10^{-4}$ in order to reduce contribution from the off-resonant tail of the noise peak at 4.6 MHz (which limits applicability of Eq. (6.2.42) to damping rates less than 200 kHz) which is due to thermal motion of the in-plane mode of the nanobeam. The effective damping rate was controlled by changing the magnitude of the electronic gain, leaving all other parameters (e.g. laser power) unaffected. Fitting the closed loop noise spectrum (see Fig. 6.10, inset) to a standard Lorentzian noise squashing model (Eq. (6.2.23)), we estimate the phonon occupancy of the mechanical mode from the formula $n_{\mathrm{m}} + 0.5 \approx \Gamma_{\mathrm{eff}} \cdot (\bar{S}_\omega[\Omega_{\mathrm{m}}] + \bar{S}_\omega^{\mathrm{imp}})/2\bar{S}_\omega^{\mathrm{zp}}$, where $\bar{S}_\omega^{\mathrm{imp}}$ denotes the off-resonant background. Accounting for extraneous back-action, we infer a minimum occupation of $n_{\mathrm{m}} \approx 5.3 \pm 0.6$ at an optimal damping rate of 52 kHz, corresponding to a fractional ground state population of $1/(1 + n_{\mathrm{m}}) \approx 16\%$.

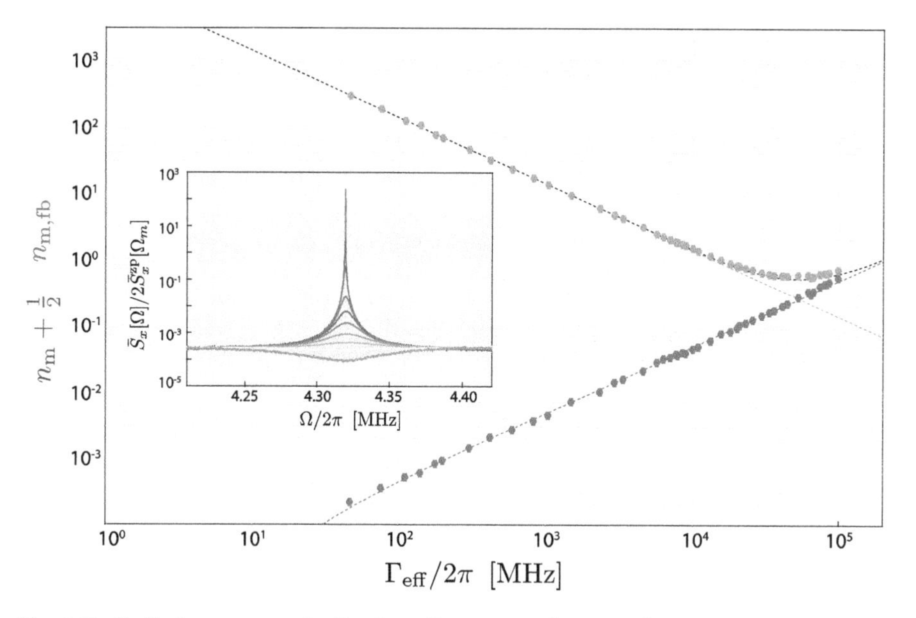

Fig. 6.10 Radiation pressure feedback cooling to near the ground state. Blue and red points correspond to measurements of the phonon occupancy of the mechanical mode, n_m (plus a phonon-equivalent zero-point energy of 1/2) and its component due to feedback of measurement noise $n_\mathrm{m,fb} = n_\mathrm{imp} g_\mathrm{fb}^2/(1+g_\mathrm{fb})$, respectively, as a function of measured damping rate, $\Gamma_\mathrm{eff} = (1+g_\mathrm{fb})\Gamma_\mathrm{m}$. Red, blue, and black dashed lines correspond to models of components in Eq. (6.2.42): $n_\mathrm{m}/(1+g_\mathrm{fb})$, $n_\mathrm{m,fb}$, and $n_\mathrm{m}+1/2$, respectively, using experimental parameters $\Gamma_\mathrm{m}/2\pi = 5.7$ Hz, $n_\mathrm{m} = 2.4 \cdot 10^5$, and $n_\mathrm{imp} = 2.8 \cdot 10^{-4}$, respectively. Inset: in-loop mechanical noise spectra for various feedback gain settings; fits to these spectra were used to infer blue and red points

6.3 Conclusion

Collectively, the results reported in this chapter establish new benchmarks for linear measurement and control of a mechanical oscillator. The enabling advance is a displacement imprecision 39.7 ± 0.3 dB below that at the SQL, a 100-fold improvement over results reported to date, combined with imprecision-back-action product within a factor of 5.0 of the uncertainty limit, on par with state-of-the-art optomechanical systems. At a moderate cryogenic temperature of 4.4 K, this amounts to the ability to resolve the zero-point motion of our 4.3 MHz oscillator at a measurement rate within an order comparable to its intrinsic thermal decoherence rate. To illustrate the utility of this advance, we have actively cooled the nanomechanical beam to a mean phonon occupancy of 5.3 ± 0.6 using radiation-pressure cold-damping; this represents a 50-fold improvement over previous active feedback cooling applied to mechanical oscillators [70]. Most importantly, and in contrast to all prior work on the feedback control of a mechanical oscillator [36, 70–83], since $n_\mathrm{m} \ll n_\mathrm{m,BA}$ in our experiment, feedback has suppressed $2 \cdot 10^3$ quanta of measurement back-action [10, 11], satisfying the basic premise of *quantum feedback*.

References

1. C.M. Caves, Phys. Rev. Lett **45**, 75 (1980)
2. A. Abramovici, W.E. Althouse, R.W.P. Drever, Y. Gürsel, S. Kawamura, F.J. Raab, D. Shoemaker, L. Sievers, R.E. Spero, K.S. Thorne, R.E. Vogt, R. Weiss, S.E. Whitcomb, M.E. Zucker, Science **256**, 325 (1992)
3. D.E. McClelland, N. Mavalvala, Y. Chen, R. Schnabel, Laser Photon. Rev. **677** (2011)
4. V. Braginsky, A. Manukin, Sov. Phys. JETP **25**, 653 (1967)
5. M. Evans et al., Phys. Rev. Lett. **114**, 161102 (2015)
6. C.J. Hood, T.W. Lynn, A.C. Doherty, A.S. Parkins, H.J. Kimble, Science **287**, 1447 (2000)
7. K.C. Schwab, M.L. Roukes, Phys. Today **58**, 36 (2005)
8. D.J. Wilson, V. Sudhir, N. Piro, R. Schilling, A. Ghadimi, T.J. Kippenberg, Nature **524**, 325 (2015)
9. V.B. Braginsky, F.Y. Khalili, *Quantum Measurement* (Cambridge University Press, 1992)
10. H. Wiseman, Phys. Rev. A **51**, 2459 (1995)
11. J.M. Courty, A. Heidmann, M. Pinard, Europhys. Lett. **63**, 226 (2003)
12. M. Hatridge, S. Shankar, M. Mirrahimi, F. Schackert, K. Geerlings, T. Brecht, K. Sliwa, B. Abdo, L. Frunzio, S. Girvin et al., Science **339**, 178 (2013)
13. M.L. Gorodetsky, A.D. Pryamikov, V.S. Ilchenko, J. Opt. Soc. Am. B **17**, 1051 (2000)
14. T.J. Kippenberg, S.M. Spillane, K.J. Vahala, Opt. Lett. **27**, 1669 (2002)
15. T.J. Kippenberg, K.J. Vahala, Science **321**, 1172 (2008)
16. M. Aspelmeyer, T.J. Kippenberg, F. Marquardt, Rev. Mod. Phys. **86**, 1391 (2014)
17. A. Gillespie, F. Raab, Phys. Lett. A **178**, 357 (1993)
18. O. Arcizet, P.-F. Cohadon, T. Briant, M. Pinard, A. Heidmann, J.-M. Mackowski, C. Michel, L. Pinard, O. Français, L. Rousseau, Phys. Rev. Lett. **97**, 133601 (2006)
19. W.H. Glenn, IEEE, J. Quant. Elec. **25**, 1218 (1989)
20. V.B. Braginsky, M.L. Gorodetsky, S.P. Vyatchanin, Phys. Lett. A **264**, 1 (1999)
21. V.B. Braginsky, M.L. Gorodetsky, S.P. Vyatchanin, Phys. Lett. A **271**, 303 (2000)
22. A.B. Matsko, A.A. Savchenkov, N. Yu, L. Maleki, J. Opt. Soc. Am. B **24**, 1324 (2007)
23. K. Numata, A. Kemery, J. Camp. Phys. Rev. Lett. **93**, 250602 (2004)
24. G. Anetsberger, E. Gavartin, O. Arcizet, Q.P. Unterreithmeier, E.M. Weig, M.L. Gorodetsky, J.P. Kotthaus, T.J. Kippenberg, Phys. Rev. A **82**, 061804 (2010)
25. C.E. Wieman, L. Hollberg, Rev. Sci. Instru. **62**, 1 (1991)
26. P.R. Saulson, Phys. Rev. D **42**, 2437 (1990)
27. T. Uchiyama, S. Miyoki, S. Telada, K. Yamamoto, M. Ohashi, K. Agatsuma, K. Arai, M.-K. Fujimoto, T. Haruyama, S. Kawamura, O. Miyakawa, N. Ohishi, T. Saito, T. Shintomi, T. Suzuki, R. Takahashi, D. Tatsumi, Phys. Rev. Lett. **108**, 141101 (2012)
28. M.L. Gorodetsky, I.S. Grudinin, J. Opt. Soc. Am. B **21**, 697 (2004)
29. O. Arcizet, R. Rivière, A. Schliesser, G. Anetsberger, T. Kippenberg, Phys. Rev. A **80**, 021803(R) (2009)
30. R.O. Pohl, X. Liu, E. Thompson, Rev. Mod. Phys **74**, 991 (2002)
31. R. Zeller, R. Pohl, Phys. Rev. B **4**, 2029 (1971)
32. R. Shelby, M.D. Levenson, P.W. Bayer, Phys. Rev. B **31**, 5244 (1985)
33. E. Verhagen, S. Deleglise, S. Weis, A. Schliesser, T.J. Kippenberg, Nature **482**, 63 (2012)
34. G. van Lear, G.E. Uhlenbeck, Phys. Rev **38**, 1583 (1931)
35. A. Yariv, *Quantum Electronics* (John Wiley, 1989)
36. P. Cohadon, A. Heidmann, M. Pinard, Phys. Rev. Lett. **83**, 3174 (1999)
37. T.P. Purdy, R.W. Peterson, C. Regal, Science **339**, 801 (2013)
38. J. Teufel, F. Lecocq, R. Simmonds, Phys. Rev. Lett. **116**, 013602 (2016)
39. D.G. Cahill, W.K. Ford, K.E. Goodson, G.D. Mahan, A. Majumdar, H.J. Maris, R. Merlin, S.R. Phillpot, J. Appl. Phys. **93**, 793 (2002)
40. J.P. Dowling, G.J. Milburn, Phil. Trans. Roy. Soc. A **361**, 1655 (2003)
41. D. D'Alessandro, *Introduction to Quantum Control and Dynamics* (Chapman & Hall, 2008)

42. L. Viola, E. Knill, S. Lloyd, Phys. Rev. Lett. **82**, 2417 (1999)
43. F. Verstraete, M.M. Wolf, J. Ignacio Cirac, Nat. Phy. **5**, 633 (2009)
44. C. Altafini, J. Math. Phys **44**, 2357 (2003)
45. D. Lidar, I. Chuang, K. Whaley, Phys. Rev. Lett. **81**, 2594 (1998)
46. H.M. Wiseman, Phys. Rev. A **49**, 2133 (1994)
47. H.M. Wiseman, G.J. Milburn, *Quantum Measurement and Control* (Cambridge University Press, 2010)
48. S. Lloyd, Phys. Rev. A **62**, 022108 (2000)
49. C. Sayrin, I. Dotsenko, X. Zhou, B. Peaudecerf, T. Rybarczyk, S. Gleyzes, P. Rouchon, M. Mirrahimi, H. Amini, M. Brune, J.-M. Raimond, S. Haroche, Nature **477**, 73 (2011)
50. B. Peaudecerf, C. Sayrin, X. Zhou, T. Rybarczyk, S. Gleyzes, I. Dotsenko, J. Raimond, M. Brune, S. Haroche, Phys. Rev. A **87**, 042320 (2013)
51. R. Vijay, C. Macklin, D.H. Slichter, S.J. Weber, K.W. Murch, R. Naik, A.N. Korotkov, I. Siddiqi, Nature **490**, 77 (2012)
52. Y. Makhlin, G. Schön, A. Shnirman, Rev. Mod. Phys. **73**, 357 (2001)
53. A.C. Doherty, A. Szorkovszky, G.I. Harris, W.P. Bowen, Phil. Trans. Roy. Soc. A **370**, 5338 (2012)
54. S. Mancini, D. Vitali, P. Tombesi, Phys. Rev. Lett **80**, 688 (1998)
55. J. Courty, A. Heidmann, M. Pinard, Eur. Phys. J. D **17**, 399 (2001)
56. N. Wiener, *Extrapolation, Interpolation and Smoothing of Stationary Time Series* (MIT Technology Press, 1949)
57. T. Kailath, A.H. Sayed, B. Hassibi, *Linear Estimation* (Prentice-Hall, 2000)
58. V.P. Belavkin, Rep. Math. Phys. **43**, A405 (1999)
59. H.M. Wiseman, A.C. Doherty, Phys. Rev. Lett. **94**, 070405 (2005)
60. L. Bouten, R. van Handel, M. James, SIAM J. Control Optim. **46**, 2199 (2007)
61. M. James, H. Nurdin, I. Petersen, I.E.E.E. Trans, Auto. Control **53**, 1787 (2008)
62. S.G. Hofer, K. Hammerer, Phys. Rev. A **91**, 033822 (2015)
63. H.W. Bode, *Network Analysis and Feedback Amplifier Design* (Van Nostrand Company, 1945)
64. N.G. van Kampen, Phys. Rev. **91**, 1267 (1953)
65. J. Giambiagi, I. Saavedra, Nucl. Phys. B **46**, 413 (1963)
66. H.M. Nussenzveig, Phys. Rev. **177**, 1848 (1969)
67. F.W. King, *Hilbert Transforms*, vol 2 (Cambridge University Press, 2009)
68. I.M. Gelfand, S.V. Fomin, *Calculus of Variations* (Prentice-Hall, 1963)
69. A.A. Clerk, M.H. Devoret, S.M. Girvin, F. Marquardt, R.J. Schoelkopf, Rev. Mod. Phys. **82**, 1155 (2010)
70. B. Abbott et al., New J. Phys **11**, 073032 (2009)
71. J.M.W. Milatz, J.J. Van Zolingen, B.B. Van Iperen, Physica **19**, 195 (1953)
72. J.A. Sirs, J. Sci. Instrum. **36**, 223 (1959)
73. P.G. Roll, R. Krotkov, R.H. Dicke, Ann. Phys **26**, 442 (1964)
74. R.L. Forward, J. Appl. Phys. **50**, 1 (1979)
75. D. Möhl, G. Petrucci, L. Thorndahl, S. van der Meer, Phys. Rep. **58**, 73 (1980)
76. B. D'Urso, B. Odom, G. Gabrielse, Phys. Rev. Lett **90**, 043001 (2003)
77. D. Kleckner, D. Bouwmeester, Nature **444**, 75 (2006)
78. M. Poggio, C. Degen, H. Mamin, D. Rugar, Phys. Rev. Lett. **99**, 017201 (2007)
79. T. Corbitt, C. Wipf, T. Bodiya, D. Ottaway, D. Sigg, N. Smith, S. Whitcomb, N. Mavalvala, Phys Rev. Lett **99**, 160801 (2007)
80. A. Vinante, M. Bignotto, M. Bonaldi, M. Cerdonio, L. Conti, P. Falferi, N. Liguori, S. Longo, R. Mezzena, A. Ortolan, G.A. Prodi, F. Salemi, L. Taffarello, G. Vedovato, S. Vitale, J.-P. Zendri, Phys. Rev. Lett. **101**, 033601 (2008)
81. A. Kubanek, M. Koch, C. Sames, A. Ourjoumtsev, P.W.H. Pinkse, K. Murr, G. Rempe, Nature **462**, 898 (2009)
82. T. Li, S. Kheifets, M.G. Raizen, Nat. Phys. **7**, 527 (2011)
83. J. Giesler, B. Deutsch, R. Quidant, L. Novotny, Phys. Rev. Lett **109**, 103603 (2012)

Chapter 7
Observation of Quantum Correlations Using Feedback

Correlations have physical reality; that which they correlate does not. And that's all there is to it, the rest is commentary.

David Mermin

Measurements proceed by establishing correlations between a system and a meter. In a quantum description of this process [1, 2], the effect of measurement persists in the system in the form of measurement back-action. For continuous linear measurements, where the meter couples linearly and weakly to the system, correlations between the system and meter manifest as imprecision-back-action correlations in the measurement record. Following the abstract discussion in Chap. 2, any continuous observable $\hat{y}$, of a system variable $\hat{x}$, must satisfy $[\hat{y}(t), \hat{y}(t')] = 0$. However, since $\hat{x}$ need not generally commute with itself, it must be that $\hat{y}$ must be contaminated by noises arising from the measurement chain. In interferometric position measurement, where $\hat{x}$ is the position of a mechanical oscillator, and the measurement chain is composed of an optical meter, and a phase-sensitive (typically, homodyne) detector, the contamination arises due to vacuum fluctuations of the meter. In this case, as illustrated in Sect. 4.3, $\hat{y} = (\hat{x} + \hat{x}_{\mathrm{BA}}) + \hat{x}_{\mathrm{imp}}$, where, $\hat{x}_{\mathrm{BA}}$ is the physical back-action driven motion, and $\hat{x}_{\mathrm{imp}}$ is the apparent motion. The symmetrised spectrum of the output of the measurement chain, i.e. of the continuous observable $\hat{y}$,

$$\bar{S}_{yy}[\Omega] = \bar{S}_{xx}^{\mathrm{imp}}[\Omega] + (\bar{S}_{xx}[\Omega] + \bar{S}_{xx}^{\mathrm{BA}}[\Omega]) + 2\mathrm{Re}\,\bar{S}_{x_{\mathrm{BA}}x_{\mathrm{imp}}}[\Omega], \tag{7.0.1}$$

generically features correlations between back-action and imprecision in the measurement record.

These correlations may be assigned physical reality independent of the final detection process that the meter—the optical field—is subjected to. The meter possesses two degrees of freedom (quadratures): amplitude and phase. Back-action arises from

V. Sudhir, *Quantum Limits on Measurement and Control of a Mechanical Oscillator*, Springer Theses, https://doi.org/10.1007/978-3-319-69431-3_7

vacuum fluctuations of the amplitude quadrature, which are imprinted onto the phase of the outgoing field. Thus the phase and amplitude quantum noises, by being causally connected via the mechanical response, get correlated. These quadrature correlations manifest as ponderomotive squeezing in homodyne detection of an appropriately chosen field quadrature [3–6], or as motional sideband asymmetry [6–10] in heterodyne detection. Differences between these effects arise from the details of how meter fluctuations are converted to a classical signal by the detection process [6–8, 11].

7.1 Quantum Correlations Due to Light-Motion Interaction

In the (ideal) case where an optomechanical system in the bad-cavity regime $\kappa \gg \Omega_\mathrm{m}$, is probed in the over-coupled regime ($\eta_c = 1$) on resonance with an optical field $\hat{a}_\mathrm{in}(t)$, the cavity transmission (see Eq. (4.3.2)),

$$\delta\hat{a}_\mathrm{out}[\Omega] = -\delta\hat{a}_\mathrm{in}[\Omega] - i\sqrt{C\Gamma_\mathrm{m}}\left(\frac{\delta\hat{x}^{(0)}[\Omega] + \delta\hat{x}_\mathrm{BA}[\Omega]}{x_\mathrm{zp}}\right), \tag{7.1.1}$$

carries information regarding the mechanical motion in its phase quadrature. Here, we have introduced the multi-photon cooperativity,

$$C := \frac{4g^2}{\kappa\Gamma_\mathrm{m}} = C_0 n_c,$$

that describes the transduction of mechanical motion onto the phase quadrature.

Equation (7.1.1) might be naively misunderstood to imply that fluctuations in $\delta\hat{a}_\mathrm{in}$ sets the measurement imprecision on top of which the total mechanical motion $\delta\hat{x}^{(0)} + \delta\hat{x}_\mathrm{BA}$ is resolved. This perspective is true if the field is detected in homodyne detector tuned to the phase quadrature, where the tradefoff between back-action and imprecision sets the standard quantum limit (Sect. 4.3). However, the back-action motion $\delta\hat{x}_\mathrm{BA}$, given by (see Eq. (4.3.4)),

$$\delta\hat{x}_\mathrm{BA}[\Omega] = \sqrt{2C\Gamma_\mathrm{m}}\frac{\hbar\chi_x^{(0)}[\Omega]}{x_\mathrm{zp}}\,\delta\hat{q}_\mathrm{in}[\Omega] \tag{7.1.2}$$

caused by quantum fluctuations in the input amplitude quadrature, creates correlations between the transmitted phase and amplitude, so that the naive expectation is false in general. To see this, note that the transmitted phase quadrature, according to Eq. (7.1.1), is given by,

$$\delta\hat{p}_\mathrm{out}[\Omega] = -\delta\hat{p}_\mathrm{in}[\Omega] - \sqrt{2C\Gamma_\mathrm{m}}\left(\frac{\delta\hat{x}^{(0)}[\Omega]}{x_\mathrm{zp}} - \sqrt{2C\Gamma_\mathrm{m}}\frac{\hbar\chi_x^{(0)}[\Omega]}{x_\mathrm{zp}^2}\delta\hat{q}_\mathrm{out}[\Omega]\right), \tag{7.1.3}$$

where we have used the expression for $\delta\hat{x}_{\text{BA}}$ in Eq. (7.1.2) and the fact that $\delta\hat{q}_{\text{in}} = -\delta\hat{q}_{\text{out}}$. The output phase and amplitude quadratures, by being related to each other, are correlated; in terms of the (unsymmetrised) cross-correlation spectrum,

$$S_{pq}^{\text{out}}[\Omega] = S_{pq}^{\text{in}}[\Omega] + 2C\Gamma_{\text{m}} \frac{\hbar\chi_x^{(0)}[\Omega]}{x_{\text{zp}}^2} S_{qq}^{\text{out}}[\Omega] = -\frac{i}{2} + \frac{2C\,\Omega_{\text{m}}\Gamma_{\text{m}}}{\Omega^2 - \Omega_{\text{m}}^2 - i\Omega\Gamma_{\text{m}}} \tag{7.1.4}$$

Since the intrinsic mechanical motion $\delta\hat{x}^{(0)}$ is uncorrelated with the optical beam, these optical quadrature correlations arise from two sources: correlations between the input amplitude and phase necessitated by the field commutation relations (see Sect. 3.2.2), and those due to interaction of the field with the mechanical oscillator. Correlations observed using linear measurements (linear optomechanical interaction followed by linear detection of the optical field) do not involve any contribution from the vacuum fluctuations of the mechanical oscillator [8, 12].

The two canonical types of linear detection schemes available for optical fields: optical homodyning and heterodyning, reveal phase-amplitude correlations differently. In the following two sub-section we briefly treat either case.

7.1.1 Manifestation as Ponderomotive Squeezing

An obvious way to probe correlations between the amplitude and phase of the output field is to detect a quadrature that is a linear superposition of the two, and hope to see the correlations via an interference. This strategy is realised by submitting the cavity transmission, $\delta\hat{a}_{\text{out}}$ in Eq. (7.1.1), to a homodyne detector [13, 14].

The observable relevant to homodyne detection is the general quadrature,

$$\hat{q}_{\text{out}}^{\theta}[\Omega] := \delta\hat{q}_{\text{out}}[\Omega]\cos\theta + \delta\hat{p}_{\text{out}}[\Omega]\sin\theta,$$

whose spectrum,

$$\bar{S}_{qq}^{\theta,\text{out}}[\Omega] = \bar{S}_{qq}^{\text{out}}[\Omega]\cos^2\theta + \bar{S}_{pp}^{\text{out}}[\Omega]\sin^2\theta + \text{Re}\ \bar{S}_{pq}^{\text{out}}[\Omega]\sin 2\theta,$$

is directly proportional to the homodyne photocurrent spectrum, $\bar{S}_I^{\theta,\text{hom}}[\Omega]$. When $\theta = \pi/2$, corresponding to conventional phase-quadrature homodyne detection, the correlation term ($\propto \sin 2\theta$) is absent, lending credence to the interpretation of the homodyne signal as being composed of a signal proportional to the oscillator position, contaminated by measurement imprecision due to vacuum fluctuations of the phase quadrature. However, when $0 < \theta < \pi/2$, the correlation term cannot be neglected, and such an interpretation is not tenable.

For resonant probing, using the expression for $\delta\hat{p}_{\text{out}}$ in Eq. (7.1.3) and noting that $\delta\hat{q}_{\text{out}} = -\delta\hat{q}_{\text{in}}$, the homodyne spectrum is (normalised to shot noise),

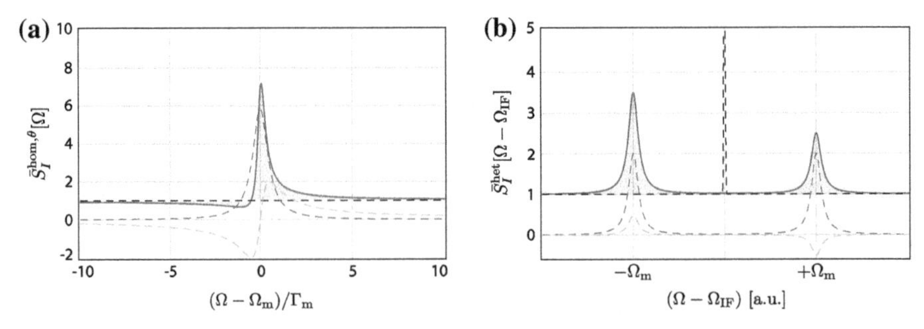

Fig. 7.1 Anatomy of ponderomotive squeezing and sideband asymmetry. a The observed homodyne spectrum (red) can be partitioned into three contributions as written in Eq. (7.1.5): a shot-noise contribution (black dashed), the contribution from the total motion of the oscillator (blue dashed), and that due to the real part of the correlations between amplitude and phase (green dashed). The anti-symmetric correlation term leads to a suppression of the photocurrent spectrum below shot-noise at a range of frequencies. The oscillator is here assumed to have $n_{\mathrm{m,th}} = 1$, and measured with a an effective cooperativity $\eta C = 1$; the spectrum depicted here is at a homodyne angle $\theta = \pi/8$. **b** The observed heterodyne spectrum (red) can be partitioned into three contributions as written in Eq. (7.1.6): a strong beat note and associated shot-noise from the heterodyne local oscillator (black dashed), the symmetrised double-sided spectrum of the total motion (blue dashed), and the imaginary part of the amplitude phase correlation (green dashed). The effect of correlations is to lead to a one-phonon asymmetry between the two sidebands. The oscillator is here assumed to have $n_{\mathrm{m,th}} = 2$

$$\begin{aligned}\bar{S}_I^{\theta,\mathrm{hom}}[\Omega] &= 1 + \frac{4\eta C\Gamma_\mathrm{m}}{x_\mathrm{zp}^2}\left(\bar{S}_{xx}[\Omega]\sin^2\theta + \frac{\hbar}{2}\mathrm{Re}\,\chi_x^{(0)}[\Omega]\,\sin 2\theta\right)\\ &= 1 + 16\eta C\frac{\left(n_\mathrm{m}+\frac{1}{2}\right)(\Omega_\mathrm{m}\Gamma_\mathrm{m})^2}{(\Omega^2-\Omega_\mathrm{m}^2)^2+(\Omega\Gamma_\mathrm{m})^2}\sin^2\theta + 4\eta C\frac{(\Omega_\mathrm{m}\Gamma_\mathrm{m})(\Omega^2-\Omega_\mathrm{m}^2)}{(\Omega^2-\Omega_\mathrm{m}^2)^2+(\Omega\Gamma_\mathrm{m})^2}\sin 2\theta\\ &\approx 1 + 16\eta C\frac{\left(n_\mathrm{m}+\frac{1}{2}\right)}{1+(2(\Omega-\Omega_\mathrm{m})/\Gamma_\mathrm{m})^2}\sin^2\theta + 4\eta C\frac{(2(\Omega-\Omega_\mathrm{m})/\Gamma_\mathrm{m})}{1+(2(\Omega-\Omega_\mathrm{m})/\Gamma_\mathrm{m})^2}\sin 2\theta.\end{aligned} \tag{7.1.5}$$

The first term, representing the vacuum fluctuations of the measured quadrature, sets the scale for photocurrent shot-noise, and is interpreted as the measurement imprecision in phase-quadrature homodyne detection (i.e. $\theta = \pi/2$). The second term, representing the total motion of the oscillator, is positive and symmetric about the mechanical resonance frequency, while the third term is anti-symmetric, and therefore negative on one side of the mechanical frequency (depending on the sign of θ); see Fig. 7.1a.

When the photocurrent variance in any frequency interval falls below the variance due to shot-noise, i.e. when $\bar{S}_I^{\theta,\mathrm{hom}}[\Omega] < 1$, the photocurrent is said to be *squeezed*. Since the homodyne photocurrent is directly proportional to the quadrature, $\hat{q}_\theta$, of the optical field exiting the signal arm of the interferometer, the spectrum of photocurrent fluctuations may be refereed back to the spectrum of the optical quadrature fluctuations; thus, photocurrent squeezing may be interpreted as squeezing of the optical quadrature fluctuations [15, 16]. From Eq. (7.1.5), squeezing occurs at frequencies

Ω, and detection angles θ, that satisfy,

$$0 < \frac{\Omega_{\rm m} - \Omega}{\Gamma_{\rm m}} \cot\theta < n_{\rm m} + \tfrac{1}{2}.$$

Physically, this condition corresponds to tuning to a quadrature sufficiently close to amplitude quadrature ($\theta = 0$) such that the transduction of the (typically large) thermal motion is suppressed, and to frequencies sufficiently far away from mechanical resonance such that the magnitude of the correlation term is comparable to shot-noise. The maximum squeezing of the photocurrent spectrum is quantified by the bound,

$$\bar{S}_I^{\theta,\rm hom}[\Omega] \gtrsim 1 - \eta \frac{n_{\rm m,BA}}{n_{\rm m,BA} + n_{\rm m,th}};$$

the largest possible squeezing is limited by the ratio of back-action to thermal motion, and the detection efficiency.

7.1.2 *Manifestation as Sideband Asymmetry*

In lieu of homodyning the output field, with the intent of measuring interference between its quadratures, it is possible to simultaneously measure both quadratures in a heterodyne detector. In this case, with a local oscillator field that is frequency-shifted by $\Omega_{\rm IF}$ (see discussion of heterodyne detection in Sect. 3.2.3.4), the spectrum of the photocurrent centred around $\Omega_{\rm IF}$ is,

$$\begin{aligned}
\bar{S}_I^{\rm het}[\Omega - \Omega_{\rm IF}] &\propto \frac{1}{2}\left(\bar{S}_{qq}^{\rm out}[\Omega] + \bar{S}_{pp}^{\rm out}[\Omega]\right) + \frac{i}{2}\left(S_{qp}^{\rm out}[\Omega] - S_{pq}^{\rm out}[\Omega]\right) \\
&= \frac{1}{2}\left(\bar{S}_{qq}^{\rm in}[\Omega] + \bar{S}_{pp}^{\rm in}[\Omega]\right) + \frac{2C\Gamma_{\rm m}}{x_{\rm zp}^2}\bar{S}_{xx}[\Omega] + \mathrm{Im}\left(S_{qp}^{\rm out}[\Omega] - S_{pq}^{\rm out}[\Omega]\right) \\
&= 1 + \frac{2C\Gamma_{\rm m}}{x_{\rm zp}^2}\left(\bar{S}_{xx}[\Omega] + \bar{S}_{xx}^{\rm zp}[\Omega] - \bar{S}_{xx}^{\rm zp}[-\Omega]\right). \qquad (7.1.6)
\end{aligned}$$

Here $\bar{S}_{xx} = \bar{S}_{xx}^{(0)} + \bar{S}_{xx}^{\rm BA}$ is the total mechanical motion. The additional contribution on either sideband, numerically equal to half of a zero-point motion, arises from the second term in the amplitude-phase correlation in Eq. (7.1.4). Ultimately, these correlations conspire to add (subtract) the equivalent of half a phonon of noise power on the upper (lower) sideband in heterodyne detection, leading to a one-phonon asymmetry between the two sidebands, as shown in Fig. 7.1b.

Unlike squeezing, the asymmetry between the two sidebands is one phonon, irrespective of detection efficiency, and the fractional contribution due to back-action. The reason is that the sideband asymmetry is a measure of correlations with respect to the thermal motion of the oscillator, and not to the detected shot-noise as in squeezing. The experimental challenge in observing the asymmetry is thus the absolute magnitude of thermal noise atop which the single phonon asymmetry must be resolved.

7.2 Observation of Quantum Correlations

In a laboratory setting, quantum correlations in interferometric position measurements, such as those studied in cavity optomechanics, are usually obscured by classical noise. An important example is thermal motion of the mechanical element.

In the case of ponderomotive squeezing, the contribution of thermo-mechanical noise is qualitatively different from the contribution due to imprecision-back-action correlations: the latter is anti-symmetric in frequency, and extends away from mechanical resonance, as shown by the green trace in Fig. 7.1a. Thus, by working away from mechanical resonance, and at homodyne detection angles close to amplitude quadrature, thermo-mechanical noise can be arbitrarily well-evaded.

In the case of sideband asymmetry, both the correlations, and the thermo-mechanical noise are qualitatively similar; quantitatively, the latter is $1/n_{\mathrm{m}}$ smaller than the former, i.e., thermal noise obscures quantum correlations. Two complementary approaches may be employed to increase the visibility of the correlations, and "distill" it from thermal noise. Coupling of the mechanical oscillators to an optical cavity mode serving as a cold bath, effectively realizing an autonomous feedback loop, has enabled thermal noise reduction to the level of the zero-point motion, thus increasing the visibility of the sideband asymmetry [8–10, 17]. In this case, back-action imposes a fundamental limit, which must be mitigated by operating in an appropriate parameter regime (the resolved-sideband regime [18, 19]). A second approach, described in Sect. 7.2.2, relies on feedback of an efficient auxiliary measurement to suppress thermal motion. As discussed in the previous chapter, measurement back-action can be suppressed by active feedback. The tradeoff in this case is between the efficiency of the measurement and the strength of an additional *feedback back-action* associated with the conversion of meter fluctuations into a classical signal. In this chapter, we show how cold damping can be used to increase the visibility of quantum-correlation-induced motional sideband asymmetry in an out-of-loop heterodyne measurement. Further we study how these quantum correlations are obscured in the regime where quantum noise in the in-loop detector causes the dominant force noise on the oscillator (termed *feedback back-action*); a regime giving rise to "squashing" of the in-loop photocurrent [20]. This demonstrates the complementary scenario where feedback is detrimental to the observation of quantum correlations. Conceptually, this feedback back-action dominated regime is analogous to the quantum back-action limit of sideband cooling [21].

7.2.1 *Observation of Ponderomotive Squeezing*

In our experiment we monitor the light transmitted after it has interacted with an optomechanical device operated in a cryogenic environment ($T \approx 6\,\mathrm{K}$). The device is nominally identical to the one used in the experiments reported in the previous chapter, and consists of a nanomechanical string coupled dispersively to an optical microcavity [22]. The fundamental mode of the string forms the oscillator (frequency $\Omega_{\mathrm{m}} = 2\pi \cdot 4.3$ MHz, damping rate $\Gamma_{\mathrm{m}} = 2\pi \cdot 7\,\mathrm{Hz}$). The meter is a laser field passing resonantly through the cavity (wavelength, $\lambda \approx 774\,\mathrm{nm}$).

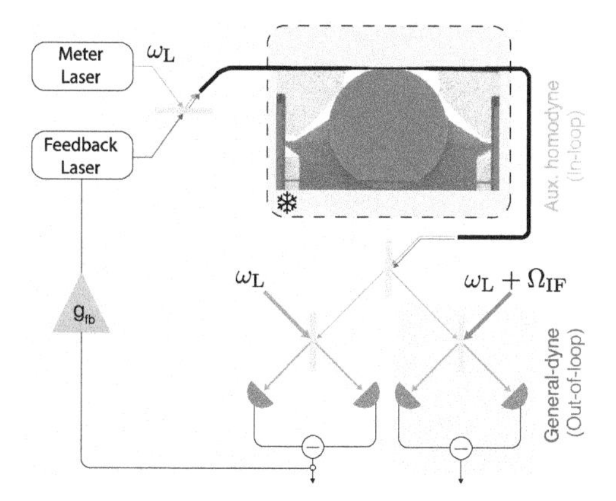

Fig. 7.2 Scheme to detect quantum correlations in the meter field. At the heart of the experiment is an optomechanical system consisting of a nanobeam (red) near-field-coupled to a whispering gallery mode optical micro-cavity (blue), placed in a cryostat (dashed black). The meter field is taper-coupled into the cavity, and is locked on resonance. The transmitted meter field can be directed to one of two detectors as shown. To detect ponderomotive squeezing, described in Sect. 7.2.1, the entirety of the meter laser is directed onto the ("out-of-loop") homodyne detector. The local oscillator in this case is degenerate in frequency with the meter (i.e., $\Omega_{IF} = 0$). To detect sideband asymmetry, described in Sect. 7.2.2, a portion of the transmitted meter field is directed onto the out-of-loop detector, now configured to operate in heterodyne mode (with $\Omega_{IF} = 2\pi \cdot 78\,\text{MHz}$). A remaining part of the meter is directed onto an (auxiliary) in-loop homodyne detector, whose signal is used to feedback cool the oscillator

The basic scheme of the experiment is depicted in Fig. 7.2 (see Fig. 5.9 for the detailed layout). In order to detect both manifestations of quantum correlations in the transmitted meter field, it is directed onto a (length- and power-) balanced general-dyne detector, which can be configured into a homodyne or a heterodyne configuration by simply changing the local oscillator frequency. In this sub-section we focus on the homodyne configuration.

As described above, in Sect. 7.2.1, phase-amplitude correlations in the meter field manifest as photocurrent squeezing when the homodyne detector is tuned to the amplitude quadrature. In contrast to the conventional operation of homodyne detector tuned to the phase quadrature (as detailed in Sect. 3.2.3.3), operation in the amplitude quadrature requires an error signal that is $\pi/2$ phase-shifted from the one used in the conventional case. We obtain this by demodulating a phase modulation tone injected using an EOM at the entrance of the homodyne interferometer (see Fig. 5.9). (In reality, we use exactly the same tone used to generate the Pound-Drever-Hall error signal for the lock of the meter laser.) By tuning the DC offset of this error signal, we can deterministically move to any quadrature, except precisely the phase quadrature. Figure 7.3c shows the suppression in the transduction of the thermo-mechanical signal ($\propto \sin^2\theta$ in Eq. (7.1.5)) as the homodyne detector is tuned away from phase quadrature and towards the amplitude quadrature. The loss of transduction may be equivalently viewed as an increased measurement imprecision, quoted as an effective quanta (read off from Eq. (7.1.5)),

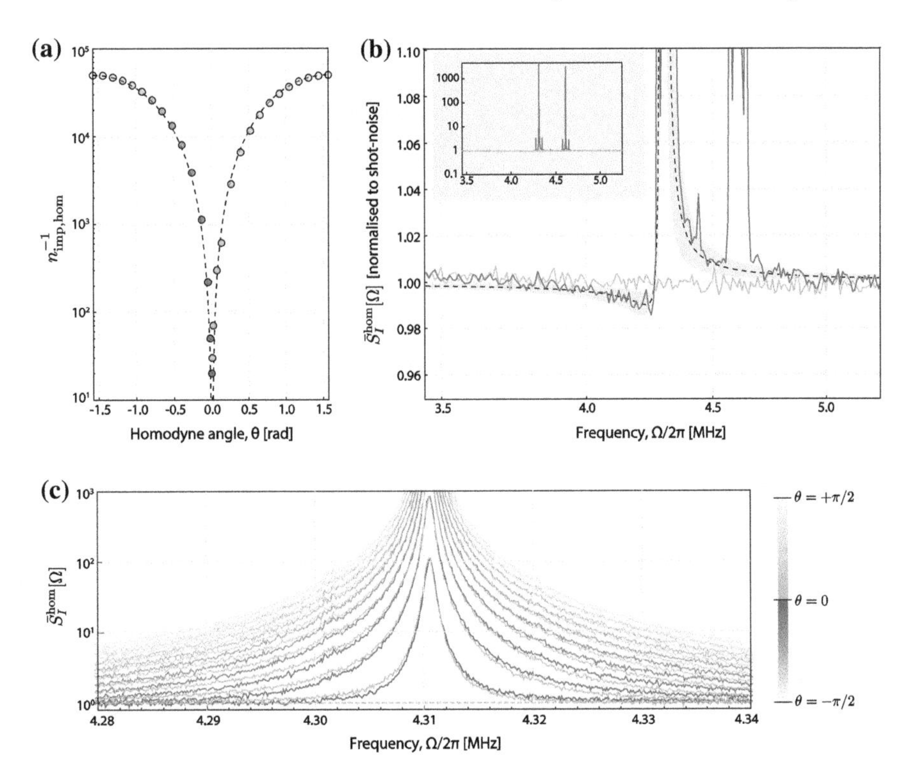

Fig. 7.3 Homodyne detection of quantum correlations. **a** Signal-to-noise ratio of the homodyne detected thermomechanical signal, quoted as an inverse imprecision quanta. Near the phase quadrature ($\theta = \pm\pi/2$), sensitivity to thermal motion is best; near the amplitude quadrature ($\theta \approx 0$), where squeezing is expected, thermal motion is suppressed by about 30 dB. **b** Squeezing spectrum probed using a homodyne detector. Red shows the homodyne photocurrent spectrum of the transmitted meter field, normalised to shot noise (orange), recorded at $\theta \approx 0.15$ rad. Black dashed shows prediction from theory using system parameters inferred from independent measurements. Gray bands show the model in Eq. (7.1.5), incorporating uncertainties in detection efficiency, and cooperativity. Inset shows zoom-out of the frequency landscape around the mechanical mode. The peak at 4.6 MHz is the in-plane mechanical mode, whose thermal noise contributes negligibly to the imprecision at $\Omega = 2\pi \cdot 4.3$ MHz. **c** A sequence of the homodyne spectra zoomed-in to near the mechanical frequency, as the homodyne angle is varied. Ponderomotive squeezing is not observed in these frequency interval because of being overwhelmed by thermomechanical noise. However, the effect of quantum correlations that are responsible for squeezing is nevertheless visible as a characteristic asymmetry in the spectra taken at two detection angles symmetric about $\theta = 0$ (red, and, blue). (Figure adapted with permission from Ref. [6]. Copyrighted by the American Physical Society)

$$n^{\theta}_{\text{imp,hom}} := \frac{1}{16\eta C \sin^2\theta} = \frac{n^{\pi/2}_{\text{imp,hom}}}{\sin^2\theta}. \tag{7.2.1}$$

Figure 7.3a shows the change in the (inverse of the) imprecision quanta as a function of the homodyne angle, in agreement with the expected $\sin^2\theta$ dependence.

The homodyne spectra corresponding to various quadratures are shown in Fig. 7.3c, overlaid on top of each other. Note that pairs of spectra corresponding to homodyne angles symmetric and opposite about amplitude quadrature exhibit an asymmetry at Fourier frequency detunings which are much less than the decoherence rate. This asymmetry confirms with the theoretical prediction of the spectra in Eq. (7.1.5), and is a clear signature of the presence of quantum correlations in the meter field. The suppression, or enhancement, of noise that leads to the asymmetry in the spectra are due to cancellation of quantum noise, both imprecision and back-action, vis-a-vis correlations between them.

At Fourier frequency offsets comparable to the oscillator decoherence rate, the magnitude of the imprecision-back-action correlations is sufficiently large so as to result in significant cancellation of the imprecision noise. As shown in Fig. 7.3b, this leads to a suppression of the photocurrent noise below its vacuum level, i.e. squeezing. The observation of ponderomotive squeezing provides bona-fide proof of the presence of quantum correlations in the meter field.

7.2.1.1 Effects Due to Decoherence in Homodyne Detection

Two sources of noise typically obscure (or reduce) the level of observed photocurrent squeezing, for a given amount of optical squeezing. The first is detection inefficiency: the addition of excess vacuum noise via various (primarily, optical) loss channels leads to optical fluctuations uncorrelated with the squeezed field. The second is noise in homodyne angle: the detection of quadratures other than the specific one containing correlations leads to excess vaccum noise contaminating the photocurrent signal [23]. The former has already been implicitly accounted for in the discussion given in Sect. 7.1.1; here we briefly outline the effect of the latter.

The homodyne photocurrent operator is nominally proportional to the quadrature to be measured, i.e.,

$$\delta\hat{I}_\theta(t) \propto \delta\hat{q}_\theta(t) = \delta\hat{q}(t)\cos\theta + \delta\hat{p}(t)\sin\theta.$$

Noise in homodyne angle—due to various technical sources, such as uncompensated interferometer path-length fluctuations, or uncompensated low-frequency phase fluctuations—can be modelled as a fluctuation in the quadrature angle, i.e.

$$\theta \mapsto \theta + \delta\theta(t).$$

Assuming that $\delta\theta$ is zero-mean normally distributed, and that, $\langle\delta\theta^2\rangle \ll 1$, the photocurrent operator takes the form,

$$\delta\hat{I}_\theta(t) \approx \delta\hat{q}_\theta(t) + \delta\hat{q}_{\theta+\frac{\pi}{2}}(t)\delta\theta(t). \tag{7.2.2}$$

Thus, the quadrature to be measured, $\delta\hat{q}_\theta$, is contaminated by its conjugate quadrature. Evaluating the spectrum of the photocurrent,[1]

$$\bar{S}_I^{\text{hom},\theta}[\Omega] \propto \bar{S}_{qq}^{\theta}[\Omega] + \int \bar{S}_{qq}^{\theta+\frac{\pi}{2}}[\Omega - \Omega']\bar{S}_{\theta\theta}[\Omega']\,\frac{\mathrm{d}\Omega'}{2\pi}, \tag{7.2.3}$$

it is apparent that an additional imprecision (second term) has been transduced into the photocurrent via angle fluctuations.

The spectrum of homodyne angle fluctuations, $\bar{S}_{\theta\theta}$, can in principle be directly inferred (see for example Fig. 3.4). In our experiment, these fluctuations are limited to low frequencies, so that the approximation,

$$\bar{S}_{\theta\theta}[\Omega] \approx \text{Var}\,[\theta] \cdot 2\pi\,\delta[\Omega],$$

is valid; this is essentially equivalent to assuming that the homodyne angle is a white Gaussian random variable with mean θ and variance $\text{Var}\,[\theta]$ (which can be estimated from the low-frequency integral of the fluctuations). In this case, the homodyne photocurrent spectrum in Eq. (7.2.3) takes the form,

$$\bar{S}_I^{\text{hom},\theta}[\Omega] \propto \bar{S}_{qq}^{\theta}[\Omega] + \text{Var}\,[\theta]\,\bar{S}_{qq}^{\theta+\frac{\pi}{2}}[\Omega]. \tag{7.2.4}$$

For amplitude quadrature measurements, corresponding to $\theta = 0$, this equation suggests two effects that contaminate the observed squeezing spectrum: firstly, imprecision noise from the vacuum fluctuations in the phase quadrature get transduced by a factor $\text{Var}\,[\theta]$ that contaminates squeezing away from mechanical resonance, secondly, thermo-mechanical noise in the phase quadrature is transduced by the same factor near mechanical resonance. Thus, the suppression of thermo-mechanical noise on amplitude quadrature, shown in Fig. 7.3a, allows an upper-bound[2] to be placed on $\text{Var}\,[\theta]$. Averaging over the homodyne angle fluctuations, the inverse of the imprecision in Eq. (7.2.1),

$$\left\langle (n_{\text{imp,hom}}^{\theta})^{-1} \right\rangle = (n_{\text{imp,hom}}^{\pi/2})^{-1} \left\langle \sin^2\theta \right\rangle = (n_{\text{imp,hom}}^{\pi/2})^{-1} \frac{1}{2}\left(1 - e^{-2\text{Var}[\theta]}\cos 2\theta\right),$$

takes a non-zero value even at $\theta = 0$, due to angle fluctuations. When the mean angle is set to amplitude quadrature ($\theta = 0$), and fluctuations are small ($\text{Var}\,[\theta] \ll 1$),

$$\frac{\left\langle (n_{\text{imp,hom}}^{\theta})^{-1} \right\rangle}{(n_{\text{imp,hom}}^{\pi/2})^{-1}} \approx \text{Var}\,[\theta]\,.$$

[1] We have omitted a term proportional to the correlation between the two conjugate quadratures; such a term is strictly zero when the angle fluctuations are uncorrelated with the quadrature fluctuations, which is indeed the typical case.

[2] This is only an upper-bound because in practice, thermo-mechanical motion could be transduced into the amplitude quadrature by dissipative optomechanical coupling [24].

From Fig. 7.3a, this ratio is ≈40 dB, implying a conservative upper-bound, $\sqrt{\mathrm{Var}[\theta]} < 0.01$ rad, for the homodyne angle fluctuations. This in turn implies that the excess imprecision due to vacuum noise in the phase quadrature transduced by homodyne angle noise is at the level of 1% of amplitude quadrature vacuum noise.

7.2.2 *Observation of Sideband Asymmetry Using Feedback*

When the meter field is incident on a heterodyne detector, both its quadratures are measured simultaneously (see Sect. 3.2.3.4 for details), giving access to $\bar{S}_{yy}^{\mathrm{het}}[\Omega > 0]$, where $\bar{S}_{yy}^{\mathrm{het}}[\Omega_{\mathrm{IF}} \pm \Omega_{\mathrm{m}}]$ is the position-equivalent heterodyne photocurrent spectrum corresponding to the upper (+) and lower (−) motional sidebands (displaced by the heterodyne intermediate frequency, Ω_{IF}). Quantum correlations between the phase and amplitude of the meter field are converted to imprecision-back-action correlations at the detector, that manifest as an asymmetry in the motional sidebands.

Motional sideband asymmetry can be understood from the three terms in the expression for the position-equivalent heterodyne spectrum (Eq. 7.0.1),

$$\bar{S}_{yy}^{\mathrm{het}}[\Omega] = \bar{S}_{xx}^{\mathrm{imp,het}}[\Omega] + (\bar{S}_{xx}[\Omega] + \bar{S}_{xx}^{\mathrm{BA}}[\Omega]) + 2\mathrm{Re}\,\bar{S}_{x_{\mathrm{BA}}x_{\mathrm{imp,het}}}[\Omega]$$

illustrated in Fig. 7.4a. Detector imprecision (gray)—arising from the vacuum fluctuations in the phase and amplitude quadrature of the meter—contributes a phonon-equivalent noise of,

$$n_{\mathrm{imp}}^{\mathrm{het}} := \frac{\bar{S}_{yy}^{\mathrm{het,imp}}[\Omega_{\mathrm{IF}} \pm \Omega_{\mathrm{m}}]}{\bar{S}_{xx}^{\mathrm{zp}}[\Omega_{\mathrm{m}}]}. \tag{7.2.5}$$

Physical motion—arising from a combination of thermal force and meter back-action—contributes $n_{\mathrm{m}} + \frac{1}{2}$ phonons to each sideband. Imprecision-back-action correlations—arising from amplitude-phase correlations in the meter—contribute $\pm\frac{1}{2}$ phonons to the lower/upper sideband (green dashed). The resulting asymmetry of the sidebands (red traces),

$$R := \frac{\bar{S}_{yy}^{\mathrm{het}}[\Omega_{\mathrm{het}}^{+}] - \bar{S}_{yy}^{\mathrm{het,imp}}[\Omega_{\mathrm{het}}^{+}]}{\bar{S}_{yy}^{\mathrm{het}}[\Omega_{\mathrm{het}}^{-}] - \bar{S}_{yy}^{\mathrm{het,imp}}[\Omega_{\mathrm{het}}^{-}]} \approx \frac{n_{\mathrm{m}}}{n_{\mathrm{m}} + 1}, \tag{7.2.6}$$

is commensurate with one phonon and arises purely from quantum correlations in the meter (here $\Omega_{\mathrm{het}}^{\pm} \equiv \Omega_{\mathrm{IF}} \pm \Omega_{\mathrm{m}}$). This asymmetry corresponds directly to the visibility of imprecision-back-action correlations with respect to the total resonant noise power, i.e.,

$$\frac{2\mathrm{Re}\,\bar{S}_{x_{\mathrm{BA}}x_{\mathrm{imp,het}}}[\Omega_{\mathrm{het}}^{+}]}{\bar{S}_{xx}^{\mathrm{imp}}[\Omega_{\mathrm{het}}^{+}] + S_{xx}^{\mathrm{tot}}[\Omega_{\mathrm{het}}^{+}]} \approx \frac{1 - R}{1 + R} = \frac{1}{2n_{\mathrm{m}} + 1}, \tag{7.2.7}$$

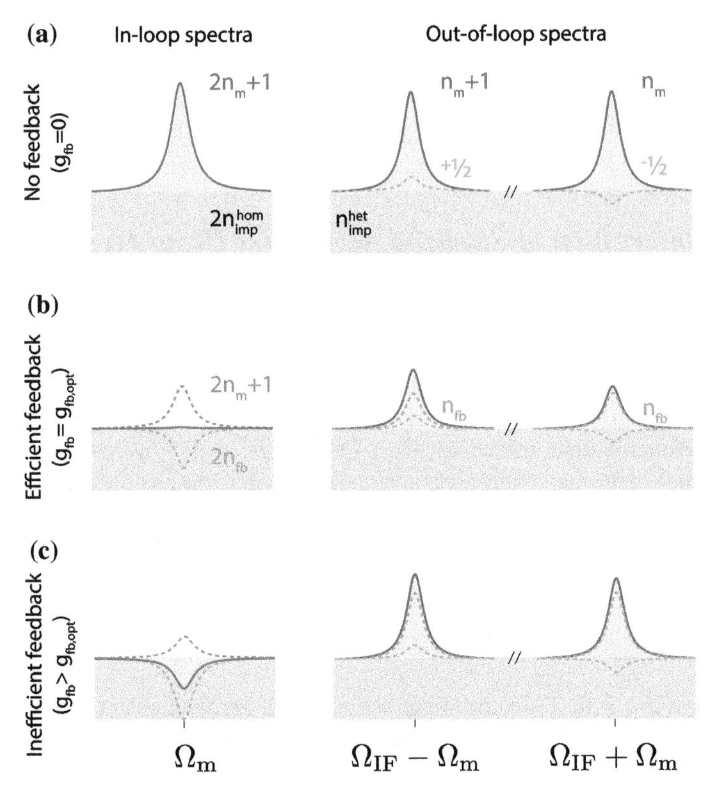

Fig. 7.4 Using homodyne feedback to increase the visibility of quantum-correlation-induced motional sideband asymmetry. Panels show the various components that constitute the in-loop homodyne (left) and out-of-loop heterodyne (right) spectrum, as the feedback gain is increased. **a** With no feedback, the homodyne signal (red left) is proportional to the total thermal occupation, while the heterodyne signal (red right) is asymmetric due to $\pm\frac{1}{2}$ phonon equivalent contributions (green dashed) from quantum correlations. **b** At optimal feedback, the homodyne signal coincides with the measurement imprecision (grey) due to classical correlations (orange dashed) from feedback back-action exactly cancelling the physical motion (blue dashed). In this case, visibility of heterodyne sideband asymmetry is maximum. **c** Further increase in the feedback gain leads to squashing of the homodyne signal, and a decrease in the visibility of sideband asymmetry in the heterodyne detector due to feedback back-action that is large compared to the thermal occupation. (Figure adapted with permission from Ref. [6]. Copyrighted by the American Physical Society)

where, $\bar{S}_{xx}^{\mathrm{tot}} = \bar{S}_{xx} + \bar{S}_{xx}^{\mathrm{BA}}$, is the total motion. Clearly, the visibility of correlations, and therefore, the fractional magnitude of sideband asymmetry, is obscured by the thermal and back-action occupation of the oscillator.

Active feedback cooling can suppress both the thermal and the back-action occupation of the oscillator, thus increasing the visibility of quantum correlations. We use the measurement record of an auxiliary homodyne measurement, as shown in Fig. 7.2, as an error signal for feedback. Briefly, the homodyne signal is imprinted onto the amplitude quadrature of an independent *feedback* laser resonant with an auxiliary cavity mode ($\lambda \approx 840$ nm). The loop delay is tuned in order to produce a purely viscous radiation pressure feedback force, effectively coupling the oscillator at a rate $\Gamma_{\mathrm{fb}} \approx g_{\mathrm{fb}}\Gamma_{\mathrm{m}}$ to a cold bath with an occupation equal to the phonon-equivalent homodyne imprecision $n_{\mathrm{imp}}^{\mathrm{hom}} = \bar{S}_{xx}^{\mathrm{imp,hom}}[\Omega_{\mathrm{m}}]/2\bar{S}_{xx}^{\mathrm{zp}}[\Omega_{\mathrm{m}}]$; here g_{fb} is the dimensionless feedback gain. The occupation of the oscillator is thereby reduced to,

$$n_{\mathrm{m}} + \frac{1}{2} \approx \frac{n_{\mathrm{tot}}}{g_{\mathrm{fb}}} + g_{\mathrm{fb}} n_{\mathrm{imp}}^{\mathrm{hom}} \geq 2\sqrt{n_{\mathrm{tot}} n_{\mathrm{imp}}^{\mathrm{hom}}}, \tag{7.2.8}$$

with the minimum achieved at an optimal gain of $g_{\mathrm{fb}}^{\mathrm{opt}} = \sqrt{n_{\mathrm{tot}}/n_{\mathrm{imp}}^{\mathrm{hom}}}$. Here, $n_{\mathrm{tot}} = n_{\mathrm{m,th}} + n_{\mathrm{m,BA}}$ is the effective bath occupation of the mechanical oscillator, including measurement back-action. Two regimes may be identified:

1. An *efficient* feedback regime, characterised by $g_{\mathrm{fb}} < g_{\mathrm{fb}}^{\mathrm{opt}}$, in which the motion of the oscillator is efficiently suppressed.
2. An *inefficient* feedback regime, in which the total motion is overwhelmed by *feedback back-action* $n_{\mathrm{fb}} = g_{\mathrm{fb}}^2 n_{\mathrm{imp}}^{\mathrm{hom}}$, arising from feedback of homodyne imprecision noise, resulting in an increase of n_{m} for feedback gains larger than the optimal value, i.e. $g_{\mathrm{fb}} > g_{\mathrm{fb,opt}}$.

An experimental demonstration of efficient feedback cooling, where feedback back-action is weak ($n_{\mathrm{fb}} < n_{\mathrm{tot}}$), is shown in Fig. 7.5. Here $n_{\mathrm{tot}} \approx 7 \cdot 10^4$, corresponding to an effective bath temperature of 13 K (arising partly due to photo-absorption, $n_{\mathrm{BA}} \approx 4 \cdot 10^4$). From the perspective of the heterodyne measurement, the objective is to "distill" a motional sideband asymmetry of one phonon out of n_{tot}. This is made possible by a low shot-noise-limited homodyne imprecision of $n_{\mathrm{imp}}^{\mathrm{hom}} \approx 1.2 \cdot 10^{-4}$. To trace out the cooling curve in Fig. 7.5, the feedback gain is tuned electronically while keeping all other experimental parameters (such as mean optical power and laser-cavity detuning) fixed. Sideband ratio R is extracted from fitting a Lorentzian to each heterodyne sideband and taking the ratio of the fitted areas. The phonon occupation n_{m} is inferred from R as well as the area beneath the lower sideband. In-loop (homodyne) and out-of-loop (heterodyne) noise spectra are shown in Fig. 7.5. As a characteristic of the efficient feedback regime, the area under the left sideband decreases linearly with g_{fb}, corresponding to $n_{\mathrm{m}} \propto g_{\mathrm{fb}}^{-1}$ (red circles in Fig. 7.5). As the optimal gain is approached, the in-loop spectrum is reduced to the imprecision noise floor (black trace in Fig. 7.5). This transition coincides with the appearance of a sideband asymmetry of $1 - R \approx 12\%$, corresponding to $n_{\mathrm{m}} \approx 7.3$.

To confirm the faithfulness of these measurements, two major sources of error were investigated:

1. Drift over the course of measurement can introduce small changes in the relative magnitude of $\bar{S}_{yy}^{\mathrm{het}}[\Omega_{\mathrm{het}}^{\pm}]$. In our experiment, this effect is mitigated by recording both heterodyne sidebands simultaneously. Augmented by operating in the bad cavity regime ($\Omega_{\mathrm{m}}/\kappa \sim 10^{-3}$), and the exceptionally low imprecision of the heterodyne measurement, $n_{\mathrm{imp}}^{\mathrm{het}} = (4\eta_{\mathrm{het}} C_0 n_c)^{-1} \approx 3 \cdot 10^{-3}$ (see Fig. 7.6a), statistical fluctuations of R over the course of a typical measurement set can

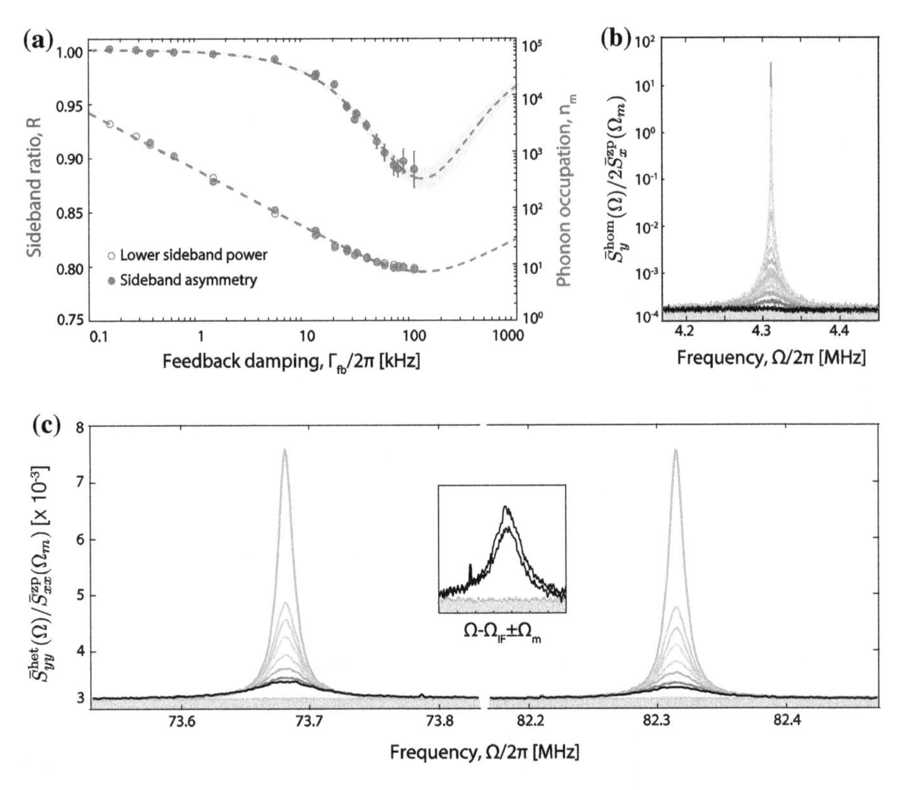

Fig. 7.5 Motional sideband asymmetry in the regime of efficient feedback. a Heterodyne sideband asymmetry (R, blue) and inferred mechanical mode occupation (n_m, red) versus closed-loop mechanical damping rate (Γ_fb) for various feedback gains. A maximum asymmetry of $1-R\approx 12\%$ ($n_\mathrm{m}\approx 7.3$) appears as the feedback gain approaches its optimal value. Dashed lines correspond to models $R=\frac{n_\mathrm{m}}{n_\mathrm{m}+1}$ (Eq. 7.2.6, blue line) and $n_\mathrm{m}+\frac{1}{2}\approx\frac{\Gamma_\mathrm{m}}{\Gamma_\mathrm{fb}}n_\mathrm{tot}+\frac{\Gamma_\mathrm{fb}}{\Gamma_\mathrm{m}}n_\mathrm{imp}^\mathrm{hom}$ (Eq. (7.2.8), red line). Solid blue band is a confidence interval based on uncertainties in estimates of n_tot, $n_\mathrm{imp}^\mathrm{hom}$, and Γ_m. Open red circles are independent estimates of n_m based on the area beneath the left heterodyne sideband. **b, c** Homodyne (**b**) and heterodyne (**c**) spectra used to obtain (**a**). Black traces correspond to lowest occupation; asymmetry is highlighted in the inset. Only a subset of heterodyne spectra are shown, for low n_m, with colours matching the corresponding homodyne spectra. An important feature of these spectra are their low imprecision, $n_\mathrm{imp}^\mathrm{hom}=(16\eta_\mathrm{hom}C_0n_\mathrm{c})^{-1}=1.2\cdot 10^{-4}$ and $n_\mathrm{imp}^\mathrm{het}=(4\eta_\mathrm{het}C_0n_c)^{-1}=2.9\cdot 10^{-3}$. This is made possible by the high photon collection efficiency $\eta\sim 0.2$, single photon cooperativity $C_0=4g_0^2/\kappa\Gamma_\mathrm{m}=0.3$, and power handling capacity of the microcavity-based sensor (allowing for intracavity photon numbers of $n_\mathrm{c}\sim 10^4$). (Figure adapted with permission from Ref. [6]. Copyrighted by the American Physical Society)

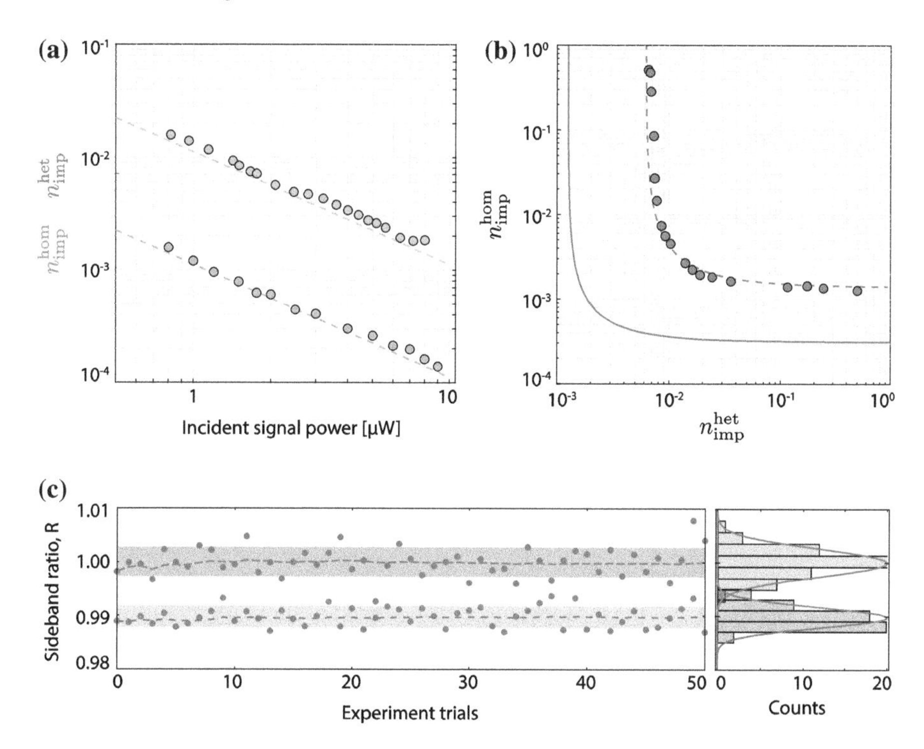

Fig. 7.6 Experimental sensitivity and precision. a Deep sub-SQL measurement sensitivity in both the in-loop (homodyne) detector and the out-of-loop (heterodyne) detector. The dashed lines show expected behaviour for quantum-limited detection. **b** Plot shows sensitivity tradeoff between the in-loop and out-of-loop detector; the relative sensitivity is changed by using a waveplate and a polarizing beam-splitter to distribute the signal between the two detectors. The solid line shows expected model for unit detection efficiency; dashed line corresponds to realistic efficiencies of $\eta_{\rm hom} \approx 20\%$ and $\eta_{\rm het} \approx 15\%$. **c** Statistical fluctuations of R for low feedback gain, indicating the ability to discriminate a 0.5% asymmetry, corresponding to $n_{\rm m} \approx 100$. (Figure adapted with permission from Ref. [6]. Copyrighted by the American Physical Society)

be as small as 0.5% (see Fig. 7.5c). Error bars for R in Fig. 7.5a are derived from the standard deviation of similar data sets (shown in Fig. 7.6c), in addition to a small contribution from the fit covariance matrix. At the largest damping rates, the reduced heterodyne signal-to-noise results in insufficient convergence of the periodogram estimate of the spectra (keeping acquisition time and analysis bandwidth fixed), leading to larger error bars, $\delta R = \pm 2\%$.

2. Excess laser noise affects R by producing additional imprecision-back-action correlations as discussed in Sect. 7.2.2.1. Assuming a mean thermal photon occupation of $C_{qq(pp)}$ for the amplitude (phase) quadrature of the injected meter field, the correlator in Eq. (7.0.1) becomes (see Eq. 7.2.6),

$$\frac{2\text{Re}\,\bar{S}^{\rm het}_{x_{\rm ba}x_{\rm imp}}[\Omega^{\pm}_{\rm het}]}{\bar{S}^{\rm zp}_{xx}[\Omega_{\rm m}]} = \mp\eta_{\rm het}\left(\frac{1}{2} + C_{qq} \pm \frac{4\bar{\Delta}\Omega_{\rm m}}{\kappa^2}C_{pp}\right), \tag{7.2.9}$$

where η_{het} is the heterodyne detection efficiency, and $\bar{\Delta}$ is the mean laser-cavity detuning. In our experiment, independent measurements reveal that $C_{qq} < 0.01$ and $C_{pp} < 30$ (owing partly to excess cavity frequency noise) for typical meter powers of $P_{\text{in}} < 5\,\mu\text{W}$. Operating on resonance ($\bar{\Delta} \approx 0$) and in the bad-cavity regime substantially reduces sensitivity to C_{pp}. Using a typical value of $\bar{\Delta} = 0.01 \cdot \kappa$, we estimate that $\frac{4\bar{\Delta}\Omega_{\text{m}}}{\kappa^2} C_{pp} < 0.005$ negligibly to Eq. (7.2.9).

Having established that our measurements of motional sideband asymmetry are not contaminated by classical artefacts, the results shown in Fig. 7.5 may be interpreted as a 'distillation' of quantum correlations using efficient feedback. We now explore the complementary regime of inefficient feedback, where feedback back-action is stronger than the thermal force and measurement back-action ($n_{\text{fb}} > n_{\text{tot}}$). We access this regime by changing the homodyne/heterodyne splitting ratio, thereby increasing the homodyne imprecision to $n_{\text{imp}}^{\text{hom}} \approx 10^{-3}$. As shown in Fig. 7.7, increasing the gain beyond its optimum value (corresponding to $n_{\text{m}} \approx 13.4$ and $1-R \approx 7\%$), results in a reduction of the homodyne signal below the shot-noise level (Fig. 7.7b left panel). Simultaneously, the areas of the heterodyne sidebands increase, while their asymmetry ($1 - R$) decreases. The discrepancy between "squashing" [20, 25] of the in-loop signal and the disappearance of sideband asymmetry relates to a basic difference between feedback back-action and meter back-action, namely, feedback back-action is correlated with the in-loop imprecision and not with the out-of-loop imprecision [20].

Squashing of the in-loop signal is caused by correlations between the feedback back-action driven motion x_{fb} and the in-loop measurement imprecision,

$$\frac{2\text{Re}\bar{S}^{\text{hom}}_{x_{\text{fb}}x_{\text{imp}}}[\Omega_{\text{m}}]}{2\bar{S}^{\text{zp}}_{xx}[\Omega_{\text{m}}]} = -n^{\text{hom}}_{\text{imp}}\, g_{\text{fb}}. \tag{7.2.10}$$

represented by the negative-valued green trace in Fig. 7.4c (right panel). Interestingly, these classical correlations, in conjunction with the generalised Heisenberg uncertainty principle [1, 26] can be used to predict the transition from efficient to inefficient feedback; viz.

$$\bar{S}_{FF} \cdot \bar{S}^{\text{imp,hom}}_{xx} \geq \frac{\hbar^2}{2} + (2\text{Re}\,\bar{S}_{Fx_{\text{imp,hom}}})^2, \tag{7.2.11}$$

is saturated for $g_{\text{fb}}^{\text{opt}} = \sqrt{n_{\text{tot}}/n_{\text{imp}}^{\text{hom}}}$ (using $F_{\text{fb}} \propto g_{\text{fb}} x_{\text{imp}}^{\text{hom}}$ and Eq. (7.2.10)). The limits of feedback cooling, and the prospects for feedback-based enhancement of quantum correlations, is related to the detection of meter fluctuations and the choice of feedback strategy—optimisation of either seems pertinent (see Chap. 8 for some ideas).

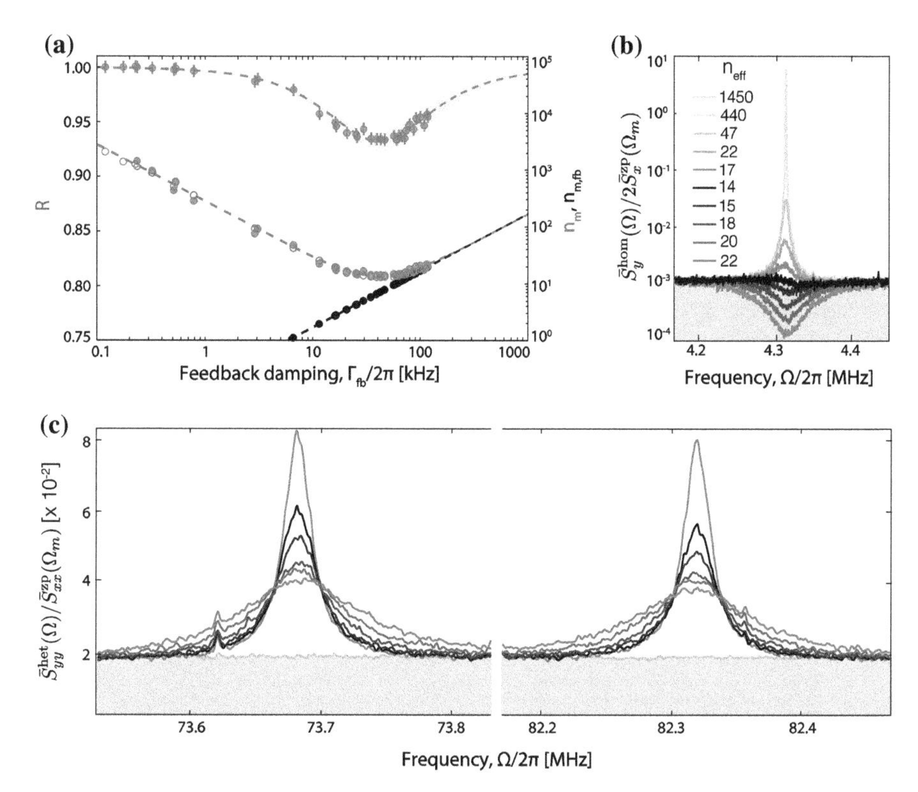

Fig. 7.7 Appearance and disappearance of sideband asymmetry. a Repeat of the experiment shown in Fig. 7.5 with lower homodyne detection efficiency. Feedback with the same range of gain results in lower asymmetry ($R \approx 6\%$) and access to the 'strong feedback' regime in which feedback back-action (n_{fb}) dominates physical motion, resulting in reduced R. Black points are an estimate of the mechanical occupation due to feedback back-action, $n_{\text{m,fb}} = \frac{\Gamma_{\text{m}}}{\Gamma_{\text{fb}}} n_{\text{fb}} = g_{\text{fb}} n_{\text{imp}}^{\text{hom}}$, based on the noise floor of the homodyne spectra. **b** In-loop homodyne spectra. In the strong feedback regime, noise is 'squashed' (reduced below the open-loop imprecision), corresponding to in-loop squeezing. **c** Out-of-loop heterodyne spectra. Inefficient feedback manifests as an increase in the off-resonant noise power and reduced asymmetry. (Figure adapted with permission from Ref. [6]. Copyrighted by the American Physical Society)

7.2.2.1 Classical and Quantum Contribution to Sideband Asymmetry

The cross-correlation spectrum, S_{pq}^{out}, by being directly related to S_{qq}^{out} (see Eq. (7.1.4)), can get contaminated by classical contribution to S_{qq}^{out}. These contributions may arise from classical contributions to S_{qq}^{in} (input amplitude noise), or from classical contributions to S_{pp}^{in} (input phase noise) that get transduced by the cavity into S_{qq}^{out}. In the following, a formal treatment of these two contributions is provided for the experimentally relevant case of simultaneous measurement of the sidebands in the bad-cavity regime.[3] In Sect. 7.2.2.2, we present measurements confirming the negligible contribution of laser noise to the results reported above. We also treat the optical cavity and the mechanical oscillator on equal footing so as to identify the contribution to sideband asymmetry from the vacuum fluctuations of either—we show that only the vacuum fluctuations from the optical cavity, vis-a-vis back-action, gives rise to sideband asymmetry.

In our experiment, we probe the optomechanical system using a resonant laser at frequency ω_ℓ. The photon flux amplitude operator of the laser, $a_{\text{in}}(t)$, is assumed to have the form (as in Eq. (3.2.51)),

$$\hat{a}_{\text{in}}(t) = e^{-i\omega_L t}(\bar{a}_{\text{in}} + \delta\hat{a}_{\text{in}}(t)),$$

where $\bar{a}_{\text{in}} = \sqrt{P_{\text{in}}/\hbar\omega_\ell}$ is the mean photon flux and the fluctuations $\delta a_{\text{in}}(t)$ satisfy,

$$[\delta\hat{a}_{\text{in}}(t), \delta\hat{a}_{\text{in}}^\dagger(t')] = \varepsilon_a\,\delta(t-t').$$

Note that we explicitly "tag" the commutator so as to follow its contribution to the measured quantities [8]; in reality $\varepsilon_a = 1$. The canonically conjugate quadratures corresponding to the fluctuations are defined by,

$$\delta\hat{q}_{\text{in}}(t) := \frac{\delta\hat{a}_{\text{in}}(t) + \delta\hat{a}_{\text{in}}^\dagger(t)}{\sqrt{2}}, \qquad \delta\hat{p}_{\text{in}}(t) := \frac{\delta\hat{a}_{\text{in}}(t) - \delta\hat{a}_{\text{in}}^\dagger(t)}{i\sqrt{2}},$$

so that,

$$[\delta\hat{q}_{\text{in}}(t), \delta\hat{p}_{\text{in}}(t')] = i\varepsilon_a\,\delta(t-t'). \tag{7.2.12}$$

Following the ansatz of optical fluctuations adopted in Sect. 3.2.2, excess noise in the laser is modelled as Gaussian fluctuations, for which (see Eq. 3.2.13),

$$\begin{pmatrix} \langle\delta\hat{q}_{\text{in}}(t)\delta\hat{q}_{\text{in}}(t')\rangle & \langle\delta\hat{q}_{\text{in}}(t)\delta\hat{p}_{\text{in}}(t')\rangle \\ \langle\delta\hat{p}_{\text{in}}(t)\delta\hat{q}_{\text{in}}(t')\rangle & \langle\delta\hat{p}_{\text{in}}(t)\delta\hat{p}_{\text{in}}(t')\rangle \end{pmatrix} = \frac{1}{2}\begin{pmatrix} \varepsilon_a + 2n_{qq} & i\varepsilon_a + 2n_{qp} \\ -i\varepsilon_a + 2n_{qp} & \varepsilon_a + 2n_{pp} \end{pmatrix}\delta(t-t'). \tag{7.2.13}$$

The terms n_{ij} $(i = q, p)$ represent the noise in excess of the fundamental vacuum fluctuations in the field quadratures, distributed uniformly (i.e. "white") in frequency. We henceforth omit the cross-correlation n_{qp} and attempt to bound its effect via an appropriate inequality[4] (see Eq. 3.2.14). Thus,

---

[3]The effect of laser noise on sideband asymmetry measurements is well-studied for cavity optomechanical systems in the resolved sideband regime [27, 28]. In this case sidebands have been observed separately by scattering them into the cavity with a probe laser red/blue detuned.

[4] In addition, it is known that for semiconductor lasers, phase-amplitude correlations are limited to frequencies close to their relaxation oscillation frequency [29, 30]; the latter is typically at a few GHz from the carrier [31, 32]—irrelevant for our experiment.

$$\begin{pmatrix} \langle \delta\hat{a}_{\rm in}(t)\delta\hat{a}_{\rm in}(t')\rangle & \langle \delta\hat{a}_{\rm in}(t)\delta\hat{a}_{\rm in}^{\dagger}(t')\rangle \\ \langle \delta\hat{a}_{\rm in}^{\dagger}(t)\delta\hat{a}_{\rm in}(t')\rangle & \langle \delta\hat{a}_{\rm in}^{\dagger}(t)\delta\hat{a}_{\rm in}^{\dagger}(t')\rangle \end{pmatrix} = \frac{1}{2}\begin{pmatrix} n_{qq}-n_{pp} & 2\varepsilon_a+n_{qq}+n_{pp} \\ n_{qq}+n_{pp} & n_{qq}-n_{pp} \end{pmatrix}. \tag{7.2.14}$$

We now consider an optomechanical system where the optical cavity is driven by a noisy input field described by Eq. (7.2.14). Fluctuations of the intracavity field amplitude (δa) and the mechanical oscillator amplitude (δb) around their stable steady states satisfy (Eq. 4.2.7)

$$\begin{aligned} \delta\dot{\hat{a}} &= +i\Delta\delta\hat{a} - \frac{\kappa}{2}\delta\hat{a} + ig(\delta\hat{b}+\delta\hat{b}^{\dagger}) + \sqrt{\kappa}\,\delta\hat{a}_{\rm in} \\ \delta\dot{\hat{b}} &= -i\Omega_m\delta\hat{b} - \frac{\Gamma_m}{2}\delta\hat{b} + i(g^*\delta\hat{a}+g\delta\hat{a}^{\dagger}) + \sqrt{\Gamma_m}\,\delta\hat{b}_{\rm in}. \end{aligned} \tag{7.2.15}$$

Here $\Delta = \omega_\ell - \omega_c$ is the laser detuning, $g = g_0\bar{a}$ is the dressed optomechanical coupling rate, and $\bar{a} = \frac{\sqrt{\kappa}\bar{a}_{\rm in}}{\frac{\kappa}{2}-i\Delta}$ is the mean intracavity field amplitude. We have also assumed here that the cavity decay rate is dominated by its external coupling, i.e., $\kappa = \kappa_0 + \kappa_{\rm ex} \approx \kappa_{\rm ex}$. The mechanical Langevin noise correlators are

$$\begin{aligned} \langle \delta\hat{b}_{\rm in}(t)\delta\hat{b}_{\rm in}^{\dagger}(t')\rangle &= (n_{\rm m,th}+\varepsilon_b)\delta(t-t') \\ \langle \delta\hat{b}_{\rm in}^{\dagger}(t)\delta\hat{b}_{\rm in}(t')\rangle &= n_{\rm m,th}\,\delta(t-t'), \end{aligned}$$

where $n_{\rm th}$ is the ambient mean thermal phonon occupation of the oscillator. Note that we also "tag" the contribution due to the zero-point fluctuation of the thermal bath to determine its role in the observables; in reality $\varepsilon_b = 1$.

Equation (7.2.15) can be solved in the Fourier domain,

$$\begin{aligned} \delta\hat{a}[\Omega] &= \chi_a[\Omega]\left[\sqrt{\kappa}\,\delta\hat{a}_{\rm in}[\Omega] + ig(\delta\hat{b}[\Omega]+\delta\hat{b}^{\dagger}[\Omega])\right] \\ \delta\hat{a}^{\dagger}[\Omega] &= \delta a[-\Omega]^{\dagger} = \chi_a^*[-\Omega]\left[\sqrt{\kappa}\,\delta\hat{a}_{\rm in}^{\dagger}[\Omega] - ig^*(\delta\hat{b}[\Omega]+\delta\hat{b}^{\dagger}[\Omega])\right] \end{aligned}$$

and

$$\begin{aligned} \begin{pmatrix} \delta\hat{b}[\Omega] \\ \delta\hat{b}^{\dagger}[\Omega] \end{pmatrix} &= \frac{\sqrt{\Gamma_{\rm m}}}{\mathcal{N}[\Omega]}\begin{pmatrix} \chi_b^{*-1}[-\Omega]-i\Sigma[\Omega] & -i\Sigma[\Omega] \\ +i\Sigma[\Omega] & \chi_b^{-1}[\Omega]+i\Sigma[\Omega] \end{pmatrix}\begin{pmatrix} \delta\hat{b}_{\rm in}[\Omega] \\ \delta\hat{b}_{\rm in}^{\dagger}[\Omega] \end{pmatrix} \\ &+ \frac{i\sqrt{\kappa}}{\mathcal{N}[\Omega]}\begin{pmatrix} g^*\chi_b^{*-1}[-\Omega]\chi_a[\Omega] & g\chi_b^{*-1}[-\Omega]\chi_a^*[-\Omega] \\ -g^*\chi_b^{-1}[\Omega]\chi_a[\Omega] & -g\chi_b^{-1}[\Omega]\chi_a^*[-\Omega] \end{pmatrix}\begin{pmatrix} \delta\hat{a}_{\rm in}[\Omega] \\ \delta\hat{a}_{\rm in}^{\dagger}[\Omega] \end{pmatrix}. \end{aligned} \tag{7.2.16}$$

Here χ_b and χ_a are the bare mechanical and cavity response functions, respectively, given by,

$$\chi_b[\Omega] := [\Gamma_{\rm m}/2 - i(\Omega-\Omega_{\rm m})]^{-1}, \qquad \chi_a[\Omega] := [\kappa/2 - i(\Omega+\Delta)]^{-1}.$$

$\Sigma[\Omega]$ is the mechanical "self-energy",

$$\Sigma[\Omega] = -i|g|^2(\chi_a[\Omega] - \chi_a^*[-\Omega]) = \Sigma^*[-\Omega], \tag{7.2.17}$$

which describes the modification to the mechanical response due to radiation pressure, and

$$\mathcal{N}[\Omega] = \chi_b^{-1}[\Omega]\chi_b^{*-1}[-\Omega] + 2\Omega_\mathrm{m}\Sigma[\Omega] = \mathcal{N}^*[-\Omega].$$

The input-output relation [33], $\delta a_\mathrm{out} = \delta a_\mathrm{in} - \sqrt{\kappa}\,\delta a$, gives the fluctuations of the output fields in terms of the fluctuations of the input fields:

$$\begin{aligned}
\delta\hat{a}_\mathrm{out} &= A[\Omega]\delta\hat{a}_\mathrm{in} + B[\Omega]\delta\hat{a}_\mathrm{in}^\dagger + C[\Omega]\delta\hat{b}_\mathrm{in} + D[\Omega]\delta\hat{b}_\mathrm{in}^\dagger \\
\delta\hat{a}_\mathrm{out}^\dagger &= A^*[-\Omega]\delta\hat{a}_\mathrm{in}^\dagger + B^*[-\Omega]\delta\hat{a}_\mathrm{in} + C^*[-\Omega]\delta\hat{b}_\mathrm{in}^\dagger + D^*[-\Omega]\delta\hat{b}_\mathrm{in}
\end{aligned}$$

where,

$$\begin{aligned}
A[\Omega] &= 1 - \kappa\chi_a[\Omega] - \frac{2i|g|^2\kappa\Omega_\mathrm{m}\chi_a[\Omega]^2}{\mathcal{N}[\Omega]} \approx -\left(1 + 4i\frac{\Delta}{\kappa}\right)\left(1 + C_0 n_c\frac{2i\Omega_\mathrm{m}\Gamma_\mathrm{m}}{\mathcal{N}[\Omega]}\right) \\
B[\Omega] &= -\frac{2ig^2\kappa\Omega_m\chi_a[\Omega]\chi_a^*[-\Omega]}{\mathcal{N}[\Omega]} \approx -C_0 n_c\frac{2i\Omega_\mathrm{m}\Gamma_\mathrm{m}}{\mathcal{N}[\Omega]} \\
C[\Omega] &= -\frac{ig\sqrt{\kappa\Gamma_\mathrm{m}}}{\mathcal{N}[\Omega]}\chi_a[\Omega]\chi_b^{*-1}[-\Omega] \approx -i\sqrt{C_0 n_c}\left(1 + 2i\frac{\Delta}{\kappa}\right)\Gamma_\mathrm{m}\chi_b[\Omega] \\
D[\Omega] &= -\frac{ig\sqrt{\kappa\Gamma_\mathrm{m}}}{\mathcal{N}[\Omega]}\chi_c[\Omega]\chi_b^{-1}[\Omega] \approx -i\sqrt{C_0 n_c}\left(1 + 2i\frac{\Delta}{\kappa}\right)\Gamma_\mathrm{m}\chi_b^*[-\Omega].
\end{aligned}$$

Here approximate expressions are given for the case of interest, namely, resonant probing ($|\Delta| \ll \kappa$), small sideband resolution ($\Omega_\mathrm{m} \ll \kappa$), and weak coupling ($|g| \ll \kappa$). We have also introduced the single-photon cooperativity, $C_0 = 4g_0^2/(\kappa\Gamma_\mathrm{m})$, and the mean intracavity photon number, $n_c = |\bar{a}|^2$.

Balanced heterodyne detection of the cavity output is used to measure motional sideband asymmetry. We assume, as in the experiment, that the local oscillator and signal paths are balanced in length; together with a balance of power beyond the combining beam-splitter, this ensures suppression of common-mode excess noise [11]. Following standard arguments for heterodyne detection with a LO frequency shifted by Ω_IF (see Sect. 3.2.3.4), the photocurrent spectrum normalised to the local oscillator shot noise is given by,

$$\begin{aligned}
\bar{S}_{II}^\mathrm{het}[\Omega - \Omega_\mathrm{IF}] \approx \varepsilon_a + 4C_0 n_c \Bigg[&\frac{\Gamma_\mathrm{m}^2}{4}|\chi_b[-\Omega]|^2\left(n_\mathrm{m} + \frac{\varepsilon_b}{2} - \left(\frac{\varepsilon_a}{2} + n_{qq}\right) + \frac{4\Delta\Omega_\mathrm{m}}{\kappa^2}n_{pp}\right) \\
&+ \frac{\Gamma_\mathrm{m}^2}{4}|\chi_b[\Omega]|^2\left(n_\mathrm{m} + \frac{\varepsilon_b}{2} + \left(\frac{\varepsilon_a}{2} + n_{qq}\right) + \frac{4\Delta\Omega_\mathrm{m}}{\kappa^2}n_{pp}\right)\Bigg].
\end{aligned} \tag{7.2.18}$$

This represents the heterodyne spectrum measured in the experiment and depicted later in Figs. 7.5 and 7.7. Here the total bath occupation, arising from the ambient thermal bath and the measurement back-action due to the meter beam, is given by,

$$n_{\mathrm{m}} = n_{\mathrm{m,th}} + \underbrace{C_0 n_c \left(\varepsilon_a + 2n_{qq} + \left(\frac{4\Delta\Omega_{\mathrm{m}}}{\kappa^2} \right)^2 2n_{pp} \right)}_{n_{\mathrm{m,BA}}}.$$

The *sideband ratio* extracted from such the heterodyne spectrum is,

$$R := \frac{\int_{0^+}^{+\infty} \left(\bar{S}_{II}^{\mathrm{het}}[\Omega - \Omega_{\mathrm{IF}}] - \bar{S}_{II}^{\mathrm{het}}[\Omega = \Omega_{\mathrm{IF}}^+] \right) \frac{d\Omega}{2\pi}}{\int_{-\infty}^{0^-} \left(\bar{S}_{II}^{\mathrm{het}}[\Omega - \Omega_{\mathrm{IF}}] - \bar{S}_{II}^{\mathrm{het}}[\Omega = \Omega_{\mathrm{IF}}^-] \right) \frac{d\Omega}{2\pi}} = \frac{n_{\mathrm{m}} + \frac{\varepsilon_b - \varepsilon_a}{2} - n_{qq} + \frac{4\Delta\Omega_{\mathrm{m}}}{\kappa^2} n_{pp}}{n_{\mathrm{m}} + \frac{\varepsilon_b + \varepsilon_a}{2} + n_{qq} + \frac{4\Delta\Omega_{\mathrm{m}}}{\kappa^2} n_{pp}};$$

inserting the values of the optical and mechanical commutators ($\varepsilon_{a,b} = 1$),

$$R = \frac{n_{\mathrm{m}} + \left(\frac{4\Delta\Omega_{\mathrm{m}}}{\kappa^2} n_{pp} - n_{qq} \right)}{n_{\mathrm{m}} + 1 + \left(\frac{4\Delta\Omega_{\mathrm{m}}}{\kappa^2} n_{pp} + n_{qq} \right)} \tag{7.2.19}$$

Firstly, characteristic of linear detection, deviation of R from unity in the ideal case ($n_{qq} = 0 = n_{pp}$) is due to vacuum fluctuations in the optical field, leading to a $\pm\frac{1}{2}\varepsilon_a$ contribution to the lower/upper sideband; physically, this is due to correlations developed between the quantum-back-action driven mechanical motion and the measurement imprecision of the detection process [8, 28]. When n_{qq} and n_{pp} are finite, classical correlations are established that affect R. The response of the cavity (for $\Delta/\kappa \approx 0$) ensures that excess classical correlations due to input amplitude noise lead to an enhanced asymmetry, whereas those arising from input phase noise lead to a common increase in the sideband noise power.

7.2.2.2 Measurement of Excess Laser Noise

Excess amplitude noise

In order to measure the noise in the amplitude quadrature, we employ direct photodetection of the probe laser. The measurement is made at the output of the tapered fiber, with the fiber retracted from the cavity. Analysis of the resulting photocurrent reveals the single-sided spectrum of the incident optical intensity (referred here for convenience to the incident optical power $\hat{P} = \hbar\omega_\ell \hat{n}$, where $\hat{n}$ is the photon flux),

$$\bar{S}_P[\Omega] = (\hbar\omega_\ell)^2 \cdot 2\bar{S}_{nn}[\Omega] = (\hbar\omega_\ell)^2 \cdot 2\langle \hat{n} \rangle (1 + 2n_{qq}).$$

A convenient characterisation of the intensity noise is via the relative intensity noise (RIN) spectrum,

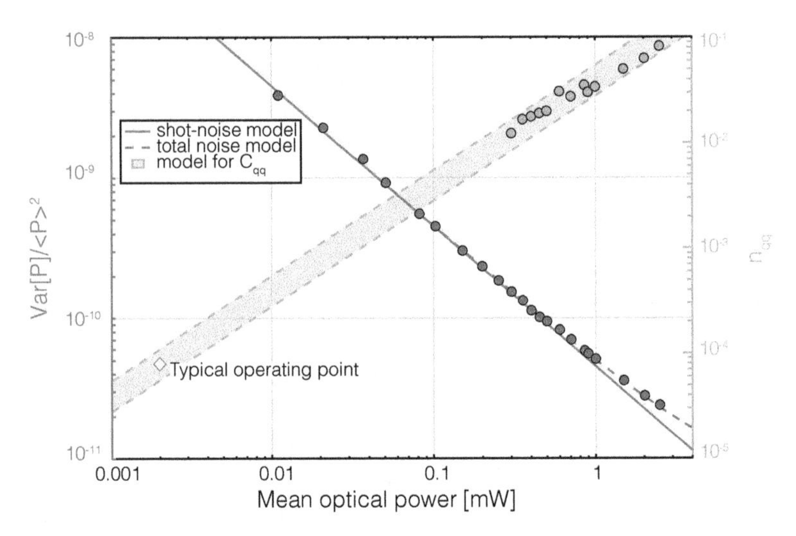

Fig. 7.8 Measurement of laser amplitude noise. Integrated (in a 100 kHz band) relative intensity noise, $\bar{S}_{\mathrm{RIN}}$, versus mean optical power. Using Eq. (7.2.20), the integral gives the relative variance in power, $\mathrm{Var}/\langle\hat{P}\rangle^2$, which should scale as $\langle\hat{P}\rangle^{-1}$ for shot noise. Deviation from shot-noise scaling is evident for $\langle\hat{P}\rangle \gtrsim 1\,\mathrm{mW}$, attributed to classical amplitude noise; using Eq. (7.2.21), this deviation can be used to infer n_{qq}, the average thermal phonon occupation in the amplitude quadrature. (Figure adapted with permission from Ref. [6]. Copyrighted by the American Physical Society)

$$\bar{S}_{\mathrm{RIN}}[\Omega] := \frac{\bar{S}_P[\Omega]}{\langle\hat{P}\rangle^2}, \tag{7.2.20}$$

for which, excess amplitude noise manifests as a deviation from the shot-noise scaling $\propto \frac{1}{\langle P\rangle}$; more precisely,

$$n_{qq} = \frac{1}{2}\left(\frac{\langle\hat{n}\rangle}{2}\bar{S}_{\mathrm{RIN}}[\Omega] - 1\right), \tag{7.2.21}$$

at given incident photon flux. Figure 7.8 shows an inference of n_{qq} using Eq. (7.2.21) and a measurement of $\bar{S}_{\mathrm{RIN}}[\Omega]$ versus mean optical power. For typical experimental conditions ($\langle\hat{P}\rangle = 1 - 5\,\mu\mathrm{W}$), $n_{qq} \ll 0.01$, so that its contribution to sideband asymmetry is negligible.

Excess phase noise

Noise in the phase quadrature of the field leaking from the cavity is measured using balanced homodyne detection. This signal reveals phase noise originating from the input laser as well as apparent phase noise from the cavity. Referred to cavity frequency noise, the homodyne photocurrent spectral density is given by,

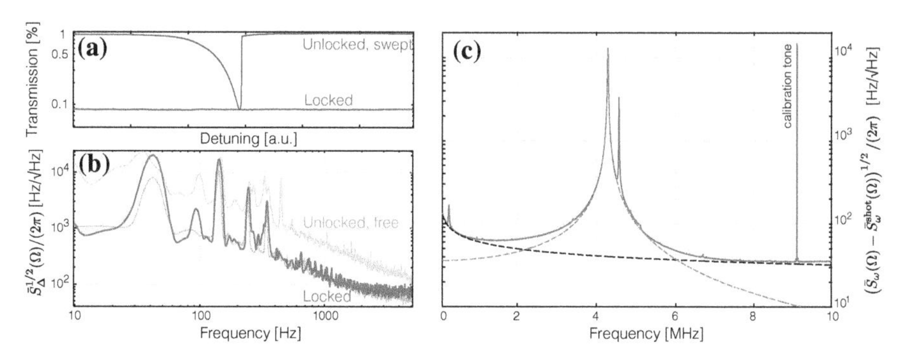

Fig. 7.9 Estimate of laser-cavity detuning noise. a Residual detuning offset at DC estimated from transmission signal when the laser is locked to cavity. **b** Spectrum analysis of the lock error signal, generated via frequency-modulation spectroscopy (see Sect. 3.2.4), reveals low frequency detuning jitter; when locked (red), apparent detuning noise is limited by electronic noise (gray) in the feedback loop, predominantly from the photodetector. **c** Excess frequency noise around the mechanical frequency inferred from a balanced homodyne measurement of the cavity output on resonance. The shot-noise-subtracted signal (red) is composed of the thermomechanical motion of the mechanical mode (blue dashed) and a contribution from excess frequency noise in the laser and cavity substrate (black dashed). (Figure adapted with permission from Ref. [6]. Copyrighted by the American Physical Society)

$$\bar{S}_{\omega}[\Omega] = \Omega^2 \bar{S}_{\phi}[\Omega] = \Omega^2 \left(\bar{S}_{\phi}^{\text{in,shot}}[\Omega] + \bar{S}_{\phi}^{\text{in,ex}}[\Omega] + \bar{S}_{\phi}^{\text{cav,ex}}[\Omega] + \bar{S}_{\phi}^{\text{cav,mech}}[\Omega] \right). \tag{7.2.22}$$

$\bar{S}_{\omega}$ contains contributions from laser phase noise (shot and excess), cavity substrate noise, including thermorefractive [34] and thermomechanical noise [35]. The total excess noise in the phase quadrature is modeled by n_{pp}, which allows us to infer the latter using,

$$\frac{n_{pp}}{\langle \hat{n} \rangle} = \bar{S}_{\phi}^{\text{in,ex}}[\Omega_{\text{m}}] + \bar{S}_{\phi}^{\text{cav,ex}}[\Omega_{\text{m}}]. \tag{7.2.23}$$

Figure 7.9c shows a homodyne measurement made with 3 mW of local oscillator power, whose shot-noise has been subtracted. The spectrum is calibrated by referencing it against a known phase modulation tone injected at the input of the homodyne interferometer. The total excess frequency noise (red) is dominated by thermal motion of the in-plane and out-of-plane modes, both of which are gas damped for this measurement. A joint fit to (a) a model of a velocity-damped oscillator (blue, dashed) and, (b) a model combining thermorefractive [34, 36] and white frequency noise (black, dashed), gives an estimate of $\bar{S}_{\omega}^{\text{ex}}(\Omega)$. Near the mechanical frequency, $\bar{S}_{\omega}^{\text{ex}}(\Omega_{\text{m}}) \approx 2\pi \cdot (35\,\text{Hz}/\sqrt{\text{Hz}})^2$, implying (via Eq. (7.2.23)), $n_{pp} \approx 30$.

From this estimate of n_{pp} we are able to bound two quantities. First, in conjunction with $n_{qq} \ll 0.01$, the excess noise cross-correlation is bounded as $n_{qp} \ll 1$ (using the inequality in Eq. (3.2.14)). Secondly, referring to Eq. (7.2.18), we are able to estimate the contribution of phase noise to the heterodyne sideband. This contribution, characterised as an equivalent phonon occupation (since it adds positive noise power to either sideband),

$$n_\phi := \frac{\Delta}{\kappa}\frac{4\Omega_\mathrm{m}}{\kappa} n_{pp}, \tag{7.2.24}$$

has a mean value determined by the mean offset in the detuning $\bar{\Delta}$. Figure 7.9a allows an estimate, $\bar{\Delta} \approx 0.01 \cdot \kappa$, giving,

$$\bar{n}_\phi = \frac{\bar{\Delta}}{\kappa}\frac{4\Omega_\mathrm{m}}{\kappa} n_{pp} = 0.0052 \cdot \left(\frac{\bar{\Delta}/\kappa}{0.01}\right)4\left(\frac{\Omega_\mathrm{m}/2\pi}{4.3\,\mathrm{MHz}}\right)\left(\frac{1\,\mathrm{GHz}}{\kappa/2\pi}\right)\left(\frac{n_{pp}}{30}\right). \tag{7.2.25}$$

Low frequency detuning noise $\delta\Delta$ (Fig. 7.9b) causes deviations from this mean, which are significant if their effect is comparable to $\bar{n}_\phi$. We bound the probability for such "large" statistical excursions using Chebyshev's inequality [37],

$$\Pr(|n_\phi - \bar{n}_\phi| > \bar{n}_\phi) \leq \frac{\mathrm{Var}[n_\phi]}{\bar{n}_\phi^2} = \left(\frac{4\Omega_\mathrm{m}}{\kappa}\frac{n_{pp}}{\bar{n}_\phi}\right)^2 \frac{\mathrm{Var}[\Delta]}{\kappa^2} \approx 10^{-6}. \tag{7.2.26}$$

We thus estimate that mean residual detuning is the leading contribution to phase noise contamination; the contamination, characterised as a phonon-equivalent noise power $\bar{n}_\phi = 0.005$ is however an insignificant contribution to the sideband ratio Eq. (7.2.19).

Together with the bounds, $n_{qq} \ll 0.01$ and $n_{qp} \ll 1$, this implies that sources of classical noise may be excluded in the interpretation of the experimental data.

7.3 Conclusion

The experiments reported in this chapter probe several distinct and unique features of quantum measurements and feedback. Firstly, quantum correlations between the phase and amplitude of the meter field is shown to manifest in one of two different fashions—optical squeezing, or sideband asymmetry—depending on the nature of the detection process. Both these manifestations are observed using a quantum-noise-limited interferometer operating with an imprecision deeply below the standard quantum limit. Secondly, building on the capability of quantum feedback reported in Chap. 6, feedback is used to distil quantum correlations without destroying them. Feedback control of a mechanical oscillator thus joins the exclusive collective of a handful of platforms where manipulation of non-classical resources using feedback has been demonstrated [38–41]. Finally, the fundamental limit of linear feedback control is elucidated: feedback, though capable of suppressing in-loop measurement back-action, is limited by quantum fluctuations amplified by the in-loop detector.

References

1. V.B. Braginsky, F.Y. Khalili, *Quantum Measurement* (Cambridge University Press, Cambridge, 1992)
2. H.M. Wiseman, G.J. Milburn, *Quantum Measurement and Control* (Cambridge University Press, Cambridge, 2010)
3. D.W.C. Brooks, T. Botter, S. Schreppler, T.P. Purdy, N. Brahms, D.M. Stamper-Kurn, Nature **488**, 476 (2012)
4. A.H. Safavi-Naeini, S. Gröblacher, J.T. Hill, J. Chan, M. Aspelmeyer, O. Painter, Nature **500**, 185 (2013)
5. T.P. Purdy, R.W. Peterson, C. Regal, Science **339**, 801 (2013)
6. V. Sudhir, D.J. Wilson, R. Schilling, H. Schütz, A. Ghadimi, A. Nunnenkamp, T.J. Kippenberg, Phys. Rev. X **7**, 011001 (2017)
7. F. Khalili, H. Miao, H. Yang, A. Safavi-Naeini, O. Painter, Y. Chen, Phys. Rev. A **86**, 033840 (2012)
8. A. Weinstein, C. Lei, E. Wollman, J. Suh, A. Metelmann, A. Clerk, K. Schwab, Phys. Rev. X **4**, 041003 (2014)
9. T.P. Purdy, P.-L. Yu, N.S. Kampel, R.W. Peterson, K. Cicak, R.W. Simmonds, C.A. Regal, Phys. Rev. A **92**, 031802 (2015)
10. M. Underwood, D. Mason, D. Lee, H. Xu, L. Jiang, A.B. Shkarin, K. Børkje, S.M. Girvin, J.G.E. Harris, Phys. Rev. A **92**, 061801 (2015)
11. J.H. Shapiro, IEEE, J. Quant. Elec. **21**, 237 (1985)
12. V.B. Braginsky, M.L. Gorodetsky, F.Y. Khalili, A.B. Matsko, K.S. Thorne, S.P. Vyatchanin, Phys. Rev. D **67**, 082001 (2003)
13. C. Fabre, M. Pinard, S. Bourzeix, A. Heidmann, E. Giacobino, S. Reynaud, Phys. Rev. A **49**, 1337 (1994)
14. S. Mancini, P. Tombesi, Phys. Rev. A **49**, 4055 (1994)
15. H.J. Carmichael, J. Opt. Soc. Am. B **4**, 1588 (1987)
16. R. Loudon, P.L. Knight, J. Mod. Opt. **34**, 709 (1987)
17. A.H. Safavi-Naeini, J. Chan, J.T. Hill, T.P.M. Alegre, A. Krause, O. Painter, Phys. Rev. Lett. **108**, 033602 (2012)
18. I. Wilson-Rae, N. Nooshi, W. Zwerger, T.J. Kippenberg, Phys. Rev. Lett. **99**, 093901 (2007)
19. F. Marquardt, J. Chen, A.A. Clerk, S.M. Girvin, Phys. Rev. Lett. **99**, 093902 (2007)
20. M.S. Taubman, H. Wiseman, D.E. McClelland, H.-A. Bachor, J. Opt. Soc. Am. B **12**, 1792 (1995)
21. R.W. Peterson, T.P. Purdy, N.S. Kampel, R.W. Andrews, K.W. Yu, P.-L. Lehnert, C.A. Regal, Phys. Rev. Lett **116**, 063601 (2016)
22. R. Schilling, H. Schütz, A. Ghadimi, V. Sudhir, D. Wilson, T. Kippenberg, Phys. Rev. Appl. **5**, 054019 (2016)
23. S. Steinlechner, B.W. Barr, A.S. Bell, S.L. Danilishin, A. Gläfke, C. Gräf, J.-S. Hennig, E.A. Houston, S.H. Huttner, S.S. Leavey, D. Pascucci, B. Sorazu, A. Spencer, K.A. Strain, J. Wright, S. Hild, Phys. Rev. D **92**, 072009 (2015)
24. A. Sawadsky, H. Kaufer, R. Moghadas-Nia, S. Tarabrin, F. Khalili, K. Hammerer, R. Schnabel, Phys. Rev. Lett. **114**, 043601 (2015)
25. H.M. Wiseman, J. Opt. B **1**, 459 (1999)
26. A.A. Clerk, M.H. Devoret, S.M. Girvin, F. Marquardt, R.J. Schoelkopf, Rev. Mod. Phys. **82**, 1155 (2010)
27. A.M. Jayich, J.C. Sankey, K. Børkje, D. Lee, C. Yang, M. Underwood, L. Childress, A. Petrenko, S.M. Girvin, J.G.E. Harris, New J. Phys. **14**, 115018 (2012)
28. A.H. Safavi-Naeini, J. Chan, J.T. Hill, S. Gröblacher, H. Miao, Y. Chen, M. Aspelmeyer, O. Painter, New J. Phys. **15**, 035007 (2013)
29. K. Vahala, C. Harder, A. Yariv, Appl. Phys. Lett. **42**, 211 (1983)
30. M. van Exeter, W. Hamel, J.P. Woerdman, B. Zeijlmans, IEEE, J. Quant. Elec. **28**, 1470 (1992)

31. C.E. Wieman, L. Hollberg, Rev. Sci. Instru. **62**, 1 (1991)
32. T.J. Kippenberg, A. Schliesser, M.L. Gorodetsky, New J. Phys **15**, 015019 (2013)
33. C.W. Gardiner, A.S. Parkins, M.J. Collett, J. Opt. Soc. Am. B **4**, 1617 (1987)
34. M.L. Gorodetsky, I.S. Grudinin, J. Opt. Soc. Am. B **21**, 697 (2004)
35. A. Gillespie, F. Raab, Phys. Lett. A **178**, 357 (1993)
36. V.B. Braginsky, M.L. Gorodetsky, F.Y. Khalili, K.S. Thorne, Phys. Rev. D **61**, 044002 (2000)
37. H. Cramer, *Mathematical Methods of Statistics* (Princeton University Press, Princeton, 1946)
38. C. Sayrin, I. Dotsenko, X. Zhou, B. Peaudecerf, T. Rybarczyk, S. Gleyzes, P. Rouchon, M. Mirrahimi, H. Amini, M. Brune, J.-M. Raimond, S. Haroche, Nature **477**, 73 (2011)
39. D. Ristè, M. Dukalski, C.A. Watson, G. de Lange, M.J. Tiggelman, Y.M. Blanter, K.W. Lehnert, R.N. Schouten, L. DiCarlo, Nature **502**, 350 (2013)
40. Y. Liu, S. Shankar, N. Ofek, M. Hatridge, A. Narla, K. Sliwa, L. Frunzio, R. Schoelkopf, M. Devoret, Phys. Rev. X **6**, 011022 (2016)
41. K.C. Cox, G.P. Greve, J.M. Weiner, J.K. Thompson, Phys. Rev. Lett. **116**, 093602 (2016)

Chapter 8
Epilogue

It's a magical world Hobbes ol' buddy... ...let's go exploring!

Bill Watterson, Calvin and Hobbes

The work reported in this thesis broaches a qualitatively new regime of cavity optomechanics, one where the measurement of the motion of a mechanical oscillator conforms to the predictions of the Heisenberg uncertainty principle. We have accessed this regime by making a measurement of the mechanical oscillator's position fluctuations with an imprecision 40 dB below that at the standard quantum limit, so that the concomitant back-action is in excess of the intrinsic motion [1]. We have demonstrated that being in this regime offers the possibility of performing measurement-based feedback control of the oscillator's quantum state. This may be interpreted as a heuristic principle of information economy—if the state of the oscillator, including random back-action, can be measured with high fidelity, then that measurement record is informationally complete with respect to the state of the oscillator [2]. Once this is true, the record may be used to perform feedback on the oscillator state. We have demonstrated this capability by cooling the oscillator over four orders of magnitude in temperature, resulting in a final average phonon occupation of 5.

A salient feature of being in the measurement back-action dominated regime is that it allows the study of the subtle nature of quantum measurements. In this thesis, we investigate correlations that arise due to the measurement and its relation to measurement-base feedback [3]. In particular, correlations between the amplitude and phase quadratures of the meter beam can be distilled using measurement-based feedback to suppress classical contamination. We also demonstrate the fundamental limitation of this technique, which arises from quantum noise in the detection of the meter state.

V. Sudhir, *Quantum Limits on Measurement and Control of a Mechanical Oscillator*, Springer Theses, https://doi.org/10.1007/978-3-319-69431-3_8

8.1 Quantum Correlations for Metrology and Control

These studies open doors to newer and richer possibilities for the immediate future. The primary resource that enables a host of such experiments is the quantum correlation generated in the optical field after it has interacted strongly with the mechanical oscillator.

A generic feature of linear measurements is that they produce quantum correlations in the meter. For interferometric position measurements (described in Chap. 7), quantum correlations are generated between the amplitude and phase of the optical field used for the measurement. Homodyne detection of this field produces the (shot-noise normalised) photocurrent spectrum (Eq. 7.15),

$$\bar{S}_I^{\theta,\mathrm{hom}}[\Omega] = 1 + \frac{4\eta C \Gamma_\mathrm{m}}{x_\mathrm{zp}^2}\left(\bar{S}_{xx}[\Omega]\sin^2\theta + \frac{\hbar}{2}\mathrm{Re}\,\chi_x^{(0)}[\Omega]\sin 2\theta\right), \tag{8.1.1}$$

consisting of the motion of the mechanical oscillator—optimally measured at phase quadrature ($\theta = \pi/2$)—and a contribution due to quantum correlations—absent in the phase quadrature.

The presence of correlations at other quadratures motivates the question of whether they can be employed for a better estimation of the intrinsic motion of the oscillator. The discussion in Sect. 4.3 provides the answer in the specific case of phase quadrature detection—the SQL that arises therein is due to a trade-off between detector imprecision and measurement back-action. In the general setting to be treated here, we will show that quantum correlations can be used to cancel back-action in the measurement, allowing better estimation precision than that dictated by the SQL for phase quadrature detection.

To analyse this idea, we start from the observed photocurrent spectrum in (Eq. 8.1.1),

$$\bar{S}_I^{\theta,\mathrm{hom}}[\Omega] = 1 + \frac{4\eta C \Gamma_\mathrm{m}}{x_\mathrm{zp}^2}\left(\left(\bar{S}_{xx}^{(0)}[\Omega] + \bar{S}_{xx}^{\mathrm{BA}}[\Omega]\right)\sin^2\theta + \frac{\hbar}{2}\mathrm{Re}\,\chi_x^{(0)}[\Omega]\sin 2\theta\right),$$

where the total motion has been split into the intrinsic motion $\bar{S}_{xx}^{(0)}$, which is to be estimated, and the back-action motion (see Eq. 4.3.5),

$$\bar{S}_{xx}^{\mathrm{BA}}[\Omega] = C\frac{\hbar^2 \Gamma_\mathrm{m}}{x_\mathrm{zp}^2}\left|\chi_x^{(0)}[\Omega]\right|^2.$$

Denoting by $\delta\hat{x}_{\mathrm{est},\theta}$ the unbiased estimator for the position based on the photocurrent record $\delta\hat{I}_\theta$, its spectrum is given by,

$$\bar{S}_{xx}^{\mathrm{est},\theta}[\Omega] = \frac{\bar{S}_I^{\theta,\mathrm{hom}}[\Omega]}{(4\eta C\Gamma_\mathrm{m}/x_\mathrm{zp}^2)\sin^2\theta} = \bar{S}_{xx}^{(0)}[\Omega] + \bar{S}_{xx}^{\mathrm{BA}}[\Omega] + \bar{S}_{xx}^{\mathrm{imp},\theta}[\Omega] + \hbar\cot\theta\,\mathrm{Re}\,\chi_x^{(0)}[\Omega], \tag{8.1.2}$$

where the phase-dependent detection imprecision is given by,

$$\bar{S}_{xx}^{\mathrm{imp},\theta}[\Omega] = \frac{x_\mathrm{zp}^2}{4\eta C\Gamma_\mathrm{m}\sin^2\theta} = \frac{\bar{S}_{xx}^{\mathrm{imp},\pi/2}[\Omega]}{\sin^2\theta}.$$

In Eq. (8.1.2), the objective of the estimation—the intrinsic motion $\bar{S}_{xx}^{(0)}$—is contaminated by three sources of noise: measurement back-action, imprecision, and measurement-induced correlations between the two. Both back-action and imprecision, by being positive, increase the uncertainty in the estimate of the intrinsic position; however, there are frequency intervals where the correlation term can be negative, and thus can be used to reduce the uncertainty. At these frequency intervals, the negative correlations can be thought of as leading to a coherent cancellation of imprecision and back-action.

To make this precise, consider the uncertainty in the estimate:

$$\varepsilon_x(C,\theta) := \bar{S}_{xx}^{\mathrm{est},\theta}[\Omega] - \bar{S}_{xx}^{(0)}[\Omega] = \bar{S}_{xx}^{\mathrm{BA}} + \frac{\bar{S}_{xx}^{\mathrm{imp},\pi/2}}{\sin^2\theta} + \hbar\cot\theta\,\mathrm{Re}\,\chi_x^{(0)}$$

here, we make explicit that the uncertainty depends on the choice of measurement strength (i.e. cooperativity C) and detection angle (θ). For a fixed measurement strength, this uncertainty is minimised for a frequency-dependent detection angle,

$$\theta_\mathrm{opt}[\Omega] = 4\eta C\frac{\Omega_\mathrm{m}\Gamma_\mathrm{m}(\Omega^2-\Omega_\mathrm{m}^2)}{(\Omega^2-\Omega_\mathrm{m}^2)^2+(\Omega\Gamma_\mathrm{m})^2}$$

which coincides with the optimal angle for the measurement of ponderomotive squeezing. At this optimal detection quadrature,

$$\varepsilon_x(C,\theta_\mathrm{opt}) = \bar{S}_{xx}^{\mathrm{BA}}[\Omega]\left[1-\eta\left(\mathrm{Re}\,\chi_x^{(0)}/|\chi_x^{(0)}|\right)^2\right] + \bar{S}_{xx}^{\mathrm{imp},\pi/2}[\Omega] < \varepsilon_x(C,\tfrac{\pi}{2}),$$

i.e. at every measurement strength (limited by the detection efficiency, and the technical challenge of realising a stable frequency-dependent detection angle), the ability to estimate the intrinsic motion is enhanced by the cancellation of back-action in the measurement record.

Further optimisation of measurement strength achieves the ultimate bound on the achievable estimation uncertainty. However, broadband enhancement across all Fourier frequencies is not possible [4, 5]; at desired frequency intervals, the minimum uncertainty, $\min_C \varepsilon_x(C,\theta_\mathrm{opt})$, satisfies,

$$\min_C \varepsilon_x(C,\tfrac{\pi}{2}) \geq \min_C \varepsilon_x(C,\theta_\mathrm{opt}) = \left[1-\eta\left(\mathrm{Re}\,\chi_x^{(0)}/|\chi_x^{(0)}|\right)^2\right]^{1/2}\cdot\frac{\hbar}{\sqrt{\eta}}\left|\chi_x^{(0)}\right|$$

i.e. an uncertainty lower than what is predicted by the SQL for phase quadrature detection.

The ideas described above can be directly transposed to the problem of force estimation. In that context, proper choice of detection angle cancels the back-action force in the measurement record—a technique termed variational measurement [4, 6, 7]. Defining an uncertainty for the estimation of the thermal force,

$$\varepsilon_F(C,\theta) := \left|\chi_x^{(0)}\right|^{-2} \varepsilon_x(C,\theta),$$

it can be shown that [8], an enhancement due to quantum correlations is achieved for (frequency-independent) detection angles away from phase quadrature. Figure 8.1 show the enhancement achieved in a recent experiment where the system described in this thesis was deployed at room-temperature [8]. (This experiment was also the first to demonstrate quantum correlations developed due to a room-temperature mechanical oscillator.)

Ultimately, these ideas illustrate a general perspective on quantum metrology. To date, the multitude of experiments that report quantum metrology, do so by following one of two strategies [9]:

(a) the state of the meter is prepared in some non-classical state [10]; for example, squeezed states of light for interferometric position measurements [11, 12], "NOON" states for super-resolved optical phase measurements [13–15], entangled states of atoms/ions for spectroscopy [16, 17],
(b) the system-meter coupling is engineered to achieve reduced back-action [18]; for example, by performing a non-demolition measurement [19–21], or measuring only a single quadrature of an oscillator [22].

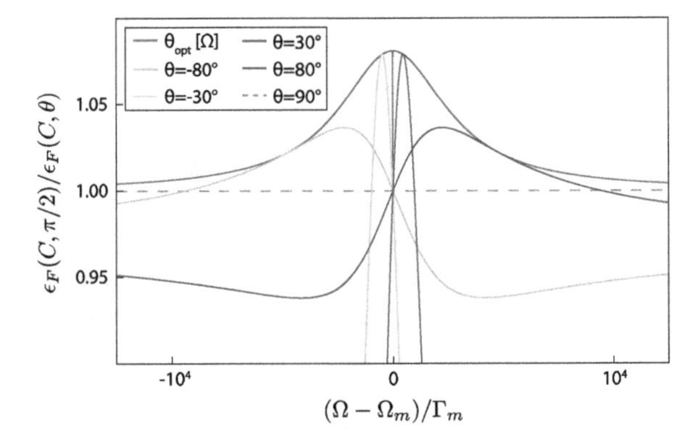

Fig. 8.1 **Quantum-enhanced force metrology.** Plot shows the reduction in uncertainty in the estimation of thermal force, quoted as the enhancement over conventional phase quadrature detection. An enhancement of 7% would be realised with current device parameters at room-temperature, practically limited by an overall detection efficiency of 25%

A third strategy, that neither requires a priori non-classical states of the meter, or a precisely tuned system-meter interaction, is what has been described above. Such schemes rely on a cunning choice of meter-detector coupling so as to harness non-classical correlations that are developed in-situ during the act of measurement.

In fact, these optimal measurements, in addition to shedding light on the inner workings of quantum measurements, may be envisioned to provide quantum-enhanced record of a state's trajectory for better feedback control. For example, in feedback cooling of a mechanical oscillator using a phase-quadrature measurement of the oscillator position (described in Chap. 6), the final phonon occupation is lower-bounded by the Heisenberg uncertainty product (see Eq. 6.2.28),

$$n_{\mathrm{m,min}}^{\pi/2} \approx \frac{1}{\hbar}\sqrt{\bar{S}_{FF}^{\mathrm{tot}} \cdot \bar{S}_{xx}^{\mathrm{imp},\pi/2}}. \tag{8.1.3}$$

By choosing to estimate the motion using a variational measurement, this limit can be naturally evaded as described above. It can be shown that the lowest achievable phonon occupation is bounded by [23],

$$n_{\mathrm{m,min}}^{\theta} \approx \frac{1}{\hbar}\sqrt{\bar{S}_{FF}^{\mathrm{tot}} \cdot \bar{S}_{xx}^{\mathrm{imp},\theta} - |\bar{S}_{Fx_{\mathrm{imp}}^{\theta}}|^2} \gtrsim n_{\mathrm{m,min}}^{\pi/2}\sqrt{1-\eta}; \tag{8.1.4}$$

here, the last bound is for the optimal measurement quadrature θ_{opt}. Similar to the quantum enhancement for position and force metrology, quantum correlations can be used to cancel back-action in the measurement record to enhance the performance of feedback cooling.

References

1. D.J. Wilson, V. Sudhir, N. Piro, R. Schilling, A. Ghadimi, T.J. Kippenberg, Nature **524**, 325 (2015)
2. M. Hatridge, S. Shankar, M. Mirrahimi, F. Schackert, K. Geerlings, T. Brecht, K. Sliwa, B. Abdo, L. Frunzio, S. Girvin et al., Science **339**, 178 (2013)
3. V. Sudhir, D.J. Wilson, R. Schilling, H. Schütz, A. Ghadimi, A. Nunnenkamp, T.J. Kippenberg, Phys. Rev. X **7**, 011001 (2017)
4. A. Buonanno, Y. Chen, Phys. Rev. D **64**, 042006 (2001)
5. H. Miao, H. Yang, R.X. Adhikari, Y. Chen, Class. Quant. Grav. **31**, 165010 (2014)
6. S.P. Vyatchanin, E.A. Zubova, Phys. Lett. A **201**, 269 (1995)
7. H.J. Kimble, Y. Levin, A.B. Matsko, K.S. Thorne, S.P. Vyatchanin, Phys. Rev. D **65**, 022002 (2001)
8. V. Sudhir, R. Schilling, S. Fedorov, H. Schütz, D.J. Wilson, T.J. Kippenberg, (2016)
9. V. Giovannetti, S. Lloyd, L. Maccone, Phys. Rev. Lett. **96**, 010401 (2006)
10. G.M. D'Ariano, P. Lo Presti, M.G.A. Paris, Phys. Rev. Lett. **87**, 270404 (2001)
11. J. Aasi et al., Nature Phot. **7**, 613 (2013)
12. J.B. Clark, F. Lecocq, R.W. Simmonds, J. Aumentado, J.D. Teufel, Nature Phys. **12**, 683 (2016)
13. M.W. Mitchell, J.S. Lundeen, A.M. Steinberg, Nature **429**, 161 (2004)
14. T. Nagata, R. Okamoto, J.L. O'Brien, K. Sasaki, S. Takeuchi, Science **316**, 726 (2007)
15. I. Afek, O. Ambar, Y. Silberberg, Science **328**, 879 (2010)

16. D. Leibfried, M.D. Barrett, T. Schaetz, J. Britton, J. Chiaverini, W.M. Itano, J.D. Jost, C. Langer, D.J. Wineland, Science **304**, 1476 (2004)
17. C.F. Roos, M. Chwalla, K. Kim, M. Riebe, R. Blatt, Nature **443**, 316 (2006)
18. V.B. Braginsky, Y.I. Vorontsov, K.S. Thorne, Science **209**, 547 (1980)
19. W. Nagourney, J. Sandberg, H. Dehmelt, Phys. Rev. Lett. **56**, 2797 (1986)
20. C. Guerlin, J. Bernu, S. Deléglise, C. Sayrin, S. Gleyzes, S. Kuhr, M. Brune, J.-M. Raimond, S. Haroche, Nature **448**, 889 (2007)
21. B.R. Johnson, M.D. Reed, A.A. Houck, D.I. Schuster, L.S. Bishop, E. Ginossar, J.M. Gambetta, L. DiCarlo, L. Frunzio, S.M. Girvin, R.J. Schoelkopf, Nature Phys. **6**, 663 (2010)
22. J. Suh, A.J. Weinstein, C.U. Lei, E.E. Wollman, S.K. Steinke, P. Meystre, A.A. Clerk, K.C. Schwab, Science **344**, 1262 (2014)
23. H. Habibi, E. Zeuthen, M. Ghanaatshoar, K. Hammerer, J. Opt. **18**, 084004 (2016)

Appendix A
Uncertainty Inequalities

For a set of observables $\{\hat{X}_i\}_{i=1,\ldots,N}$ of a quantum mechanical system, the fact that their expectation values are determined by an underlying quantum state, determine a set of fundamental bounds to be satisfied by their outcome statistics.

The experiment where an identical quantum state of the system, $\hat{\rho}$, is independently prepared and any of the observables is measured once per preparation, determines a distribution of outcomes for each of the observables. Denoting by $x_i \in \mathbb{R}$ the continuous eigenvalues of the observable $\hat{X}_i$ corresponding to the eigenstate $|x_i\rangle$, the probability distribution of the outcomes of $\hat{X}_i$ is given by [1–3],

$$\Pr\left[x_i\right] = \mathrm{Tr}\left[|x_i\rangle\langle x_i|\,\hat{\rho}\right]. \tag{A.0.1}$$

Broadly, uncertainty relations are general statements that describe the constraints satisfied by the set of these probability distributions.

In the case of a large number of experimental trials, each of these distributions tend to a gaussian distribution, in which case, a convenient measure of measurement uncertainty is the deviation from the mean outcome, represented by the operators,

$$\delta\hat{X}_i := \hat{X}_i - \left\langle \hat{X}_i \right\rangle. \tag{A.0.2}$$

The uncertainty in each observable may be characterized as the variance of the distribution,

$$\mathrm{Var}\left[\hat{X}_i\right] := \left\langle \delta\hat{X}_i^{\,2} \right\rangle, \tag{A.0.3}$$

while the mutual correlations between the observables described by the covariance,

$$\mathrm{Cov}\left[\hat{X}_i, \hat{X}_j\right] := \left\langle \frac{1}{2}\left\{\delta\hat{X}_i, \delta\hat{X}_j\right\} \right\rangle, \tag{A.0.4}$$

where the possible non-commutativity of observables necessitates the symmetrization. Note that, $\mathrm{Var}\left[\hat{X}_i\right] = \mathrm{Cov}\left[\hat{X}_i, \hat{X}_i\right]$. In a general setting, the observables may

V. Sudhir, *Quantum Limits on Measurement and Control of a Mechanical Oscillator*, Springer Theses, https://doi.org/10.1007/978-3-319-69431-3

commute amongst themselves according to,

$$\left[\hat{X}_j, \hat{X}_k\right] = i\hat{C}_{jk}, \tag{A.0.5}$$

where $\hat{C}_{jk}$ are some operators encoding the commutation structure. Note that $\hat{C}_{jk}$ are necessarily hermitian, and satisfy, $\hat{C}_{jk} = -\hat{C}_{kj}$.

Theorem 1 (Uncertainty principle) *In the setting described herein, the covariance matrix satisfies the matrix inequality,*

$$\mathrm{Cov}\left[\hat{X}_j, \hat{X}_k\right] + \frac{i}{2}\left\langle \hat{C}_{jk}\right\rangle \geq 0. \tag{A.0.6}$$

Proof Note that a general operator, defined by,

$$\hat{M} := \sum_j \alpha_j \,\delta\hat{X}_j,$$

for some arbitrary complex numbers α_i, satisfies the identity, $\mathrm{Tr}\left[\hat{M}^\dagger \hat{M}\hat{\rho}\right] \geq 0$, for any state $\hat{\rho}$ (see Lemma 2.1). Working out the trace explicitly,

$$\begin{aligned}
\mathrm{Tr}\left[\hat{M}^\dagger \hat{M}\hat{\rho}\right] &= \sum_{j,k} \alpha_j^* \alpha_k \,\mathrm{Tr}[\delta\hat{X}_j \delta\hat{X}_k\,\hat{\rho}] \\
&= \sum_{j,k} \alpha_j^* \alpha_k \,\mathrm{Tr}\left[\left(\frac{1}{2}\left\{\delta\hat{X}_j, \delta\hat{X}_k\right\} + \frac{1}{2}\left[\delta\hat{X}_j, \delta\hat{X}_k\right]\right)\hat{\rho}\right] \\
&= \sum_{j,k} \alpha_j^* \alpha_k \left(\mathrm{Cov}\left[\hat{X}_i, \hat{X}_j\right] + \frac{i}{2}\left\langle \hat{C}_{jk}\right\rangle\right) \\
&= \boldsymbol{\alpha}^H \mathbf{M} \boldsymbol{\alpha},
\end{aligned}$$

where, $\boldsymbol{\alpha} := [\alpha_1, \ldots, \alpha_N]^T$, is the vector of the arbitrary complex numbers α_i, $\boldsymbol{\alpha}^H = (\alpha^*)^T$ is its hermitian conjugate, and $\mathbf{M}$ is a complex matrix whose elements are given by,

$$M_{jk} := \mathrm{Cov}\left[\hat{X}_i, \hat{X}_j\right] + \frac{i}{2}\left\langle \hat{C}_{jk}\right\rangle.$$

The identity $\mathrm{Tr}\left[\hat{M}^\dagger \hat{M}\hat{\rho}\right] \geq 0$ implies that the quadratic form,

$$\boldsymbol{\alpha}^H \mathbf{M} \boldsymbol{\alpha} \geq 0, \quad \text{for any } \alpha_i.$$

This implies that the matrix $\mathbf{M}$ must itself be positive, giving the desired result. □

Corollary 1 (Robertson-Schrodinger [4, 5]) *For the case of two observables, $\hat{X}_1, \hat{X}_2$,*

$$\mathrm{Var}[\hat{X}_1]\mathrm{Var}[\hat{X}_2] \geq \frac{1}{4}\left|\left\langle\left\{\delta\hat{X}_1, \delta\hat{X}_2\right\}\right\rangle\right|^2 + \frac{1}{4}\left|\left\langle\left[\delta\hat{X}_1, \delta\hat{X}_2\right]\right\rangle\right|^2. \tag{A.0.7}$$

Proof The $N = 2$ case of Eq. (A.0.6) gives,

$$\begin{pmatrix} \mathrm{Var}\left[\hat{X}_1\right] & \mathrm{Cov}\left[\hat{X}_1, \hat{X}_2\right] + \frac{i}{2}\left\langle\left[\hat{X}_1, \hat{X}_2\right]\right\rangle \\ \mathrm{Cov}\left[\hat{X}_1, \hat{X}_2\right] - \frac{i}{2}\left\langle\left[\hat{X}_1, \hat{X}_2\right]\right\rangle & \mathrm{Var}\left[\hat{X}_2\right] \end{pmatrix} \geq 0.$$

The sufficient condition for this to be true is that its lowest eigenvalue be positive, i.e.

$$\left(\mathrm{Var}\left[\hat{X}_1\right] + \mathrm{Var}\left[\hat{X}_2\right]\right) - \sqrt{\left(\mathrm{Var}\left[\hat{X}_1\right] - \mathrm{Var}\left[\hat{X}_2\right]\right)^2 + 4\mathrm{Cov}\left[\hat{X}_1, \hat{X}_2\right]^2 + \left\langle\left[\hat{X}_1, \hat{X}_2\right]\right\rangle^2} \geq 0;$$

simplifying this gives the required result. □

References

1. P.A.M. Dirac, *The Principles of Quantum Mechanics*, 4th edn. (Clarendon Press, 1982)
2. J. von Neumann, *Mathematical Foundations of Quantum Mechanics* (Princeton University Press, 1955)
3. V.B. Braginsky, F.Y. Khalili, *Quantum Measurement* (Cambridge University Press, 1992)
4. H.P. Robertson, Phys. Rev. **46**, 794 (1934)
5. E. Schrödinger, Sitzungsberichte Preus. Akad. Wiss. **19**, 296 (1930)

Appendix B
Miscellanea on Elastodynamics

B.1 Principle of Least Action

Following from Sect. 3.1, the action for the elastodynamic field is,

$$\mathscr{S}[\![u_i]\!] = \int \mathrm{d}t \int_D \mathrm{d}^3 r\, \mathscr{L}(u_i, \dot{u}_i, \partial_j u_i), \tag{B.1.1}$$

for the set of independent displacement fields $u_i(\mathbf{r}, t)$. Note that for the sake of generality, we here retain a possible functional dependence of the Lagrangian on u_i, even though the actual Lagrangian of interest (Eq. (3.1.15)),

$$\mathscr{L} = \frac{\rho}{2}\dot{u}_i \dot{u}_i - \frac{1}{2} t_{ij} u_{ij}^{(1)}, \tag{B.1.2}$$

depends only on the derivatives of u_i. Note the constitute relation for the stress in terms of the strain (Eq. (3.1.16))

$$t_{ij} = \alpha_{ijkl} u_{kl}^{(1)}, \tag{B.1.3}$$

with the Hooke tensor given by (Eq. (3.1.17)),

$$\alpha_{ijkl} = \mu_1 \delta_{ij}\delta_{kl} + \mu_2(\delta_{ik}\delta_{jl} + \delta_{il}\delta_{jk}). \tag{B.1.4}$$

The principle of least action dictates that the field configuration $u_i(\mathbf{r}, t)$ that is realized is the one that renders the action minimum. Note that the action Eq. (B.1.1) is an example of a *functional*, i.e., a map that associates to a set of functions (here u_i), a real number (here, the value of the definite integral in Eq. (B.1.1)). Thus, it is reasonable to compare values of the action for different field configurations and determine one for which the action attains a minimum. In order to determine such a point, we are led to consider a space of test functions, so as to be able to explore

V. Sudhir, *Quantum Limits on Measurement and Control of a Mechanical Oscillator*, Springer Theses, https://doi.org/10.1007/978-3-319-69431-3

the neighbourhood of each element of this functional space in a systematic fashion. Variational calculus [1–3] provides the machinery to accomplish this task.[1]

We consider the *variation* of the functions u_i,

$$u_i(\mathbf{r}, t) \to u_i(\mathbf{r}, t) + \mathrm{D}u_i(\mathbf{r}, t),$$

where the symbol D denotes a functional variation, signifying the fact that these changes are simply a device to enable exploration of the functional neighbourhood of u_i. Since the fields u_i are independent, they maybe varied independently, and so the corresponding variations $\mathrm{D}u_i$ are also independent. The resulting variation in the action is,

$$\mathrm{D}\mathscr{S} = \int \mathrm{d}t\, \mathrm{d}^3 r \left[\frac{\partial \mathscr{L}}{\partial u_i} \mathrm{D}u_i + \frac{\partial \mathscr{L}}{\partial \dot{u}_i} \mathrm{D}\dot{u}_i + \frac{\partial \mathscr{L}}{\partial(\partial_j u_i)} \mathrm{D}(\partial_j u_i) \right].$$

The second and third terms of the integrand can be re-expressed as,

$$\begin{aligned} \frac{\partial \mathscr{L}}{\partial \dot{u}_i} \mathrm{D}\dot{u}_i &= \frac{\partial \mathscr{L}}{\partial \dot{u}_i} \partial_t(\mathrm{D}u_i) = \partial_t \left(\frac{\partial \mathscr{L}}{\partial \dot{u}_i} \mathrm{D}u_i \right) - \partial_t \left(\frac{\partial \mathscr{L}}{\partial \dot{u}_i} \right) \mathrm{D}u_i \\ \frac{\partial \mathscr{L}}{\partial(\partial_j u_i)} \mathrm{D}(\partial_j u_i) &= \frac{\partial \mathscr{L}}{\partial(\partial_j u_i)} \partial_j(\mathrm{D}u_i) = \partial_j \left(\frac{\partial \mathscr{L}}{\partial(\partial_j u_i)} \mathrm{D}u_i \right) - \partial_j \left(\frac{\partial \mathscr{L}}{\partial(\partial_j u_i)} \right) \mathrm{D}u_i \end{aligned} \tag{B.1.5}$$

and re-inserted back. Thus we arrive at,

$$\begin{aligned} \mathrm{D}\mathscr{S} = &\int \mathrm{d}t \int_D \mathrm{d}^3 r \left[\frac{\partial \mathscr{L}}{\partial u_i} - \partial_t \left(\frac{\partial \mathscr{L}}{\partial \dot{u}_i} \right) - \partial_j \left(\frac{\partial \mathscr{L}}{\partial(\partial_j u_i)} \right) \right] \mathrm{D}u_i \\ &+ \int_D \mathrm{d}^3 r \left[\frac{\partial \mathscr{L}}{\partial \dot{u}_i} \mathrm{D}u_i \right]_{t=0}^{t=\infty} + \int \mathrm{d}t \oint_{\partial D} \mathrm{d}A_j \frac{\partial \mathscr{L}}{\partial(\partial_j u_i)} \mathrm{D}u_i. \end{aligned} \tag{B.1.6}$$

Here, the second and third integrals arise from integrating the total derivatives in Eq. (B.1.5). In the third integral, this is performed through an application of the divergence theorem, resulting in an integral over the boundary ∂D of the domain D.

For the principle of least action to be implemented in the form,

$$\frac{\mathrm{D}\mathscr{S}}{\mathrm{D}u_i} = 0,$$

it is therefore required that each of the integrals in Eq. (B.1.6) vanish separately. Since the variations $\mathrm{D}u_i$ are arbitrary, this is tantamount to each of the integrands vanishing independently. This results in three conditions:

1. the Euler-Lagrange equations,

[1] Incidentally, note that the principle of least action is really a principle of stationary action [1].

$$\frac{\partial \mathscr{L}}{\partial u_i} - \partial_t \left(\frac{\partial \mathscr{L}}{\partial \dot{u}_i} \right) - \partial_j \left(\frac{\partial \mathscr{L}}{\partial (\partial_j u_i)} \right) = 0 \tag{B.1.7}$$

2. fulfilment of initial and/or final conditions,

$$\left[\frac{\partial \mathscr{L}}{\partial \dot{u}_i} \, \mathrm{D} u_i \right]_{t=0}^{t=\infty} = 0 \tag{B.1.8}$$

3. fulfilment of boundary conditions,

$$\oint_{\partial D} \mathrm{d} A_j \, \frac{\partial \mathscr{L}}{\partial (\partial_j u_i)} \, \mathrm{D} u_i = 0. \tag{B.1.9}$$

Note that the principle of least action not only furnishes the dynamical equation Eq. (B.1.7) to be satisfied by the true field configuration, but also provides a consistent set of *natural* boundary conditions Eqs. (B.1.8) and (B.1.9).

B.1.1 Equations of Motion

To implement the Euler-Lagrange equation Eq. (B.1.7), we compute the various terms in it, for the Lagrangian Eq. (B.1.2):

$$\begin{aligned} \frac{\partial \mathscr{L}}{\partial u_i} &= 0 \\ \frac{\partial \mathscr{L}}{\partial \dot{u}_i} &= \frac{\partial}{\partial \dot{u}_i} \left(\frac{\rho}{2} \dot{u}_a \dot{u}_a \right) = \frac{\rho}{2} (2 \dot{u}_a \delta_{ia}) = \rho \dot{u}_i \\ \frac{\partial \mathscr{L}}{\partial (\partial_j u_i)} &= \frac{\partial}{\partial (\partial_j u_i)} \left(-\frac{\alpha_{abcd}}{2} (\partial_b u_a)(\partial_d u_c) \right) = -\alpha_{ijcd} (\partial_d u_c). \end{aligned} \tag{B.1.10}$$

Inserting these in Eq. (B.1.7) gives,

$$\rho \ddot{u}_i - \alpha_{ijkl} \partial_j \partial_l u_k = 0. \tag{B.1.11}$$

Finally using the explicit form of the Hooke tensor (Eq. (B.1.3)) gives the Navier equations,

$$\begin{aligned} \rho \, \ddot{u}_i &= (\mu_1 + \mu_2) \partial_i \partial_j u_j + \mu_2 \, \partial_j \partial_j u_i \\ \text{or,} \quad \rho \ddot{\mathbf{u}} &= (\mu_1 + \mu_2) \nabla (\nabla \cdot \mathbf{u}) + \mu_2 \nabla^2 \mathbf{u} \\ &= (\mu_1 + 2\mu_2) \nabla (\nabla \cdot \mathbf{u}) - \mu_2 \nabla \times (\nabla \times \mathbf{u}), \end{aligned} \tag{B.1.12}$$

where the last two forms are expressed using vector operators appropriate for 3D domains. The third form is obtained by using the generally valid vector identity,

$$\nabla \times (\nabla \times \mathbf{u}) = \nabla(\nabla \cdot \mathbf{u}) - \nabla^2 \mathbf{u}. \tag{B.1.13}$$

B.1.2 Boundary Conditions

The natural boundary condition Eq. (B.1.9), applied to the Lagrangian Eq. (B.1.2) results in,

$$\oint_{\partial D} \mathrm{d}A_j \, t_{ij} \, \mathrm{D}u_i = 0,$$

where, $t_{ij} := \alpha_{ijkl} \partial_l u_k$, is the stress tensor according to Hooke's law. Using the fact that the force F_i (along the i direction) is given in terms of the stress t_{ij} acting on the area element $\mathrm{d}A_j$ (normal to the direction j), $t_{ij} \, \mathrm{d}A_j = \mathrm{d}F_i$, the boundary condition reads,

$$\oint_{\partial D} \mathrm{d}F_i \, \mathrm{D}u_i = 0.$$

Since the variation $\mathrm{D}u_i$ are independent of the force on the boundary, this is equivalent to two conditions, viz.

$$\begin{aligned} \mathrm{d}F_i|_{\partial D} &= t_{ij} \mathrm{d}A_j|_{\partial D} = 0 \\ \mathrm{D}u_i|_{\partial D} &= 0. \end{aligned} \tag{B.1.14}$$

Physically, the first is appropriate for a free boundary, on which no force impinges, whereas the second is appropriate for a fixed boundary, whose displacement is prescribed.

B.2 Transverse and Longitudinal Elastic Waves

The Navier equations Eq. (B.1.12) expressed as a single vector equation,

$$\ddot{\mathbf{u}} = \left(\frac{\mu_1 + 2\mu_2}{\rho} \right) \nabla(\nabla \cdot \mathbf{u}) - \left(\frac{\mu_2}{\rho} \right) \nabla \times (\nabla \times \mathbf{u}), \tag{B.2.1}$$

makes explicit the two kinds of excitations referred to in Sect. 3.1. In order to exhibit this claim, we make use of the fact that any vector field, here $\mathbf{u}$, in a simply connected domain D, maybe expressed uniquely in terms of *potentials*, $\phi(\mathbf{r}, t)$ and $\mathbf{\Phi}(\mathbf{r}, t)$:

$$\mathbf{u} = \nabla \phi + \nabla \times \mathbf{\Phi}. \tag{B.2.2}$$

Identifying these two terms as $\mathbf{u}_L$ and $\mathbf{u}_T$ respectively, standard vector identities imply $\nabla \cdot \mathbf{u}_T = 0$ and $\nabla \times \mathbf{u}_L = 0$; $\mathbf{u}_T$ ($\mathbf{u}_L$) is the transverse (longitudinal) component of $\mathbf{u}$. Substituting this decomposition into Eq. (B.2.1), and realizing that the transverse and longitudinal components are independent, results in two wave equations,

$$\begin{aligned}\ddot{\mathbf{u}}_L &= \left(\frac{\mu_1 + 2\mu_2}{\rho}\right) \nabla^2 \mathbf{u}_L \\ \ddot{\mathbf{u}}_T &= \left(\frac{\mu_2}{\rho}\right) \nabla^2 \mathbf{u}_T.\end{aligned} \tag{B.2.3}$$

The phase velocities of the two elastic waves can be immediately identified, viz.,

$$c_L := \sqrt{\frac{\mu_1 + 2\mu_2}{\rho}}, \quad c_T := \sqrt{\frac{\mu_2}{\rho}}. \tag{B.2.4}$$

B.3 Hermiticity of the Elastic Operator

The elasticity operator $\hat{\mathbf{L}}$, defined in Eq. (3.1.20) viz.

$$\hat{L}_{ik} = \frac{\alpha_{ijkl}}{\rho} \partial_j \partial_l, \tag{B.3.1}$$

acts on vector functions $\mathbf{u}$ defined on some finite domain D. Corresponding to some such function $\mathbf{v}$, we define a linear functional $\langle \mathbf{v}, \cdot \rangle$ that acts as,

$$\langle \mathbf{v}, \mathbf{u} \rangle := \frac{1}{\mathrm{Vol}(D)} \int_D v_i^*(\mathbf{r}) u_i(\mathbf{r})\, \mathrm{d}^3 r. \tag{B.3.2}$$

We now restrict attention to functions $\mathbf{u}$ for which $\langle \mathbf{u}, \mathbf{u} \rangle < \infty$, and satisfies one of the boundary conditions in Eq. (B.1.14) viz.

$$\begin{aligned}&\text{Type 1:} \quad \mathrm{d}F_i|_{\partial D} = t_{ij} \mathrm{d}A_j|_{\partial D} = \alpha_{ijkl} (\partial_j u_i)(\partial_l u_k)|_{\partial D} = 0 \\ &\text{Type 2:} \quad u_i|_{\partial D} = 0,\end{aligned} \tag{B.3.3}$$

where we have assumed (without loss of generality) that in the case of a fixed boundary condition, the boundary displacement is zero.

Each set of such functions—bounded and satisfying boundary condition of Type s ($s = 1, 2$)—forms a Hilbert space[2] $\mathscr{H}_s$ under the inner product $\langle \cdot, \cdot \rangle$. For every $\mathbf{u} \in \mathscr{H}_s$, there is a functional $\langle \mathbf{u}, \cdot \rangle \in \mathrm{Dual}(\mathscr{H}_s)$ in the dual of $\mathscr{H}_s$ [4].

Having identified the two distinct Hilbert spaces at play, the proof of the hermiticity of $\hat{\mathbf{L}}$ is straightforward. Using the definition of $\hat{\mathbf{L}}$ (Eq. (B.3.1)),

$$\langle \mathbf{v}, \hat{\mathbf{L}} \mathbf{u} \rangle = \frac{\rho^{-1}}{\mathrm{Vol}(D)} \int_D v_i^*(\mathbf{r})\, \alpha_{ijkl} \partial_j \partial_l u_k(\mathbf{r})\, \mathrm{d}^3 r.$$

[2]Physically the two spaces $\mathscr{H}_{1,2}$ describe the displacement fields for the physically incompatible boundary conditions of each type; mathematically, this incompatibility manifests as the fact that a function satisfying one type of boundary condition does not form a superposition with that satisfying a different boundary condition, such that the superposed function satisfies any well-defined boundary condition. Closure under superposition is necessary for a Hilbert space.

Manipulating the integral, and freely using the symmetries of the Hooke tensor (Eq. (3.1.10)) $\alpha_{ijkl} = \alpha_{jikl} = \alpha_{ijlk} = \alpha_{klij}$,

$$\begin{aligned}\int_D v_i^* \alpha_{ijkl}\partial_j\partial_l u_k \, \mathrm{d}^3r &= \int_D v_i^* \alpha_{ijkl}\partial_j\partial_k u_l \, \mathrm{d}^3r \\ &= \int_D \partial_j (v_i^* \alpha_{ijkl} \partial_k u_l) \, \mathrm{d}^3r - \int_D (\partial_j v_i^*) \alpha_{ijkl} (\partial_k u_l) \, \mathrm{d}^3r \\ &= \int_{\partial D} v_i^* \underbrace{\alpha_{ijkl} \partial_k u_l}_{t_{ij}} \, \mathrm{d}A_j - \int_D (\partial_j v_i^*) \alpha_{ijkl} (\partial_k u_l) \, \mathrm{d}^3r;\end{aligned}$$

the second equality follows by partial integration, while the third follows from Gauss' Theorem. Finally, either type of boundary condition ensures that the first term in the last line is zero. Treating the remaining integral similarly,

$$\begin{aligned}\int_D v_i^* \alpha_{ijkl}\partial_j\partial_l u_k \, \mathrm{d}^3r &= -\int_D (\partial_j v_i^*) \alpha_{ijkl} (\partial_k u_l) \, \mathrm{d}^3r \\ &= -\int_{\partial D} (\partial_j v_i^*) \alpha_{ijkl} u_l \, \mathrm{d}A_k + \int_D (\partial_k \partial_j v_i^*) \alpha_{ijkl} u_l \, \mathrm{d}^3r \\ &= -\int_{\partial D} u_i \underbrace{\alpha_{ijkl} \partial_k v_l^*}_{t_{ij}^*} \, \mathrm{d}A_k + \int_D (\alpha_{ijkl} \partial_j \partial_l v_k^*) u_i \, \mathrm{d}^3r \\ &= \int_D (\alpha_{ijkl} \partial_j \partial_l v_k^*) u_i \, \mathrm{d}^3r,\end{aligned}$$

i.e., the differential operator $\partial_j \partial_l$ can be freely commuted within the integral as long as the functions satisfy one of the boundary conditions (Eq. (B.3.3)), and the Hooke tensor is symmetric. In particular, this means that the inner product satisfies,

$$\langle \mathbf{v}, \hat{\mathbf{L}}\mathbf{u} \rangle = \langle \hat{\mathbf{L}}\mathbf{v}, \mathbf{u} \rangle, \tag{B.3.4}$$

i.e. $\hat{\mathbf{L}}$ is hermitian in either Hilbert space $\mathscr{H}_{1,2}$.

B.4 Eigensolution of the Doubly-Clamped Stressed Elastic Beam

The normalized mode functions $v_n(\zeta)$, of a 1D stressed elastic beam are given by the Euler-Bernoulli equations with stress Eq. (5.1.9), viz.

$$\epsilon \frac{\partial^4 v_n}{\partial \zeta^4} - \frac{\partial^2 v_n}{\partial \zeta^2} = \left(\frac{\Omega_n}{\Omega_0} \right)^2 v_n, \tag{B.4.1}$$

where, $\epsilon = KM/T\ell_z^2$ is the (dimensionless) ratio of bending to tensile energy, $\Omega_0 = (T/\rho\mathscr{A}\ell_z^2)^{1/2}$ is the frequency determined by the ratio of tensile energy to inertia. The equation is well-posed for the case where the beam is clamped on both ends, described by the boundary conditions,

$$v(0) = v(1) = 0, \quad \partial_\zeta v(0) = \partial_\zeta v(1) = 0. \tag{B.4.2}$$

The fourth order differential operator forming the right-hand side of Eq. (B.4.1) has four eigenfunctions, viz. $e^{\pm k_n^+ \zeta}$, $e^{\pm i k_n^- \zeta}$, where,

$$k_n^\pm := \left(\frac{1}{2\epsilon}\right)^{1/2} \left(\pm 1 + \sqrt{1 + 4\epsilon(\Omega_n/\Omega_0)^2}\right)^{1/2}, \tag{B.4.3}$$

are the normalized wave vectors of the vibration at frequency Ω_n. Note that this relation indicates a nonlinear dispersion for waves excited on the stressed beam. Indeed, the small$-\epsilon$ approximation,

$$\begin{aligned} k_n^+ &\approx \frac{1}{\sqrt{\epsilon}}\left[1 + \frac{(\Omega_n/\Omega_0)^2}{2}\epsilon + \mathscr{O}(\epsilon^3)\right] \\ k_n^- &\approx \frac{\Omega_n}{\Omega_0}\left[1 - \frac{(\Omega_n/\Omega_0)^2}{2}\epsilon + \mathscr{O}(\epsilon^2)\right] \end{aligned} \tag{B.4.4}$$

seems to suggest that the k_n^- branch describes excitations with linear dispersion—familiar from the case of the purely tensile string ($\epsilon = 0$), while the k_n^+ branch arises from corrections due to the bending term—leading to deviations from a sinusoidal mode that occupy a spatial scale approximated by $\ell_z/k_n^+ \approx \ell_z\sqrt{\epsilon}$.

In the following, exact shapes of the mode functions, and their small$-\epsilon$ approximation—describing the afore-mentioned deviations—will be presented. The general mode $v_n(\zeta)$ is that superposition of the four exponential eigenfunctions that satisfies the double-clamped boundary conditions in Eq. (B.4.2), viz. (see also [5])

$$v_n(\zeta) \propto \frac{k_n^+ \sin k_n^- \zeta - k_n^- \sinh k_n^+ \zeta}{k_n^+ \sin k_n^+ - k_n^- \sinh k_n^-} - \frac{\cos k_n^- \zeta - \cosh k_n^+ \zeta}{\cos k_n^- - \cosh k_n^+}. \tag{B.4.5}$$

Here, the proportionality indicates that an overall constant— fixed by the normalization of the mode function—is omitted. In order for the boundary conditions to be satisfied consistently, it is required that,

$$\frac{(k_n^+)^2 - (k_n^-)^2}{2k_n^- k_n^+} = \frac{\cosh k_n^+ \cos k_n^- - 1}{\sinh k_n^+ \sin k_n^-}; \tag{B.4.6}$$

an algebraic equation that, expressed in terms of Ω_n (via Eq. (B.4.3)), determines the eigenfrequencies of the beam.

Convenient approximations, relevant for the case $\epsilon \ll 1$, can be derived from noting that in this regime, $k_n^+ \gg 1$, and, $k_n^+ \gg k_n^-$. Applied to the characteristic Eq. (B.4.6),

$$k_n^+/k_n^- \approx 2\coth k_n^+ \cot k_n^- \approx 2\cot k_n^-,$$

where the second approximation follows from, $\coth k_n^+ \to 1$, for $k_n^+ \approx \Omega_n/\Omega_0 \gtrsim 1$ (and improving for higher order modes). Thus the approximate characteristic equation,

$$k_n^- \cot k_n^- \approx 2k_n^+,$$

holds. Since $k_n^+ \gg 1$, the solutions of this equation are well approximated by those values of k_n^- that make, $\cot k_n^-$, singular; i.e., $k_n^- \approx n\pi$, for $n \in \mathbb{Z}$. Finally using Eq. (B.4.3), the approximate eigenfrequencies are given by,

$$\Omega_n \approx n\pi\Omega_0\sqrt{1+(n\pi)^2\epsilon}. \tag{B.4.7}$$

For the mode functions, a similar approach may be followed, noting that for $\epsilon \ll 1$, $\sinh k_n^+ \approx \cosh k_n^+ \gg 1$. Applying these crude estimates in Eq. (B.4.5), for the case $\zeta < 1$, gives the approximate mode function, $f_n(\zeta) := v_n(0 \le \zeta \lesssim \frac{1}{2})|_{\epsilon \ll 1}$, viz.

$$\begin{aligned}
f_n(\zeta) &\approx \frac{k_n^+ \sin k_n^-\zeta - k_n^- \sinh k_n^+\zeta}{-k_n^- \sinh k_n^+} - \frac{\cosh k_n^+\zeta - \cos k_n^-\zeta}{\cosh k_n^+} \\
&\propto \sin k_n^-\zeta - \frac{k_n^-}{k_n^+}\sinh k_n^+\zeta + \frac{k_n^-}{k_n^+}\tanh k_n^+\left(\cosh k_n^+\zeta - \cos k_n^-\zeta\right) \\
&\approx \sin k_n^-\zeta + \frac{k_n^-}{k_n^+}\left(\cosh k_n^+\zeta - \sinh k_n^+\zeta - \cos k_n^-\zeta\right) \\
&= \sin k_n^-\zeta + \frac{k_n^-}{k_n^+}\left(e^{-k_n^+\zeta} - \cos k_n^-\zeta\right).
\end{aligned}$$

This approximate form indicates that the mode functions deviate slightly from the sinusoidal modes of a tensile string, by a factor proportional to $\sqrt{\epsilon}$, and the form of the deviation is an exponential correction at the boundary. Indeed the mode function, $v_n(\zeta)$, over the full domain can be approximated by the piecewise smooth function (used, for example in [6]),

$$v_n(\zeta) \approx \begin{cases} f_n(\zeta), & 0 \le \zeta \le \frac{1}{2} \\ (-1)^{n+1} f_n(1-\zeta), & \frac{1}{2} < \zeta \le 1 \end{cases}. \tag{B.4.8}$$

References

1. I.M. Gelfand, S.V. Fomin, *Calculus of Variations* (Prentice-Hall, 1963)
2. C. Lanczos, *The Variational Principles of Mechanics* (Dover, 1970)

3. W. Heitler, *The Quantum Theory of Radiation*, 3rd edn. (Clarendon Press, 1954)
4. N. Akhiezer, I. Glazman, *Theory of Linear Operators in Hilbert Space* (Dover, 1993)
5. A. Bokaian, J. Sound Vib. **142**, 481 (1990)
6. P.-L. Yu, T.P. Purdy, C.A. Regal, Phys. Rev. Lett. **108**, 083603 (2012).

Appendix C
Response of an Imbalanced Interferometer

Following the discussion in Sect. 3.2.2, assume that the amplitude flux $\hat{a}(t)$ of a coherent source of mean amplitude $\bar{a}$ undergoes classical amplitude and phase fluctuations, so that in the rotating frame (the ansatz in Eq. (3.2.6)),

$$a_{\text{in}}(t) = (\bar{a} + \delta\alpha(t))e^{i\delta\phi(t)}, \tag{C.1}$$

where $\delta\alpha(t)$ and $\delta\phi(t)$ are real-valued stochastic processes. Note that since we are interested in classical noise in the amplitude $\delta\alpha(t)$ and in phase $\delta\phi(t)$, all vacuum contributions will be ignored here.

Figure C.1 shows such a field passing through an interferometer. When the input field is split at a beam splitter of transmissivity η_1 at the input of the interferometer, each arm is fed with the fields $a_{1,\text{in}}(t)$ and $a_{2,\text{in}}(t)$, given by,

$$a_{1,\text{in}}(t) = \sqrt{\eta_1}\, a_{\text{in}}(t), \qquad a_{2,\text{in}}(t) = i\sqrt{1-\eta_1}\, a_{\text{in}}(t). \tag{C.2}$$

The first field propagates through a path containing a frequency-shifting element (for example, AOM) implementing a radio frequency shift $\Omega_{\text{IF}} \ll \Omega_{\text{det}} \ll \omega_\ell$ (where Ω_{det} is the final detection span), while the other field propagates through a relative delay (for example using a long path length) of duration τ. The two fields emerging at the end of these paths are,

$$a_{1,\text{out}}(t) = a_{1,\text{in}}(t)e^{-i\Omega_{\text{IF}}t}, \qquad a_{2,\text{out}}(t) = a_{2,\text{in}}(t-\tau). \tag{C.3}$$

Finally, the beams are combined at a beam-splitter of transmissivity η_2 and one of the outputs,

$$\begin{aligned} a_{\text{out}}(t) &= \sqrt{\eta_2}\, a_{1,\text{out}}(t) + i\sqrt{1-\eta_2}\, a_{2,\text{out}}(t) \\ &= \sqrt{\eta_1\eta_2}\, a_{\text{in}}(t)e^{-i\Omega_{\text{IF}}t} - \sqrt{(1-\eta_1)(1-\eta_2)}\, a_{\text{in}}(t-\tau), \end{aligned}$$

V. Sudhir, *Quantum Limits on Measurement and Control of a Mechanical Oscillator*, Springer Theses, https://doi.org/10.1007/978-3-319-69431-3

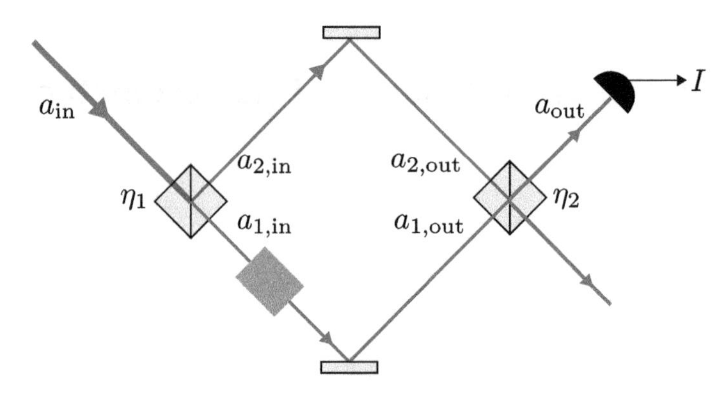

Fig. C.1 Schematic of an imbalanced interferometer. An interferometer in Mach-Zehnder configuration with a noisy input field that is possibly frequency-shifted in one of the arms, and phase delayed in the other

is photodetected. The resulting photocurrent $I(t) \propto |a_{\text{out}}(t)|^2$ is given by,

$$\begin{aligned} I(t) = \eta_1\eta_2\ |a_{\text{in}}(t)|^2 + (1-\eta_1)(1-\eta_2)\ |a_{\text{in}}(t-\tau)|^2 \\ + 2\sqrt{\eta_1\eta_2(1-\eta_1)(1-\eta_2)}\, \text{Re}\, a_{\text{in}}^*(t-\tau)a_{\text{in}}(t)e^{-i\Omega_{\text{IF}}t}. \end{aligned}$$

The last (interference) term describes fluctuations in the photocurrent,

$$\begin{aligned} \delta I(t) &:= \text{Re}\, a_{\text{in}}^*(t-\tau)a_{\text{in}}(t)e^{-i\Omega_{\text{IF}}t} \\ &= \bar{a}^2\left(1+\frac{\delta\alpha(t-\tau)}{\bar{a}}\right)\left(1+\frac{\delta\alpha(t)}{\bar{a}}\right)\cos\left[\delta\phi(t)-\delta\phi(t-\tau)-\Omega_{\text{IF}}t\right], \end{aligned}$$

that carry traces of the amplitude and phase fluctuations of the field at the input of the interferometer. Introducing the cumulative relative amplitude fluctuations,

$$\delta\text{A}(t) := \left(\delta\alpha(t)+\delta\alpha(t-\tau)\right)/\bar{a}, \tag{C.4}$$

and the differential phase fluctuations,

$$\delta\Phi(t) := \delta\phi(t)-\delta\phi(t-\tau), \tag{C.5}$$

the photocurrent fluctuations can be approximated as,

$$\delta I(t) \approx (1+\delta\text{A}(t))\cos\left[\delta\Phi(t)-\Omega_{\text{IF}}t\right]. \tag{C.6}$$

Henceforth, we assume that the amplitude ($\delta\alpha(t)$) and phase ($\delta\phi(t)$) fluctuations are stationary gaussian processes with zero mean; a property that is inherited by $\delta\text{A}(t)$, and, $\delta\Phi(t)$. However, due to the nonlinear transformation relating the phase to the photocurrent, the latter is not gaussian.

Despite this fact, useful information about the amplitude and phase fluctuations can be garnered from the lowest order correlation function of the fluctuating photocurrent. Indeed, assuming that the amplitude and phase fluctuations are uncorrelated (see in Chap. 7 footnote 4, on page 183), the two-time correlation of the photocurrent fluctuations take the form,

$$\begin{aligned}\langle \delta I(t)\delta I(0)\rangle &= \bar{a}^2\Big\langle\Big(1+\delta \mathrm{A}(t)\Big)\Big(1+\delta \mathrm{A}(0)\Big)\cos\big[\delta\Phi(t)-\Omega_{\mathrm{IF}}t\big]\cos\big[\delta\Phi(0)\big]\Big\rangle \\ &= \bar{a}^2\Big(1+\langle\delta \mathrm{A}(t)\delta \mathrm{A}(0)\rangle\Big)\Big(\big\langle\cos[\delta\Phi(t)]\cos[\delta\Phi(0)]\big\rangle\cos\Omega_{\mathrm{IF}}t \\ &\qquad +\big\langle\sin[\delta\Phi(t)]\cos[\delta\Phi(0)]\big\rangle\sin\Omega_{\mathrm{IF}}t\Big).\end{aligned} \tag{C.7}$$

Using standard techniques,[3] the expectation values of the product of the cosine/sine phase terms can be shown to be equal, and given by,

$$\begin{aligned}\langle\cos[\delta\Phi(t)]\cos[\delta\Phi(0)]\rangle &= \langle\sin[\delta\Phi(t)]\cos[\delta\Phi(0)]\rangle \\ &= \tfrac{1}{2}+\tfrac{1}{2}\exp\big[-\langle\delta\Phi(t)\delta\Phi(0)\rangle-\langle\delta\Phi(0)^2\big]\rangle.\end{aligned} \tag{C.8}$$

Finally using the Fourier representation of $\delta\Phi$, and then using its definition given in Eq. (C.5),

$$\begin{aligned}\langle\delta\Phi(t)\delta\Phi(0)\rangle &= \int\frac{\mathrm{d}\Omega\,\mathrm{d}\Omega'}{(2\pi)^2}e^{-i\Omega t}\left\langle\delta\Phi[\Omega]\delta\Phi[\Omega']\right\rangle \\ &= \int\frac{\mathrm{d}\Omega\,\mathrm{d}\Omega'}{(2\pi)^2}e^{-i\Omega t}\left\langle\delta\phi[\Omega](1-e^{i\Omega\tau})\,\delta\phi[\Omega'](1-e^{i\Omega'\tau})\right\rangle \\ &= \int\frac{\mathrm{d}\Omega\,\mathrm{d}\Omega'}{(2\pi)^2}e^{-i\Omega t}\,(1-e^{i\Omega\tau})(1-e^{i\Omega'\tau})\cdot 2\pi\,S_{\phi\phi}[\Omega]\,\delta[\Omega-\Omega'] \\ &= -4\int\frac{\mathrm{d}\Omega}{2\pi}e^{-i\Omega(t-\tau)}\sin^2\left(\frac{\Omega\tau}{2}\right)S_{\phi\phi}[\Omega];\end{aligned} \tag{C.9}$$

thus, the two-time correlators in the exponent of Eq. (C.8) can be expressed in terms of the spectrum of phase fluctuations. Similarly, the two-time correlator, $\langle\delta \mathrm{A}(t)\delta \mathrm{A}(0)\rangle$ in Eq. (C.7), can be expressed in terms of the spectrum of amplitude fluctuations, viz.

[3]Re-writing the trigonometric functions as exponentials, multiplying them out, and then using the identity $\left\langle\exp[i\delta X(t)]\right\rangle = \exp\left[-\frac{1}{2}\langle\delta X(t)\delta X(0)\rangle\right]$, on each exponential term; here δX denotes the relevant random process.

$$
\begin{aligned}
\langle \delta \mathrm{A}(t) \delta \mathrm{A}(0) \rangle &= \int \frac{d\Omega \, d\Omega'}{(2\pi)^2} e^{-i\Omega t} \langle \delta \mathrm{A}[\Omega] \delta \mathrm{A}[\Omega'] \rangle \\
&= \int \frac{d\Omega \, d\Omega'}{(2\pi)^2} e^{-i\Omega t} \left\langle \delta \mathrm{A}[\Omega](1 + e^{i\Omega\tau}) \, \delta \mathrm{A}[\Omega'](1 + e^{i\Omega'\tau}) \right\rangle \\
&= 4 \int \frac{d\Omega}{2\pi} e^{-i\Omega(t-\tau)} \cos^2\left(\frac{\Omega\tau}{2}\right) \frac{S_{\alpha\alpha}[\Omega]}{\bar{a}^2}.
\end{aligned} \tag{C.10}
$$

Inserting Eq. (C.9) in Eq. (C.8) and subsequently in Eq. (C.7), and inserting Eq. (C.10) in Eq. (C.7), taking the limit where $S_{\phi\phi} \ll 1$, and dropping irrelevant constant factors, the photocurrent correlation takes the approximate form,

$$
\begin{aligned}
\langle \delta I(t) \delta I(0) \rangle \propto \sin\left(\Omega_{\mathrm{IF}} t + \frac{\pi}{4}\right) \Bigg[1 &+ 4 \int \frac{d\Omega}{2\pi} e^{-i\Omega(t-\tau)} \cos^2\left(\frac{\Omega\tau}{2}\right) \frac{S_{\alpha\alpha}[\Omega]}{\bar{a}^2} \\
&+ 4 \int \frac{d\Omega}{2\pi} e^{-i\Omega(t-\tau)} \sin^2\left(\frac{\Omega\tau}{2}\right) S_{\phi\phi}[\Omega] \Bigg].
\end{aligned} \tag{C.11}
$$

The (symmetrised) spectrum of photocurrent fluctuations recorded by a spectrum analyser is the cosine transform of this quantity. Shifted by the heterodyne beat frequency, the photocurrent spectrum is,

$$
\bar{S}_{II}[\Omega - \Omega_{\mathrm{IF}}] \propto \delta[\Omega - \Omega_{\mathrm{IF}}] + \frac{2}{\pi} \left(\cos^2\left(\frac{\Omega\tau}{2}\right) \frac{\bar{S}_{\alpha\alpha}[\Omega]}{\bar{a}^2} + \sin^2\left(\frac{\Omega\tau}{2}\right) \bar{S}_{\phi\phi}[\Omega] \right), \tag{C.12}
$$

a result consistent with earlier treatments of phase fluctuations alone [1, 2].

Thus, an imbalanced interferometer transduces input phase and relative amplitude fluctuations, onto the output photocurrent, depending on the time delay τ between the two arms. Typically, by operating a laser far above threshold with a large photon flux $\bar{a}$, the input relative intensity noise can be made arbitrarily small, so that an imbalanced Mach-Zehnder interferometer can be used to measure input phase noise.

References

1. J.A. Armstrong, J. Opt. Soc. Am. **56**, 1024 (1966)
2. P. Gallion, G. Debarge, IEEE J. Quant. Elec. **20**, 343 (1984)

CPSIA information can be obtained
at www.ICGtesting.com
Printed in the USA
LVHW02*1315271217
560944LV00001B/54/P